DATE DUE

DEMCO, INC. 38-2931

Computing with Cells

Computing with Cells

Advances in Membrane Computing

PIERLUIGI FRISCO

Heriot-Watt University
School of Mathematical and Computer Sciences
Riccarton Campus, Edinburgh EH14 4AS
United Kingdom

OXFORD
UNIVERSITY PRESS

OXFORD
UNIVERSITY PRESS

Great Clarendon Street, Oxford OX2 6DP

Oxford University Press is a department of the University of Oxford.
It furthers the University's objective of excellence in research, scholarship,
and education by publishing worldwide in

Oxford New York

Auckland Cape Town Dar es Salaam Hong Kong Karachi
Kuala Lumpur Madrid Melbourne Mexico City Nairobi
New Delhi Shanghai Taipei Toronto

With offices in

Argentina Austria Brazil Chile Czech Republic France Greece
Guatemala Hungary Italy Japan Poland Portugal Singapore
South Korea Switzerland Thailand Turkey Ukraine Vietnam

Oxford is a registered trade mark of Oxford University Press
in the UK and in certain other countries

Published in the United States
by Oxford University Press Inc., New York

© Pierluigi Frisco 2009

British Library Cataloguing in Publication Data
Data available

Library of Congress Cataloging in Publication Data
Data available

Typeset by SPI Publisher Services, Pondicherry, India
Printed in Great Britain
on acid-free paper by the
MPG Books Group, Bodmin and King's Lynn

ISBN 978-0-19-954286-4

1 3 5 7 9 10 8 6 4 2

To Anna

Preface

How could we use living cells to perform computation? Would our definition of *computation* change as a consequence of this? Could such a *cell-computer* outperform digital computers?

These are some of the questions that Membrane Computing tries to answer and at the base of what is treated by this monograph.

Membrane Computing originates from the observation that eukaryotic cells have an internal organisation defined by membranes. In the framework provided by Membrane Computing several biological processes and phenomena occurring in eukaryotic cells have been studied from a computational point of view or have been used to develop algorithms for solving computationally hard problems.

Descriptional and computational complexity of models in Membrane Computing are the two lines of research on which we focus. In this context we recall the results of only *some* of the models present in this framework.

The models considered by us represent a very relevant part of all the models introduced so far in Membrane Computing. They are among the most studied models in the field and they cover a broad range of features (using multisets of symbols or sets of strings, based only on communications, inspired by intra- and inter-cellular processes, that may or may not have a tree as underlying structure, etc.) that gives a grasp of the enormous flexibility of this framework.

A third line of research, concerning the modelling of biological processes and also part of Membrane Computing, has not been included in this work.

Links with biology are constant through this monograph. In addition to having a chapter entirely dedicated to the required biological background, the definition of the models considered by us is preceded by a section with the specific indication of the appropriate biological reality.

Another constant element present in this monograph is the link with Petri nets. Using these nets it is possible to prove that some features present in abstract models of computation are different aspects of the same thing. Moreover, their use facilitates the study of the computational power of these models and introduces new measures of complexity.

This monograph also aims to inspire research. We suggest research topics for each of the considered models. The research topics are targeted both to the people that will use this monograph as an introduction to Membrane Computing and those who want to become researchers in this field, and to the experts. We give suggestions for research of different levels of difficulty and we clearly indicate their importance and the relevance of the possible outcomes.

Readers new to this field of research will find the provided examples particularly useful. These examples are meant to give an initial understanding of the models considered by us and to highlight their operational differences.

The first four chapters give short introductions to computing, membranes in biological cells, theoretical computer science and Petri nets, respectively. In the following five chapters we study the descriptional complexity of the abstract models of computation considered by us. In the last chapter we deal with computational complexity issues. These last six chapters are independent of each other: they can be read in any order after reading the introductory chapters.

In order to improve readability citations have been systematically collected in the last section of each chapter.

I hope the present monograph will be of valuable help to the people who want to approach Membrane Computing and a source of inspiration for further research in this exciting field.

Acknowledgements

The basis of the material of this monograph is the fruit of the efforts of many scientists who researched on membrane systems. For this reason my foremost thanks go to the whole community of people, many of whom are friends, working in the fascinating field of Membrane Computing.

It is a great privilege to have the possibility to convey this exciting body of research material to a new audience and it has been a joy for me to have this privilege. Anyhow, I could not have pursued this task without the constant encouragement and guidance of Gheorghe Păun.

I am very grateful to Christiaan Henkel for explanations and discussions about biology, to Jetty Kleijn for explanation and discussions about Petri nets, to Mario J. Pérez-Jímenez for explanation and discussions about P systems with active membranes and to Artiom Alhazov, Matteo Cavaliere, Erzsébet Csuhaj-Varjú, Marian Gheorghe, Hendrick Jan Hoogeboom, Oscar Ibarra, Vincenzo Manca, Marion Oswald, Andrei Păun, Petr Sosík, György Vaszil, Serghei Verlan, Sara Woodworth, and Claudio Zandron for their ready help many, many times during several stages of the writing process.

Heartfelt thanks to the colleagues and friends in the School of Mathematical and Computer Sciences at Heriot-Watt University for providing enough independence and freedom to prepare this work together with constant support and advice.

Finally, I thank my wife Anna, to whom this work is dedicated, for her patience in the first year of our marriage while I was writing this monograph.

Pierluigi Frisco
Edinburgh, May 2008

Contents

1 Introduction

We give a very brief introduction to computability emphasising concepts that play an important role in the remaining chapters of this monograph.

1.1 From clerks to computers

In the 1930s *Alan Turing* was interested in defining in mathematical terms the activity of what at that time were called *computers*: clerks performing computations. One of his definitions considers a read/write *tape* that is infinitely long and subdivided into *locations*, one next to the other and each location is able to record (any) one symbol from a finite set of symbols. This tape is equipped with a read/write *head* scanning one location at the time. Moreover, the machine can be in (any) one of a finite number of states.

If, for instance, one such abstract device is in one of its possible states and the head is reading one symbol from one location, then another symbol can be written on the same location, the head can move from one location either to the left or to the right and the machine can change state.

There are several kinds of these devices: with only one tape, with several tapes, working as generators or acceptors of languages, etc. and they are all known as *Turing machines*.

If such a machine starts with a certain content in its tape, changes from state to state eventually changing the content of the tape and then it halts, then it is said to *compute* the final content of the tape.

Turing also defined *universal Turing machines* (*UTM*): abstract devices able to simulate any other Turing machine. A UTM receives as input the encoding of another Turing machine, it simulates it and renders as output the encoding of the output of the simulated machine.

Alan Turing and Alonzo Church introduced the formalisation of an *algorithm* indicating the limits on what can be computed and starting in this way the study of *computability*. Computability includes, among others, *theory of computation* and *computational complexity theory*.

The *Church–Turing thesis* (also known as *Church's thesis, Church's conjecture* and *Turing's thesis*) states that any real-world computation can be translated into an equivalent computation performed by a Turing machine. In other words this thesis says that any device performing computation cannot compute more than what a Turing machine does. A device able to compute as much as a Turing machine is then said to be *computationally complete*.

1.2   From cells to computers

Clerks have not been the only models for computation. Several scientists started to define abstract computational devices inspired by biological processes. The devices that we consider in this monograph are called *membrane systems*.

In 1998 Gheorghe Păun defined the class of these devices, also called *P systems*, where 'P' stands for 'Păun', inspired by the internal organisation of eukariotic cells.

These cells have an internal hierarchical organisation defined by *membranes*. These membranes delimit *compartments* where finite sets of symbols (corresponding to the chemicals present in a cell compartment) and finite sets of evolution rules (describing the possible interactions between chemicals), can be placed; this is illustrated in Fig. 1.1.

The topological structure, the locality of interactions, the inherent parallelism, and also the capacity (in less basic models) for membrane division, represent the distinguishing hallmarks of membrane systems.

Membrane Computing is then the field of research using membrane systems to define computability models in order to study computation and computational complexity issues.

In this light, Membrane Computing is part of *Molecular Computing*, a field of research at the interface of computer science and molecular biology driven by the idea that molecular processes can be used for implementing computations or can be regarded as computations. Molecular Computing (together with *Quantum Computing*, *Evolutionary Algorithms*, and *Neural Networks*) belongs to *Natural Computing* which is concerned with computing taking place in nature and computing inspired by nature.

We make an important observation now. When we speak about Molecular Computing or Membrane Computing, we do not imply that molecules or

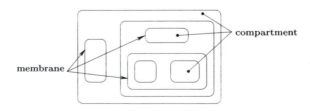

Fig. 1.1 A schematic cell structure

cells compute in a strict sense. We refer to the fact that molecular and cellular processes can be regarded as computations according to our definition of computation. Because of this, these processes and their formalisations follow the Church–Turing thesis.

Anyhow, there is a fundamental difference between Turing machines and the abstract computing devices belonging to Molecular Computing. In the latter case these devices are inspired by direct observations of a natural process, Turing machines were not. Turing did not try to mimic what happened in the brain of the clerks, he was interested in defining this process in general.

How does the approach of Molecular Computing affect our understanding of computation?

Some answers to this question are presented in the following chapters of this monograph and in the literature of Molecular Computing in general.

Moreover, this approach offers at least three further advantages:

1. Firstly, the models of computations have direct counterparts in the natural processes and phenomena by which they have been inspired. This means that if one were interested in implementing such computing devices, then this implementation would not be too far from the abstract model. This is indeed the case for one of the latest developments of Membrane Computing seeing initial attempts (at the time this section is being written) to implement a *membrane computer* in a biological laboratory.

2. The second advantage is that the formalisation of biological processes (resulting from billions of years of evolution) can help define new concepts or algorithms. The classical examples are Evolutionary Algorithms and Neural Networks. In Membrane Computing these formalisations resulted, among others, in the definition of a new class of computational complexity classes inspired by cell division.

3. The third advantage is that such abstract devices can be used as a modelling platform (not necessarily only for biological processes). Also this is applicable to membrane systems: an active and successful research direction in Membrane Computing uses membrane systems to model processes in biology, linguistic, economy, etc.

1.3 Bibliographical notes

We refer readers to classical monographs such as [113, 182, 191] in order to know more about Turing machines and computation in general.

Membrane systems were defined for the first time in a technical report [225] that was later published [207].

The ISI Essential Science Indicators in February 2003 indicated [207] as a 'fast breaking paper' and in October 2003 *Membrane Computing* was indicated by the same institute to be an 'emerging research front in computer science'.

The primary meeting for research on membrane systems is the series of Workshops on Membrane Computing. The first of these workshops was held in 2000 and yearly since then:

2000 August: Curtea de Argeş, Romania, [39];

2001 August: Curtea de Argeş, Romania, [172];

2002 August: Curtea de Argeş, Romania, [221];

2003 July: Tarragona, Spain, [170];

2004 July: Milan, Italy, [177];

2005 July: Vienna, Austria, [67];

2006 July: Leiden, The Netherlands, [110];

2007 July: Thessaloniki, Greece, [60];

2008 July: Edinburgh, UK, [53];

2009 August: Curtea de Argeş, Romania.

Moreover, an increasing number of conferences have membrane systems within their scope.

The most up-to-date source of information on Membrane Computing is the P systems webpage [238]. At the time of writing it contains around 1300 articles, 23 PhD thesis, 19 collective volumes and four monographs on the subject (counting the present one).

The initial ideas regarding the possible implementation of a membrane computer can be found in [93]. Readers interested in Molecular Computing can refer to [20, 21, 59, 92, 220].

2 Biology: an introduction to membranes in cells

The presentation of biochemical and biological concepts that we provide in this monograph is quite simplified. It is directed to computer scientists, not biologists. Our aim is to provide a sufficient background for the understanding of the biological processes and phenomena which inspired models of membrane systems.

The present chapter contains a brief introduction to membranes in living cells.

2.1 The lipid bilayer

The word *cell* was coined in 1665 by Robert Hooke who, looking at thin slices of cork through a microscope, observed tiny, hollow, room-like structures which reminded him of rooms that monks lived in. Actually what Hooke was observing were what we call now *cell walls*, that is, a fairly rigid layer surrounding plant cells, located external to the plasma membrane.

Eukaryotic cells have several membranes; each of them is a thin and pliable layer. All membranes present in a cell are based on *lipid molecules*. The lipid molecules composing membranes in cells have a *hydrophilic* ('water-loving') or *polar* end and a *hydrophobic* ('water-fearing') or *non-polar* end; Fig. 2.1 depicts a phospholipid, that is, a lipid having a polar head and two hydrophobic tails.

Lipid molecules can spontaneously assemble to form *bilayers* approximately 5 nanometres thick. Figure 2.2 illustrates a section of a bilayer composed by phospholipids.

If a bilayer of phospholipids is in a solution where water is present, then it closes on itself forming a *compartment*. The bilayer defining the compartment is a *membrane*. A section of such a compartment is depicted in Fig. 2.3. This figure has to be understood as a section of a hollow sphere (and not as a section of a cylinder). This extraordinary behaviour, fundamental to the functioning of living cells, is a direct consequence of the hydrophobic/hydrophilic nature of phospholipids.

Between 25 and 75% of the mass of membranes in cells consists of proteins. These proteins can be classified in two categories: *integral* and *peripheral*

Fig. 2.1 A phospholipid

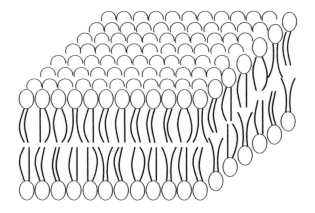

Fig. 2.2 A bilayer of phospholipids

Fig. 2.3 Section of a compartment composed of a bilayer of phospholipids immersed in water

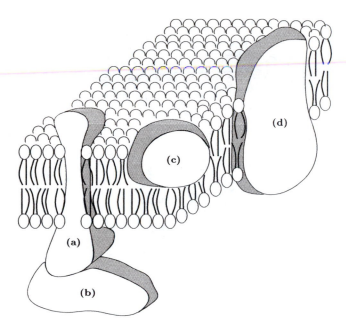

Fig. 2.4 Membrane proteins: **(a)**, **(c)** and **(d)** are integral, **(a)** and **(d)** are transmembrane and **(b)** is a peripheral protein

membrane proteins. Integral membrane proteins are embedded within the membrane. Some of these proteins, called *transmembrane proteins*, span the bilayer. Peripheral membrane proteins are associated with the membrane in an indirect way, generally by interactions with integral membrane proteins. These kinds of membrane proteins are depicted in Fig. 2.4. Phospholipids are free to move in a bilayer. This means that a bilayer is not a rigid structure, rather it is a two-dimensional fluid in which individual molecules are relatively free to rotate and move in lateral directions.

2.2 Membranes in cells

While the bilayer of phospholipids is the basic structure of the membranes present in a cell, the kind of membrane proteins present on a specific bilayer characterise the functions of a bilayer. In general, the functions performed by a membrane are:

separator: bilayers define *compartments*, that is, topological structures within a cell and in between cells, in which specific processes take place;

channel of communication: except for a small number of molecules that can diffuse in and out of a compartment depending on their relative concentration,

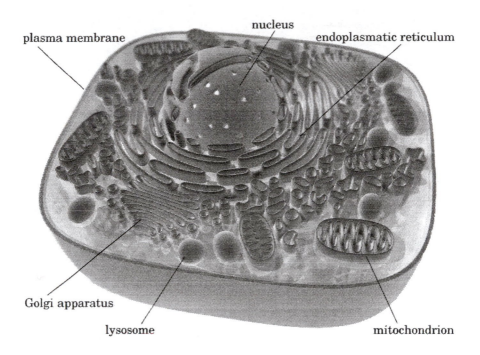

Fig. 2.5 Sketch of an animal eukaryotic cell

the vast majority of passages of molecules across bilayers is mediated by membrane proteins; this function is described in more detail in Section 5.1;

support for several chemical reactions.

In Fig. 2.5 an animal eukaryotic cell is sketched with an indication of some of the membranes and compartments (*organelles*) defined by them.

In order to give a glimpse of the variety of the functions of these organelles, in the following we briefly describe some of them.

Plasma membrane (or *cell membrane*). This physically separates the cellular components from the extracellular environment, it facilitates and regulates the transport of molecules into and out of the cell and it is a support for cell signalling (for this last function see Section 5.13.1). The plasma membrane encloses the *cytoplasm*: a fluid with three major elements in it: *cytosol*, *inclusions* and *organelles* (with the exclusion of the nucleus).

Nuclear envelope. This is formed by two concentric bilayers of phospholipids: the *inner nuclear membrane* and the *outer nuclear membrane*. The nuclear envelope defines the *nuclear compartment* (or *nucleus*) and contains the *nucleolus*, a sub-organelle surrounded by a complex of DNA and proteins called *chromatin* where ribosome components are produced. The outer nuclear membrane is perforated by *nuclear pore complexes*, that is, pores composed by different proteins regulating the traffic of substances needed and produced by the nucleus. It is through these pores that the ribosome components produced in the nucleolus pass to the cytosol where they are used.

Endoplasmatic reticulum. This has several specialised functions as lipid and protein biosynthesis performed by the molecules embedded into its membrane. It is structured in two parts: the *rough* and the *smooth* endoplasmatic reticula. The rough endoplasmatic reticulum is studded with *ribosomes* synthesising proteins and its membrane is continuous with the outer nuclear membrane.

Mitochondria. Their main function is the production of *ATP*, the cell's source of chemical energy. They are also involved in other functions such as: cell signalling, cellular differentiation, apoptosis and they have some control over the cell's cycle and growth. Mitochondria are bounded by two very specialised membranes: the *inner* and *outer* membrane. The outer membrane embeds porins (see Section 5.1) allowing the passage of ions and small polar molecules; in this way only certain kinds of the molecules present in the cytosol can enter the compartment defined by this membrane. The proteins present in the inner membrane are related to carrying specific molecules in and out and synthesising ATP.

Golgi apparatus. This mainly further processes some proteins and lipids produced by the endoplasmatic reticulum. Some of the molecules that have to be processed or that have been processed are carried into *vesicles*: small membrane compartments created by the budding of the membrane of some organelles.

Lysosomes. These are the principal sites of intracellular digestion. They contain digestive enzymes and have more acid content than the cytosol. Their membrane embeds proteins not allowing these enzymes to pass into the cytosol and proton pumps (see Section 5.1) letting protons pass from the cytosol to the lysosome.

2.3 Bibliographical notes

The biological concepts and processes considered in this monograph represent an extremely small part of the fascinating field of biology. Readers interested in knowing more about biology can refer to [1,34,51,140,163,249]. While writing the sections related to biology, in addition to the cited books, we also consulted [251].

3 Theoretical computer science: an introduction

In this chapter we survey the principal concepts in theoretical computer science necessary for the understanding of this monograph. These concepts include strings, sets, graphs, register machines, and formal grammars.

3.1 Preliminaries

A *symbol* is an abstract concept that we do not define formally. This is similar to the concepts of *point, line* and *plane* present in geometry. A *set* is a collection of symbols. We also do not give a formal definition for this concept. An *alphabet* is a nonempty finite set. Sets are defined in classical ways: for instance, indicating the symbols one by one (as $V = \{a, b, c\}$) or using a more compact way (as $A = \{i \mid 1 \le i \le 100\}$). The operations between sets are denoted in the classical way: $\cup$ union, $\cap$ intersection and $\setminus$ difference. If a is a symbol in a set V, denoted by $a \in V$, then we say that a is an *element* of V. The *empty set*, that is, the set without any symbol, is denoted by $\emptyset$. The symbol $\mathbb{N}$ denotes the set of natural numbers $\{0, 1, 2, \ldots\}$, $\mathbb{N}_+ = \mathbb{N} \setminus \{0\}$, and $\mathbb{Z} = \{-n \mid n \in \mathbb{N}_+\} \cup \mathbb{N}$. Given a nonempty finite set of numbers $A \subset \mathbb{N}, A \ne \emptyset$, the biggest element in A is denoted by $max(A)$ while the smallest is denoted by $min(A)$. Given a set V its *cardinality*, that is, the number of elements in V, is denoted with $|V|$. If $V = \{a, b, c\}$, then $|V| = |\{a, b, c\}| = 3$, while $|\emptyset| = 0$. Given a set V, $\mathcal{P}(V)$ denotes its *power set*, that is, the set of all subsets of V. If $V = \{a, b, c\}$, then $\mathcal{P}(V) = \{\emptyset, \{a\}, \{b\}, \{c\}, \{a, b\}, \{a, c\}, \{b, c\}, \{a, b, c\}\}$. If $|V| = n$, then $|\mathcal{P}(V)| = 2^n$.

A *string* is a finite sequence, that is, a finite ordered list, of symbols. Strings are defined by their sequence of symbols. For instance, $w = abc$ is a string identified by the symbol w and having a as leftmost symbol which is followed by b, which is followed by c, the rightmost symbol of w. The *empty string*, that is, the string having no symbols, is denoted by ϵ. Given a string w its *length*, that is, the number of symbols present in it, is denoted by $|w|$. If $w = abc$, then $|w| = |abc| = 3$, while $|\epsilon| = 0$. Let $A = \{a_1, \ldots, a_{|A|}\}$ be an alphabet and w a string over A. With $|w|_a$ we denote the number of occurrences of the symbol a

in w. For instance, if $w = a_1a_1a_2$, $a_1, a_2 \in A$, then $|w|_{a_1} = 2$, $|w|_{a_2} = 1$ and $|w|_{a_3} = 0$.

Given a string w a *substring* of w is any subsequence of symbols present in w, ϵ included. If $w = abc$, then substrings of w are: ϵ, a, b, c, ab, bc and abc. The *concatenation* of two strings x and y is the string w obtained by juxtaposing the symbols of x after the symbols of y. If $x = ab$ and $y = ef$, then the concatenation of x and y results in $w = abef$.

A *(formal) language* is a set of strings over a certain alphabet. The set of all finite languages is denoted by FIN. If A is an alphabet, then A^* denotes the (infinite) language of all the strings over A and is known as the *Kleene closure*, while $A^+ = A^* \setminus \{\epsilon\}$ is known as the *positive closure*. For instance, if $A = \{0\}$, then $A^* = \{\epsilon, 0, 00, 000, \ldots\}$.

A *relation* R between two sets A and B is a subset of $A \times B$, that is, its elements are pairs (a, b) with $a \in A$ and $b \in B$. Given a relation $R \subseteq A \times B$ its *inverse* (or *converse*, *transpose*) is the relation $R^{-1} \subseteq B \times A$ such that $(b, a) \in R^{-1}$ if and only if $(a, b) \in R$. A relation $R \subseteq A \times B$ is:

total (*left total*) if for each $a \in A$ there is a $b \in B$ such that $(a, b) \in R$;

surjective (*right total*) if for each $b \in B$ there is an $a \in A$ such that $(a, b) \in R$;

injective if $(a, b), (a', b) \in R$ implies $a = a'$;

bijective if R is total and for each $b \in B$ there is exactly one $a \in A$ such that $(a, b) \in R$.

A surjective relation is called a *surjection*, an injective relation is called *injection* and a bijective relation is called a *bijection*. A *function* is a relation $R \subseteq A \times B$ such that if $(a, b) \in R$, then there is no $(a, b') \in R$ with $b \neq b'$. If $R \subseteq A \times B$ is a function, then we write $R : A \to B$, and if $(a, b) \in R$, then we write $R(a) = b$.

A *vector* is an ordered k-tuple denoted by $(a_1, a_2, \ldots, a_k)$ where a_i are elements of a set A, $1 \leq i \leq k$ and k is the *dimension* of the vector. We consider also vectors with an infinite dimension. Given a vector v its dimension is denoted by $|v|$. So, for instance, $v = (4, 56, 0)$ is a vector with $|v| = |(4, 56, 0)| = 3$. In case the dimension of a vector $|v|$ is infinite, then we write $|v| = +\infty$. Given a set A and a number $k \in \mathbb{N}_+$, the set of vectors of dimension k and elements in A is denoted by A^k and $A^* = \bigcup_{i \in \mathbb{N}} A^i$. The *sum* of two vectors with the same dimension $a = (a_1, \ldots, a_k)$ and $b = (b_1, \ldots, b_k)$, $k \in \mathbb{N}_+$, is the vector $a + b = (a_1 + b_1, \ldots, a_k + b_k)$. The *product* of a number n for a vector $a = (a_1, \ldots, a_k)$ is the vector $n\,a = (na_1, \ldots, na_k)$. To indicate that all the elements of a vector a satisfy a relation we say that a satisfies the relation. So, for instance, to denote that all the elements in the vector a are bigger than zero we write $a > 0$ (0 denotes the vector with all elements 0) while $a \geq b$ denotes that $a_i \geq b_i$, $1 \leq i \leq k$.

A subset C of $\mathbb{N}^k$ is a *linear set* if there are vectors $v_0, v_1, \ldots, v_t \subseteq \mathbb{N}^k_+$ such that

$$C = \{v_0 + \sum_{i=1}^{t} p_i v_i \mid p_i \in \mathbb{N}_+,\ 1 \leq i \leq t\}.$$

The set C is *semilinear* if it is a finite union of linear sets. The class of linear sets of dimension k is denoted by LS_k, while the class of semilinear set of dimension k is denoted by SLS_k.

The empty set and every finite subset of $\mathbb{N}^k_+$ are semilinear sets. The family of semilinear sets is closed under finite union, complementation and intersection.

Let $A = \{a_1, \ldots, a_{|A|}\}$ be an alphabet and w be a string over A. We define $\Psi_A : A \to \mathbb{N}^{|A|}$ such that $\Psi_A(w) = (|w|_{a_1}, \ldots, |w|_{a_{|A|}})$. Clearly, the ordering of the symbols in A is relevant. The vector $(|w|_{a_1}, \ldots, |w|_{a_{|A|}})$ is called a *Parikh vector* and the mapping defined by the function Ψ_A is called a *Parikh mapping*. For $L \subseteq A^*$ we define $\Psi_A(L) = \{\Psi_A(w) \mid w \in L\}$ and $\Psi_A(L)$ is called the *Parikh image* of the language L.

A *multiset* (over V) is a total function $M : V \to \mathbb{N} \cup \{+\infty\}$. For $a \in V$, $M(a)$ defines the *multiplicity* of a in the multiset M. We say that an element a of a multiset M has *infinite multiplicity* if $M(a) = +\infty$.

The *support* of a multiset M is the set $supp(M) = \{a \in V \mid M(a) > 0\}$. We say that an element $a \in V$ *belongs* to a multiset M (denoted by $a \in M$) if $a \in supp(M)$. The symbol ϕ denotes the *empty multiset*, that is, the multiset whose support is the empty set $\emptyset$.

Multisets are denoted by:

multiset notation: indicating the symbols belonging to a multiset with their multiplicities; if a symbol has multiplicity one, then we only indicate the symbol;

string notation: any of the possible permutations of the elements of the multisets repeated by (or to the power of) their cardinality.

If, for instance the multiset M over $\{a, b, c, d\}$ is such that $M(a) = 3, M(b) = 1$, $M(c) = 2$ and $M(d) = 0$, then one of the multiset notations of M is $\{(a, 3), b, (c, 2)\}$, while some of the string notations of M are $aaabcc, c^2 a^3 b$ and $ba^3 cc$. In Section 3.9.1 the string notation for multisets is discussed.

The *cardinality* of a multiset M is denoted by $|M| = \sum_{a \in supp(M)} M(a)$. If, for instance, $M = \{(a, 3), b, (c, 2)\}$, then $|M| = 6$. When comparing Parikh vectors and multisets we assume a unique ordering of the symbols.

Let $M_1, M_2 : V \to \mathbb{N} \cup \{+\infty\}$ be two multisets. To say that M_1 is *included* in M_2, denoted by $M_1 \subseteq M_2$, means that $M_1(a) \leq M_2(a)$ for each $a \in V$. The *union* of M_1 and M_2 is the multiset $M_1 \cup M_2 : V \to \mathbb{N} \cup \{+\infty\}$ defined by $(M_1 \cup M_2)(a) = M_1(a) + M_2(a)$, for all $a \in V$. The *difference* $M_1 \backslash M_2$, for two multisets such that M_2 is included in M_1, is defined by $(M_1 \backslash M_2)(a) = M_1(a) - M_2(a)$ for all $a \in V$. Of course, if $M_1(a) = +\infty$ and $M_2(a)$ is finite, then $M_1(a) \backslash M_2(a) = +\infty$. The notation $M_1 \neq M_2$ means that there is a $v \in V$ such that $M_1(v) \neq M_2(v)$.

3.2 Formal grammars

Definition 3.1 *A* (Chomsky) *grammar* $G = (N, T, S, P)$ *is such that:*

N *is an alphabet whose elements are called* non-terminals;

T *is an alphabet whose elements are called* terminals *with* $N \cap T = \emptyset$;

$S \in N$, *called the* start symbol;

$P \subseteq (N \cup T)^* N (N \cup T)^* \times (N \cup T)^*$, P *finite and the elements of* P *are called* productions.

If $(\alpha, \beta) \in P$, then we write $\alpha \to \beta \in P$, indicating that α can be replaced by β in G. Let $G = (N, T, S, P)$ be a grammar and γ a string over $N \cup T$. Moreover, let $\gamma = \gamma_1 \ldots \gamma_n \gamma_{n+1} \ldots \gamma_{n+m} \gamma_{n+m+1} \ldots \gamma_{|\gamma|}$ with $\gamma_1, \ldots, \gamma_n, \gamma_{n+1}, \ldots, \gamma_{n+m}$, $\gamma_{n+m+1}, \ldots, \gamma_{|\gamma|} \in N \cup T$ and $\gamma_{n+1} \ldots \gamma_{n+m} \to \omega_1 \ldots \omega_p \in P$, then $\gamma_1 \ldots \gamma_n \gamma_{n+1}$ $\ldots \gamma_{n+m} \gamma_{n+m+1} \ldots \gamma_{|\gamma|} \Rightarrow \gamma_1 \ldots \gamma_n \omega_1 \ldots \omega_p \gamma_{n+m+1} \ldots \gamma_{|\gamma|}$ is a *derivation step* in G. A *derivation* of a grammar is a finite sequence of strings called *sentential forms* obtained via derivation steps starting from S and replacing substrings according to the productions in P. The reflexive and transitive closure of $\Rightarrow$ is denoted by $\Rightarrow^*$. The *language generated* by a grammar $G = (N, T, S, P)$, indicated by $\mathsf{L}(G)$, is given by all the strings over T which can be obtained by derivations of G. Formally: $\mathsf{L}(G) = \{\delta \in T^* \mid S \Rightarrow^* \delta\}$. Here we consider an example already present in the literature of Theoretical Computer Science.

Example 3.1 A grammar generating $\{a^n b^n \mid n \in \mathbb{N}_+\}$.

We define the grammar $G = (N, T, S, P)$ with $N = \{S\}$, $T = \{a, b\}$ and $P = \{S \to aSb, \, S \to ab\}$. One derivation of G is: $S \Rightarrow aSb \Rightarrow aaSbb \Rightarrow aaabbb$. The strings S, aSb, $aaSbb$ and $aaabbb$ are sentential forms and $aaabbb \in T^*$ belongs to the language generated by G. Section 3.9 indicates where to find the proof that $\mathsf{L}(G) = \{a^n b^n \mid n \in \mathbb{N}_+\}$. $\Diamond$

The kind of productions present in grammars classifies the language generated and the grammars themselves. Grammars $G = (N, T, S, P)$ having productions of the kind $A \to a$, $A \to aB$ and $A \to \epsilon$ with $A, B \in N$ and $a \in T$, are called *regular* grammars. Regular grammars are also known as *type-3* grammars. The set of all languages generated by type-3 grammars is denoted by REG. Each element in this set is called a *regular language*.

Grammars $G = (N, T, S, P)$ having productions of the kind $A \to \alpha$ with $A \in N$ and $\alpha \in (N \cup T)^*$ are called *context-free* or *type-2* grammars. The set of all languages generated by type-2 grammars is denoted by CF and each element of this set is called a *context-free* language.

Grammars $G = (N, T, S, P)$ having productions of the kind $\alpha A \beta \to \alpha \gamma \beta$ with $A \in N$, $\alpha, \beta \in (N \cup T)^*$ and $\gamma \in (N \cup T)^+$ are called *context sensitive* or *type-1* grammars. The set of all languages generated by type-1 grammars is denoted by CS and each element of this set is called a *context-sensitive* language.

Grammars $G = (N, T, S, P)$ having productions of the kind $\alpha \to \beta$ with $\alpha \in (N \cup T)^+$ and $\beta \in (N \cup T)^*$ are called *type-0* grammars. The set of all languages generated by type-0 grammars is denoted by RE and each element of this set is called a *recursive enumerable* language.

If a grammar $G = (N, T, S, P)$ has productions of the kind $AB \to CD$, $A \to BC$, $A \to \epsilon$ and $A \to a$ with $A, B, C, D \in N$ and $a \in T$, then it is a type-0 grammar in *Kuroda normal form*.

The classes of languages introduced until now are such that:

$$\text{FIN} \subset \text{REG} \subset \text{CF} \subset \text{CS} \subset \text{RE}$$

and this sequence of proper inclusions is known as a *Chomsky hierarchy*.

It is known that:

Theorem 3.1 *For $L \in$ CF a language over the alphabet V, $\Psi_V(L) = \text{SLS}_{|V|}$.*

Let a grammar $G = (N, T, S, P)$ be of type-τ, $0 \le \tau \le 3$. The length of the strings over T which can be obtained by derivations of G is the *set of numbers generated* by G. The respective classes of languages are denoted by: $\mathbb{N}$ REG, $\mathbb{N}$ CF, $\mathbb{N}$ CS and $\mathbb{N}$ RE. If we denote with $\mathbb{N}$ FIN the class of finite sets of numbers, then we have:

$$\mathbb{N}\,\text{FIN} \subset \mathbb{N}\,\text{REG} = \mathbb{N}\,\text{CF} \subset \mathbb{N}\,\text{CS} \subset \mathbb{N}\,\text{RE}.$$

Proposition 3.1 *For any $K \in \mathbb{N}$ REG there are finite sets K_0 and K_1 and $k \in \mathbb{N}$ such that $K = K_0 \cup \{i + jk \mid i \in K_1, \ j \in \mathbb{N}_+\}$.*

For $k \in \mathbb{N}$, $\mathbb{N}_k$ RE denotes the family of recursively enumerable sets with elements greater or equal to k, that is, $\{L \in \mathbb{N} \text{ RE} \mid \{0, \dots, k-1\} \cap L = \emptyset\}$, or equivalently, $\{k + L \mid L \in \mathbb{N} \text{ RE}\}$, where $k + L = \{k + n \mid n \in L\}$. From the point of view of computational completeness, the families $\mathbb{N}$ RE and $\mathbb{N}_k$ RE are equivalent, but here we inherit the language definition used in the literature of Membrane Computing and make this distinction.

Similarly to the universal Turing machine, there is a *universal type-0 grammar*, that is, a type-0 grammar able to simulate any other type-0 grammar.

Definition 3.2 *A* matrix grammar $G_M = (N, T, S, M)$ *is such that* N, T *and* S *are as in Definition 3.1,* M *is a finite set of* matrices $M_i, 1 \leq i \leq k$, *where each matrix is a tuple of the form* $M_i = (A_{i_1} \rightarrow \alpha_{i_1}, \ldots, A_{i_{|M_i|}} \rightarrow \alpha_{i_{|M_i|}})$ *with* $A_{i_j} \in N$, $\alpha_{i_j} \in (N \cup T)^*$ *for* $1 \leq j \leq |M_i|$.

If γ and ω are two strings over $N \cup T$ and $G_M = (N, T, S, M)$ is a matrix grammar with $M_i = (A_{i_1} \rightarrow \alpha_{i_1}, \ldots, A_{i_{|M_i|}} \rightarrow \alpha_{i_{|M_i|}}) \in M$, then we write $\gamma \Rightarrow_{M_i} \omega$, denoting a *derivation step*, if there are $\tau_i \in (N \cup T)^*$, $0 \leq i \leq |M_i|$ such that $\tau_0 = \gamma$, $\tau_{|M_i|} = \omega$ and for each $1 \leq j \leq |M_i|$ τ_j is the result of the application of $A_{i_j} \rightarrow \alpha_{i_j}$ to τ_{j-1},

A *derivation* of a matrix grammar is a finite sequence of strings called *sentential forms* obtained from S through derivations steps. Informally, each derivation step replaces a substring in a sentential form by applying the productions of a (any) matrix in sequence: initially the first production in a matrix, then the second production in the same matrix and so on until the last production in the same matrix. If in a sentential form a production in a matrix cannot be applied, then the whole matrix is not applied.

The *language generated* by a matrix grammar $G_M = (N, T, S, M)$, indicated by $\mathsf{L}(G_M)$, is given by all the strings over T which can be obtained by derivations of G_M. Formally: $\mathsf{L}(G_M) = \{\delta_k \in T^* \mid S \Rightarrow_{M_{p_1}} \delta_1 \Rightarrow_{M_{p_2}} \cdots \Rightarrow_{M_{p_k}} \delta_k, \delta_q \in (N \cup T)^*, M_{p_q} \in M$ for $1 \leq q \leq k, k \geq 1\}$.

The family of languages generated by matrix grammars is denoted by **MAT**.

Definition 3.3 *A* matrix grammar with appearance checking in Z-binary normal form $G_M = (N, T, S, M, F)$ *is such that* N, T, S *and* M *are defined as in Definition 3.2 and* F *is a set (possibly empty) of productions in matrices of* M.

The grammar G_M *is such that* $N = N_1 \cup N_2 \cup \{S, Z, \star\}$ *with these three sets mutually disjoint and the matrices in* M *of the following types:*

(i) $M_0 = (S \rightarrow S_X S_A)$ *with* $S_X \in N_1$, $S_A \in N_2$;

(ii) $M_i = (X \rightarrow Y, A \rightarrow a)$ *with* $X, Y \in N_1$, $A \in N_2$, $a \in (N \cup T)^*$, $|a| \leq 2$, $1 \leq i \leq s$;

(iii) $M_i = (X \rightarrow Y, A \rightarrow \star)$ *with* $X \in N_1$, $Y \in N_1 \cup \{Z\}$, $A \in N_2$, $s+1 \leq i \leq k-1$;

(iv) $M_k = (Z \rightarrow \epsilon)$.

There is only one matrix of type (i) and F consists of all the productions $A \to \star$ present in matrices of type (iii). Moreover, if a sentential form (see below) of G_M contains Z, then it is of the form $Z\omega$ with $\omega \in (T \cup \{\star\})^*$.

If γ and ω are two strings over $N \cup T$ and $G_M = (N, T, S, M, F)$ is a matrix grammar with appearance checking in Z-binary normal form with $M_i = (A_{i_1} \to \alpha_{i_1}, \ldots, A_{i_{|M_i|}} \to \alpha_{i_{|M_i|}}) \in M$, then we write $\gamma \Rightarrow_{M_i} \omega$, denoting a *derivation step*, if and only if there are $\tau_i \in (N \cup T)^*$, $0 \leq i \leq |M_i|$, such that $\tau_0 = \gamma$, $\tau_{|M_i|} = \omega$ and for each $1 \leq j \leq |M_i|$:

either τ_j is the result of the application of $A_{i,j} \to \alpha_{i,j}$ to τ_{j-1}; or

$A_{i,j} \to \alpha_{i,j}$ is not applicable to τ_{j-1}, $\tau_j = \tau_{j-1}$ and $A_{i,j} \to \alpha_{i,j} \in F$ (this is called *appearance checking*).

Derivations of matrix grammars with appearance checking in Z-binary normal form are similar to those of matrix grammars with the difference that if in one sentential form one production ρ in a matrix M_i cannot be applied and $\rho \in F$, then ρ is not applied and the remaining productions in M_i are applied. If instead $\rho \notin F$, then none of the productions in M_i is applied.

The *language generated* by a matrix grammar (with appearance checking in Z-binary normal form) G_M, denoted by $\mathsf{L}(G_M)$, is given by all the strings over T which can be obtained by derivations of G_M. Formally: $\mathsf{L}(G_M) = \{\delta_k \in T^* \mid S \Rightarrow_{M_{p_1}} \delta_1 \Rightarrow_{M_{p_2}} \ldots \Rightarrow_{M_{p_k}} \delta_k, \delta_q \in (N \cup T)^*, M_{p_q} \in M \text{ for } 1 \leq q \leq k, k \geq 1\}$.

The family of languages generated by matrix grammars with appearance checking (in Z-binary normal form) is denoted by $\mathsf{MAT_{ac}}$. It is known that

$$\mathsf{CF} \subset \mathsf{MAT} \subset \mathsf{MAT_{ac}} = \mathsf{RE}.$$

The family of sets of numbers generated by matrix grammars (when the length of the strings over T generated by these grammars is considered) is denoted by $\mathbb{N} \mathsf{MAT}$ and it is known that

$$\mathbb{N} \mathsf{CF} = \mathbb{N} \mathsf{MAT} \subset \mathbb{N} \mathsf{RE}.$$

3.3 Lindenmayer systems

Lindenmayer systems differ from Chomsky grammars as they work in parallel: in one derivation step all symbols in the string are replaced.

A *zero-interactions Lindenmayer system (0L system)* $R = (V, P, \alpha)$ is such that:

V is an alphabet;

$P \subseteq V \times V^*$ is a finite set of context-free productions;

$\alpha \in V^*$ is a string called an *axiom*.

The set P has to be *complete*: for each $a \in V$ there must be at least one production $a \rightarrow \beta \in P$. Given a 0L system $R = (V, P, \alpha)$ we say that γ derives ω, written by $\gamma \Rightarrow_R \omega$, if $\gamma = \gamma_1 \ldots \gamma_p$, $\omega = \omega_1 \ldots \omega_p$, where $\gamma_i \rightarrow \omega_i \in P$, $1 \leq i \leq p$.

A *tabled zero-interactions Lindenmayer system (T0L)* is a triplet $R = (V, H, \alpha)$, where V and α are defined as in the above and $H = \{H_1, \ldots, H_{|H|}\}$ is a finite set of *tables*. Each table is a complete set of context-free productions on V. Given a T0L system $R = (V, H, \alpha)$ we say that γ derives ω, written by $\gamma \Rightarrow_R \omega$, if $\gamma \Rightarrow_{R'} \omega$ for the 0L system $R' = (V, H_i, \alpha)$ and $1 \leq i \leq |H|$.

An *extended tabled zero-interactions Lindenmayer system (ET0L)* is a tuple $R = (V, H, \alpha, T)$ where $R'' = (V, H, \alpha)$ is a T0L system and $T \subseteq V$, $T \neq \emptyset$ is called a *terminal alphabet*. Given an ET0L system $R = (V, H, \alpha, T)$ we say that γ derives ω, written by $\gamma \Rightarrow_R \omega$, if $\gamma \Rightarrow_{R''} \omega$. The reflexive and transitive closure of $\Rightarrow_R$ is denoted by $\Rightarrow_R^*$.

The *language generated* by an ET0L system R is $\mathsf{L}(R) = \{\delta \in T^* \mid \alpha \Rightarrow_R^* \delta\}$.

An E0L system is one having only one table. Clearly, E0L, T0L and 0L systems are special cases of ET0L systems.

A language is said to be 0L, E0L, T0L or ET0L if there is a 0L, E0L, T0L or ET0L system, respectively, generating it. It is known that:

$$\mathsf{0L} \subset \mathsf{E0L} \subset \mathsf{T0L} \subset \mathsf{ET0L} \subset \mathsf{CS}$$

and that

$$\mathsf{CF} \subset \mathsf{E0L}.$$

Moreover:

Theorem 3.2 *For each $L \in$ ET0L there is an ET0L system R with only two tables such that $L = \mathsf{L}(R)$.*

Lemma 3.1 *For each $L \in$ ET0L there is an ET0L system $R = (V, H, \alpha, T)$, $L = \mathsf{L}(R)$, $H = (H_1, H_2)$, with only two tables and such that for each $a \in T$ if $a \rightarrow \beta \in H_1 \cup H_2$, then $\beta = a$.*

Example 3.2 A 0L system generating Fibonacci numbers.

We define the 0L system $R = (V, P, a)$ with $V = \{a, b\}$ and $P = \{a \rightarrow b, b \rightarrow ab\}$. The initial derivation steps of R are: $a \Rightarrow b \Rightarrow ab \Rightarrow bab \Rightarrow abbab \Rightarrow bababbab$. The lengths of these strings are the Fibonacci numbers.

It is possible to prove by induction that R generates all and only Fibonacci numbers. Informally, all the a's in step $n - 2$ become b's in step $n - 1$. These become ab's in step n. Since a is the axiom and no b's are generated unless they replace an a, then the statement holds for all steps. $\diamond$

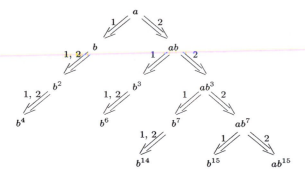

Fig. 3.1 Derivations steps for Example 3.3

Example 3.3 An ET0L system generating $\{b^{2i}, b^{j(2^i-1)} \mid i, j \in \mathbb{N}_+\}$.

We define the ET0L system $R = (\{a, b\}, H, a, \{b\})$ with:
$$H = \{H_1, H_2\};$$
$$H_1 = \{a \to b, b \to b^2\};$$
$$H_2 = \{a \to ab, b \to b^2\}.$$
The first derivation steps of R are depicted in Fig. 3.1. In this figure the number of the applied table or tables is indicated close to the arrow. $\diamond$

3.4 Register machines

The devices we define in this section use numbers to perform computations. They have *registers* (also called *counters*) each of unbounded capacity recording a natural number. Simple operations can be performed on the registers: addition of one unit and conditional subtraction of one unit. After each of these operations the machine can change state.

Formally a *register machine* with n registers ($n \in \mathbb{N}_+$), each register able to store any element in $\mathbb{N}$, is defined as $M = (S, I, s_1, s_f)$, where $S = \{s_1, \ldots, s_f\}$ is a finite set of *states*, $s_1, s_f \in S$ are respectively called the *initial* and *final* states, I is the finite set of *instructions* of the form (s, γ_i^-, v, w) or (s, γ_i^+, v) with $s, v, w \in S$, $s \neq s_f, 1 \leq i \leq n$.

A *configuration* of a register machine M with n registers is given by an element in the $(n+1)$-tuples $S \times \mathbb{N}^n$. Given two configurations $(s, val(\gamma_1), \ldots, val(\gamma_n))$, $(s', \gamma_1', \ldots, \gamma_n')$ (where $val : \{\gamma_1, \ldots, \gamma_n\} \to \mathbb{N}$ is the function returning the content of a register) we define a *computational step* as $(s, val(\gamma_1), \ldots, val(\gamma_n)) \vdash (s', \gamma_1', \ldots, \gamma_n')$ and:

if $(s, \gamma_i^-, v, w) \in I$ and $val(\gamma_i) \neq 0$, then $s' = v$, $\gamma_i' = val(\gamma_i) - 1$, $\gamma_j' = val(\gamma_j)$, $j \neq i$, $1 \leq j \leq n$.

if $val(\gamma_i) = 0$, then $s' = w$, $\gamma'_j = val(\gamma_j)$, $1 \le j \le n$.

Informally: in state s if the content of register γ_i is greater than 0, then subtract 1 from that register and change state into v, otherwise change state into w;

if $(s, \gamma_i^+, v) \in I$, then $s' = v$, $\gamma'_i = val(\gamma_i) + 1$, $\gamma'_j = val(\gamma_j)$, $j \ne i$, $1 \le j \le n$.

Informally: in state s add 1 to register γ_i and change state into v.

The reflexive and transitive closure of $\vdash$ is denoted by $\vdash^*$.

A *computation* is a sequence of computational steps of a register machine M starting from the *initial configuration* $(s_1, val(\gamma_1), 0, \ldots, 0)$. If a computation is finite, then the last configuration is called *final*. If a final configuration has s_f as state, then we say that M *halts* and it *accepts* the input $val(\gamma_1)$. For this reason γ_1 is called the *input register* and M is called an *accepting register machine*. Starting from an initial configuration $(s_1, val(\gamma_1), 0, \ldots, 0)$ a register machine M could have a finite sequence of computational steps in which the last one does not have s_f as a state. In this case we say that M *stops* and $val(\gamma_1)$ is not accepted.

The *set of numbers accepted* by M is defined as $N(M) = \{val(\gamma_1) \mid (s_1, val(\gamma_1), 0, \ldots, 0) \vdash^* (s_f, \gamma'_1, \ldots, \gamma'_n), \gamma'_1, \ldots, \gamma'_n \in \mathbb{N}\}$.

It is also possible to define *generating register machines* in which in the initial configuration some registers have a content different from 0. In this case the register machines M accepts sets of vectors defined by $W(M) = \{(val(\gamma_{i_1}), \ldots, val(\gamma_{i_k})) \mid (s_1, val(\gamma_1), \ldots, val(\gamma_n)) \vdash^* (s_f, \gamma'_1, \ldots, \gamma'_n), 1 \le i_j \le n, 1 \le j \le k, k \le n, \gamma'_1, \ldots, \gamma'_n \in \mathbb{N}\}$.

Given a register machine $M = (S, I, s_1, s_f)$ if for each state $s \in S$ at most one instruction can be applied when M is in state s, then M is *deterministic*. If, instead, for at least one state $s \in S$ more than one instruction can be applied when M is in state s, then M is *non-deterministic*.

It is also possible to describe register machines having all registers empty in the initial configuration and halting with a number stored in a specific register called the *output register*. In this case the set of numbers that can be present in the halting configuration defines the class of sets of numbers *generated* by a register machine M. This set is $N(M) = \{val(\gamma_1) \mid (s_1, 0, \ldots, 0) \vdash^* (s_f, val(\gamma_1), 0, \ldots, 0)\}$. In a similar way it is possible to define register machines having all registers empty in the initial configuration and halting with more than one register having content different from zero. In this case the register machines M generates sets of vectors defined by $W(M) = \{(val(\gamma_{i_1}), \ldots, val(\gamma_{i_k})) \mid (s_1, 0, \ldots, 0) \vdash^* (s_f, val(\gamma_1), \ldots, val(\gamma_n)), 1 \le i_j \le n, 1 \le j \le k, k \le n\}$.

The set of numbers accepted or generated by register machines is $\mathbb{N}$ RE.

It has been proved that the acceptance or generation of $\mathbb{N}$ RE can be obtained by a register machine with three registers; two registers are necessary and sufficient if one uses a specific input format (for example, 2^x instead of x). Register machines with one register can accept or generate $\mathbb{N}$ REG.

Given a register machine M' with n' registers, it is always possible to define a register machine M with $n = n' + 1$ registers such that *register 1 is only incremented*. The machine M is equivalent to M'. The instructions performed by M' on the registers $\gamma_1', \ldots, \gamma_{n'}'$ are performed by M on the registers $\gamma_2, \ldots, \gamma_{n+1}$, respectively. When the final state of M' is reached, then the content of the output register of M' is copied to γ_1.

Some authors considered register machines equipped with instructions of the kind: $(s, \gamma_i^-, v), (s, \gamma_i^{=0}, w)$ and $(s, \gamma_i^+, v), 1 \leq i \leq n$. Informally, instructions of the kind (s, γ_i^-, v) let a register machine in state s and with $val(\gamma_i) > 0$ change state into v and decrease by 1 the content of γ_i. If instead $val(\gamma_i) = 0$, then the register machine stops. Instructions of the kind $(s, \gamma_i^{=0}, w)$ let a register machine in state s and with $val(\gamma_i) = 0$ change state into w. If, instead, $val(\gamma_i) > 0$, then the register machine stops. Instructions of the kind (s, γ_i^+, v) let a register machine in state s change state into v and increase of 1 the content of γ_i.

Some authors considered register machines $M = (S, I, s_1, s_f)$ equipped also with instructions of the kind (s, v) indicating that when in state s the machine changes state into v. We do not consider this kind of operation.

One could think that a register machine equipped with instructions of the kind (s, γ_i^-, v, w) is equivalent to one equipped with pairs of instructions $(s, \gamma_i^-, v), (s, \gamma_i^{=0}, w))$. This is in part true as the indicated pair of instructions make a register machine non-deterministic (there are two instructions for the state s), while in the former case the machine could be deterministic. From a computational point of view deterministic and non-deterministic register machines are equivalent. In Section 4.3 we return to this point.

Instructions of the kind (s, γ_i^-, v, w) and $(s, \gamma_i^{=0}, w)$ are also called *test on 0* or *0-test*. This is because they allow the machine to detect if a register stores zero and perform some operations as a consequence of this. The name given to instructions of these kinds is misleading as these instructions perform more than just a test.

Reversal-bounded register machines are register machines for which the maximum numbers of alternations between increasing and decreasing the content of the registers is fixed. A register which is not reversal-bounded is called *free*.

Linear-bounded register machines are register machines whose sum of the contents of its registers is a linear function of the sum of the contents of the registers in the initial configuration. The computational equivalence of deterministic and non-deterministic linear-bounded register machines is a long-standing open problem in computational complexity theory.

Partially blind register machines are defined as register machines without test on zero. The only allowed operations are (s, γ^+, v) and (s, γ^-, v) where γ

is a register. In case the machine tries to subtract from a register having value zero it stops. They are strictly less powerful from a computational point of view than register machines.

Restricted register machines are defined as register machines restricted in their operations: they can increase the value of a register, say β, only if they decrease the value of another register, say γ at the same time.

So, restricted register machines have only one kind of instruction: $(s, \gamma^-, \beta^+, v, w)$ with s, v, w states and γ, β different registers of the restricted register machine. If when in state s the content of register γ can be decreased by 1, then that of register β is increased by 1 and the machine goes into state v, otherwise no operation is performed on the registers and the machine goes into state w. Two registers γ and β are *connected* if there is an instruction in which both registers are present.

Here is a result that we will use in Chapter 8:

Theorem 3.3 *Restricted register machines with $n+1$ registers are more powerful from a computational point of view than those with n registers.*

A consequence of this theorem is that an infinite hierarchy is induced, by means of the number of registers, among families of computed sets of numbers.

Register machines with input tape $M = (W, S, R, s_0, s_f)$ are equivalent to register machines but they are equipped with a read-only tape subdivided into locations, each location storing an element of the alphabet W. In the initial configuration the input tape contains a string $x = x_1 \ldots x_{|x|}$, $x_1, \ldots, x_{|x|} \in W$ and the reading head, able to scan one location per time, is on x_1. In addition to the instructions discussed before, register machines with input tape can use instructions of the kind

(s, a, v): if the machine is in state s and the symbol a is read from the tape, then the state is changed into v and the reading head is moved to the right of one location.

A configuration of a register machine with n registers and input tape is defined as $(s, x_i, val(\gamma_1), \ldots, val(\gamma_n))$ where $s \in S$ is the state of the register machine and $x_i \in W$, $1 \leq i \leq |x|$, indicates the symbol read by the reading head from the string x. In the initial configuration $s = s_0$ and $i = 1$.

A computational step $(s, x_i, val(\gamma_1), \ldots, val(\gamma_n)) \vdash (s', x_{i'}, val(\gamma_1'), \ldots, val(\gamma_n'))$ is defined as before with the addition of:

if $(s, a, v) \in I$, and $x_i = a$, then $s' = v$, $i' = i + 1$ and $\gamma_i' = \gamma_i$, $1 \leq i \leq n$.

A computation is defined as before. The set of numbers accepted by register machines with input tape is defined as $N(M) = \{x \in W^* \mid (s_0, x_1, 0, \ldots, 0) \vdash^* (s_f, x_i, val(\gamma_1), \ldots, val(\gamma_n)), 1 \leq i \leq |x|\}$.

Other models of register machines will be introduced in Chapter 5.

3.5 Unique-sum sets

A few proofs in this monograph require the following mathematical concepts.

Definition 3.4 *Let* $U = \{u_1, \ldots, u_p\}$ *be a set of distinct natural numbers and* $\sigma_U = \sum_{i=1}^{p} u_i$ *the sum of the elements of* U. *The set* U *is said to be a unique-sum set if the equation* $\sum_{i=1}^{p} c_i u_i = \sigma_U, c_i \in \mathbb{N}$, *has only the solutions* $c_i = 1$, $1 \leq i \leq p$.

An example of a unique-sum set is $U' = \{4, 6, 7\}$ as $4 + 6 + 7 = 17$ and 17 cannot be obtained with any other linear combination of 4, 6 and 7. The set $U'' = \{4, 5, 6\}$ is not a unique-sum set as $4 + 5 + 6 = 15 = 5 + 5 + 5$.

It should be clear that any subset of a unique-sum set is a unique-sum set, too. In particular none of the elements of a unique-sum set can be obtained as a linear combination of the remaining elements in the set.

Proposition 3.2 *The sets* $U_p = \cup_{m=1}^{p}\{2^p - 2^{p-m}\}$ *with* $p \in \mathbb{N}_+$, *are unique-sum sets.*

The sum of the elements of the sets in this family is $\sigma_{U_p} = (p-1)2^p + 1$. The first sets in this family are:

$$U_1 = \{1\};$$
$$U_2 = \{2, 3\};$$
$$U_3 = \{4, 6, 7\};$$
$$U_4 = \{8, 12, 14, 15\};$$
$$U_5 = \{16, 24, 28, 30, 31\};$$
$$U_6 = \{32, 48, 56, 60, 62, 63\}.$$

It is known that:

Theorem 3.4 *The family of sets indicated in Proposition 3.2 contains the unique-sum sets having minimal sum in function of their number of elements.*

3.6 Graphs

A *graph* $\mu = (Q, E)$ is such that:

Q is a set of *vertices*;

$E \subseteq Q \times Q$ is a set of unordered pairs of vertices called *arcs* and denoted by (q_i, q_j), $q_i, q_j \in Q$, $q_i \neq q_j$. For each pair of vertices there can be at most one arc.

As arcs are unordered pairs of vertices, then (q_i, q_j) and (q_j, q_i) denote exactly the same arc in a graph. Arcs are depicted by lines; see, for instance, Fig. 3.2(a).

A *path* from q_1 to q_k of a graph $\mu = (Q, E)$ is a vector $p = (q_1, q_2, \ldots, q_k)$ with $q_i \in Q$, $1 \leq i \leq k$, such that $(q_i, q_{i+1}) \in E$ for $1 \leq i \leq k - 1$. In this case we say that the path p *traverses* the arcs $(q_i, q_{i+1}) \in E$ for $1 \leq i \leq k - 1$. A path is called *simple* if no arc is traversed more than once. The dimension of a path is called the *length* of the path. A graph $\mu = (Q, E)$ is said to be *connected* if for each $q_1, q_2 \in Q$, $q_1 \neq q_2$, there is a path from q_1 to q_2. In this monograph we only consider connected graphs.

A *tree* is a graph μ such that the removal of any arc makes μ not connected. Formally, let $\mu = (Q, E)$ be a connected graph. If for each arc $(q_i, q_j) \in E$ the graph $\mu' = (Q, E \setminus \{(q_i, q_j)\})$ is not connected, then μ is a tree. In a tree all paths from any vertex to itself are not simple.

A tree is said to be *rooted* if one vertex is called the *root*. Let $\mu = (Q, E)$ be a rooted tree with $Q = \{0, 1, \ldots, m\}$ and let 0 be the root of μ. It should be clear that for each $q \in Q \setminus \{0\}$ there is only one simple path from the root to q; we denote such a path by $path_{0,q}$. The *depth* of μ is $max(\{path_{0,q} \mid q \in Q \setminus \{0\}\})$. Let $path_{q_1,q_k} = (q_1, q_2, \ldots, q_k)$ with $q_1 = 0$ be a path in μ. We say that q_i is the *parent* of q_{i+1} and that q_{i+1} is the *child* of q_i for $1 \leq i \leq k - 1$. If for $q \in Q$ there is no simple path $path_{0,q}$ in μ such that q is the parent of another vertex, then q is said to be a *leaf*. We also define the function $parent : Q \setminus \{0\} \rightarrow Q$, returning the parent of a vertex, and the relation $child \subseteq Q \times Q$ such that $(q_i, q_j) \in child$ if $parent(q_j) = q_i$.

It is important to notice that the definitions of child, parent, leaf, and the function $parent(\ )$ in a rooted tree depend on the choice of the root in the tree and that a vertex can have more than one child but at most one parent.

A *directed graph* (or *digraph*) $\mu = (Q, E)$ is such that:

Q is a set of *vertices*;

$E \subseteq Q \times Q$ is a set of ordered pairs of vertices called *edges* and denoted by (q_i, q_j), $q_i, q_j \in Q$, $q_i \neq q_j$. For each pair of vertices there can be at most one edge.

As edges are ordered pairs of vertices, then (q_i, q_j) and (q_j, q_i) denote two different edges in a directed graph. Edges are depicted by arrows. The definitions of *path*, *traverse*, *path*, *length* for a directed graph can be easily derived from those given for graphs. Let $\mu = (Q, E)$ be a directed graph. If the graph $\mu' = (Q, E')$ with $E' = \{(q_i, q_j) \mid (q_i, q_j) \in E\}$ is connected, then μ is *connected*. In this monograph we only consider connected directed graphs. A directed graph $\mu = (Q, E)$ is said to be *asymmetric* if for each of two different vertices $q_i, q_j \in Q$ either $(q_i, q_j) \in E$ or $(q_j, q_i) \in E$. A directed graph having more than one edge for a pair of vertices is called a *directed multigraph*. The definition of an *asymmetric directed multigraph* easily follows from the above.

A *directed tree* is a directed graph μ such that the removal of any edge makes μ not connected. A *rooted directed tree* $\mu(Q, E)$ is a directed tree having one vertex called the *root* and such that there is only one path from the root to any other vertex in Q. The definitions of *child*, *parent*, *leaf* and the function *parent*( ) for directed rooted trees can be easily derived from those given for rooted trees.

An *edge-labelled directed graph* $\mu = (Q, E, label)$ is such that:

Q and E are defined as for directed graphs;

label $: E \times E \rightarrow L$ is a *labelling function* where L is a set of *labels*.

The function *label*( ) returns the label associated to each edge in μ.

An *edge-labelled directed multigraph* $\mu = (Q, E, label)$ is such that:

Q and E are defined as for directed graphs;

label $\subseteq E \times E \times L$ is a *labelling relation* where L is a set of *labels*.

For notational convenience the edges of edge-labelled direct (multi)graphs are denoted by (q_i, q_j, l) if $(q_i, q_j) \in E$ and $label(q_i, q_j) = l$.

An *vertex-labelled directed graph* $\mu = (Q, E, label)$ is such that:

Q and E are defined as for directed graphs;

label $: E \times E \rightarrow L$ is a *labelling function* where L is a set of *labels*.

The function *label*( ) returns the label associated to each vertex in μ.

A *vertex-labelled rooted tree* $\mu = (Q, E, label)$ is such that:

Q and E are defined as for directed graphs and one vertex in Q is called the *root*;

label $: E \times E \rightarrow L$ is a *labelling function* where L is a set of *labels*.

3.7 Topology and operational modes of membrane systems

One of the hallmarks of membrane systems is their topological structure. This structure can be defined with graphs such that the vertices in a graph are associated in a unique way with the compartments present in a membrane system. Vertex i in a graph μ corresponds to compartment i present in the membrane systems having μ as underlying topological structure. For this reason in this monograph the terms vertex and compartment are used as synonymous when they refer to a membrane system.

The topological structure of the compartments present in the membrane systems considered in this monograph can be of two kinds:

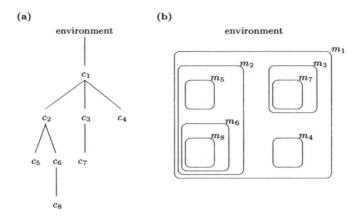

Fig. 3.2 **(a)** A cell-tree and **(b)** its compartment representation

cell-tree: a rooted tree such that the root has only one child;

cell-graph: an (edge-labelled/vertex/labelled/asymmetric/directed/multi) graph different from a cell-tree.

A cell-tree is the representation of the membrane compartments present in a cell located in an environment. In a cell-tree if i and j are two compartments and i is the parent of j, then we say that compartment i *contains* j or that j is *present in* compartment i. A cell-graph can be regarded as the network composed by different cells. Some models of membrane systems consider graphs of restricted forms.

The *degree* of a cell-graph or an asymmetric cell-graph is the number of vertices present in it, while it is the number of compartments (that is, the number of vertices minus one) in a cell-tree.

A cell-tree of degree 8 is depicted in Fig. 3.2(a) while its *membrane represen-tation* is depicted in Fig. 3.2(b). In this figure membrane m_i defines compartment c_i, $1 \leq i \leq 8$, moreover, in the membrane representation membranes defining compartments are depicted with the exception of the one defining the environ-ment. Some of the vertices in a cell-tree have specific names:

environment: the root of the tree;

skin (or external) compartment: the child of the root;

composite (or not elementary) compartment: any compartment different from the root which is not a leaf;

elementary compartment: a leaf.

In Fig. 3.2 c_1 is the external compartment; c_2, c_3 and c_6 are composite com-partments; and c_4, c_5, c_7 and c_8 are elementary compartments. The depth of the cell-tree depicted in Fig. 3.2(a) is 4 and the vertices defining it are: c_1, c_2, c_6 and c_8.

The models of membrane systems considered in this monograph operate in the following modes:

maximal parallelism: in each configuration a multiset of rules cannot be applied if there is a strictly larger multiset of rules that can be applied;

maximal strategy: in each configuration at most one rule per compartment is applied;

with priorities: priorities are assigned to rules and in each configuration a rule is applied only if no rule with a higher priority can be applied;

asynchronous: in each configuration different from the final one at least one rule is applied.

Maximal parallelism abstracts the fact that in some (bio)chemical systems all reactions occur if one waits long enough.

Similar ways to operate applied to P/T systems are studied in Section 4.2 (where the term 'sequential' instead of 'asynchronous' is used).

3.8 About simulations

The formal systems considered in this monograph are mainly membrane systems, Petri nets, register machines and formal grammars. In general a *configuration* of a formal system S can be regarded as a *snapshot* of (the relevant elements defining) S while it computes. In the following we consider that the set of configurations of S is an infinite set (the result related to the finite case can be easily deduced by the ones we indicate). If an initial configuration is given, then S can pass from one configuration to another (depending on the rules of S defining this passage). This process is called *computation* and the passage from one configuration to another is called a *transition*. A transition occurs because some operations (rules, instructions, transitions, etc.) are applied. If a computation is finite, then the last configuration is called *final*. We assume that final configurations have certain properties that make them different from the other configurations and there is no transition from a final configuration to any other configuration. If a final configuration meets certain criteria, then we say that the system *halts*, otherwise we say that the system *stops*.

In the following we formalise the simulation between formal systems.

Let S and S' be two formal systems with O and O' their respective sets of operations and $\mathbb{C} = \{c_1, c_2, \ldots\}$ and $\mathbb{C}' = \{c'_1, c'_2, \ldots\}$ their respective sets of configurations. We denote by $\overset{\sigma}{\Rightarrow}$ ($\overset{\sigma'}{\Rightarrow}$), σ a multiset over O (σ' multiset over O'), the transition from one configuration to another in a computation of S (S')

according to the application of the operations in σ (σ'). By $\overset{\sigma_1,\ldots,\sigma_n}{\Rightarrow}{}^+$ ($\overset{\sigma'_1,\ldots,\sigma'_n}{\blacktriangleright}{}^+$) we denote nonempty sequences of transitions from one configuration to another in a computation of S (S') according to the application of the operations in $\sigma_1,\ldots,\sigma_n$ $(\sigma'_1,\ldots,\sigma'_n)$ in sequence. So, for instance, if $c_1 \overset{\sigma_1}{\Rightarrow} c_2 \overset{\sigma_2}{\Rightarrow} c_3$, then we can write $c_1 \overset{\sigma_1,\sigma_2}{\Rightarrow}{}^+ c_3$.

It should be clear that, depending on the operational mode of S, the multiset σ can be a multiset of a specific kind. For instance, σ can be such that it returns at most 1.

Definition 3.5 *Let S and S' be two formal systems with O and O' their respective sets of operations, $\mathbb{C}$ and $\mathbb{C}'$ their respective sets of configurations and c_{init} and c'_{init} their respective initial configurations.*

We say that S and S' are $\alpha\beta$ strong equivalent (or that S $\alpha\beta$ strongly simulates S') if there are two relations $\alpha \subseteq \mathbb{C} \times \mathbb{C}'$ and $\beta \subseteq O \times O'$ such that:

(i) $(c_{\text{init}}, c'_{\text{init}}) \in \alpha$;

(ii) *for all $c_1, c_2 \in \mathbb{C}$, $c'_1 \in \mathbb{C}'$ and $\sigma \in O$: if $c_1 \overset{\sigma}{\Rightarrow} c_2$ and $(c_1, c'_1) \in \alpha$, then there is $c'_2 \in \mathbb{C}'$ such that $c'_1 \overset{\sigma'}{\Rightarrow} c'_2$ with $(c_2, c'_2) \in \alpha$ and $(\sigma, \sigma') \in \beta$;*

(iii) *for all $c'_1, c'_2 \in \mathbb{C}'$, $c_1 \in \mathbb{C}$ and $\sigma' \in O$: if $c'_1 \overset{\sigma'}{\blacktriangleright} c'_2$ and $(c_1, c'_1) \in \alpha$, then there is $c_2 \in \mathbb{C}$ such that $c_1 \overset{\sigma}{\blacktriangleright} c_2$ with $(c_2, c'_2) \in \alpha$ and $(\sigma, \sigma') \in \beta$.*

If α and β are bijections then we say that S strongly bisimulates S' according to α and β.

It should be clear that if S strongly bisimulates S' according to α and β, then S' strongly bisimulates S according to α and β, too.

Definition 3.6 *Let S and S' be two formal systems with O and O' their respective sets of operations, $\mathbb{C}$ and $\mathbb{C}'$ their respective sets of configurations and c_{init} and c'_{init} their respective initial configurations.*

We say that S and S' are $\alpha\beta$ weak equivalent (or that S $\alpha\beta$ weakly simulates S') if there are two relations $\alpha \subseteq \mathbb{C} \times \mathbb{C}'$ and $\beta \subseteq O^n \times O'^m$, $n, m \in \mathbb{N}_+$, such that:

(i) $(c_{\text{init}}, c'_{\text{init}}) \in \alpha$;

(ii) *for all $c_1, c_2 \in \mathbb{C}$, $c'_1 \in \mathbb{C}'$ and $\sigma \in O^n$: if $c_1 \overset{\sigma}{\Rightarrow}{}^+ c_2$ and $(c_1, c'_1) \in \alpha$, then there is $c'_2 \in \mathbb{C}'$ such that $c'_1 \overset{\sigma}{\blacktriangleright}{}^+ c'_2$ with $(c_2, c'_2) \in \alpha$ and $(\sigma, \sigma') \in \beta$;*

(iii) *for all $c'_1, c'_2 \in \mathbb{C}'$, $c_1 \in \mathbb{C}$ and $\sigma' \in O^n$: if $c'_1 \overset{\sigma}{\blacktriangleright}{}^+ c'_2$ and $(c_1, c'_1) \in \alpha$, then there is $c_2 \in \mathbb{C}$ such that $c_1 \overset{\sigma}{\Rightarrow}{}^+ c_2$ with $(c_2, c'_2) \in \alpha$ and $(\sigma, \sigma') \in \beta$.*

If α and β are bijection, then we say that S weakly bisimulates S' according to α and β.

If S $\alpha\beta$ (strongly or weakly) simulates S', then S is called the *simulating* system while S' is called the *simulated* system. It should be clear that if S strongly simulates S', then S weakly simulates S', too.

It is important to stress that in the present monograph the accepted or generated languages of simulations are mainly related to the configurations, not to the labels associated to the operations.

Several of the proofs we give do not contain the details of the simulation relations α and β. This is because the definition of these relations would be tedious and rather complex. Instead, we prefer to show how a simulating system can copy the dynamics of a simulated system. In these cases the term 'simulate' refers to weak simulation.

3.9 Bibliographical notes

As already pointed out, the concepts of theoretical computer science introduced in this chapter are those necessary to understand this monograph. Readers interested in knowing more about theoretical computer science can refer to [182], [113] (where the proof that the grammar G given in Example 3.1 generates $L(G) = \{a^n b^n \mid n \in \mathbb{N}_+\}$ can be found), [232] (where the construction of a universal type-0 grammar can be found) and [158].

We did not succeed in determining who defined register machines for the first time. For this we refer the reader to [251] under the entry 'register machine'. Nonrewriting Turing machines were studied in [181] and then reconsidered in [182] under the name of *program machines*. These machines and their variants have been studied under different names: *(multi)counter machines* in [95], *multipushdown machines* in [23], *register machines* in [138], and *counter automata* in [109]. We decided to use the name 'register machines' as it is currently more commonly used.

In [182] the results we indicated related to the computation power of register machine were proved by M.L. Minsky. From [138] we learn that in [25] (in Russian) it is proved that the use of a specific input format is necessary for a register machine with two registers to be computationally complete.

Partially blind register machines have been introduced and studied in [95] (there called *partially blind multicounter machines*). Restricted register machines were introduced in [117] (there called *restricted counter automata*). Lemma 3.1 was proved in [2] while Theorem 3.3 was proved in [117].

Unique-sum sets were studied in [84].

Readers interested in the operational modes of membrane systems can refer to [76].

Our definitions of *strong bisimulation* and *weak bisimulation* are different than the ones in [179, 180]. Similar definitions but specific for EN systems can be found in [231].

3.9.1 About the notation

In this monograph some new notation has been adopted or only some of several different notations have been used. Similar sections in each chapter highlight and motivate the chosen representations (introduced in the corresponding chapter). In general, these choices aim to give a uniform, consistent and clear notation in this monograph not too far from what readers may be accustomed to.

The string notation for multisets, common in Membrane Computing, may seem inappropriate. This is because symbols in strings are ordered while there is no order between the elements in a multiset. This notation is just a denotational convenience: no assumption on the order of the elements of multisets is ever made.

We use the terms *cell-tree* and *cell-graph* instead of *cell-like membrane system* and *tissue-like membrane system* (common in Membrane Computing), respectively, because we want to emphasise the underlying topological structure of membrane systems (tree and graph) in a more direct way. Moreover, not all membrane systems having a cell-graph as underlying topology have been inspired by (biological) tissues (see, for instance, Chapter 8).

The symbol $\star$ is used throughout this monograph to represent what in the literature of Membrane Computing is represented by *trap*, $\#$ (see, for instance, [8]) or by other symbols. As we wanted to use a symbol not otherwise adopted (the symbol $\#$ is more commonly used as is done in Chapter 9), we decided to introduce $\star$.

4 Petri nets

Petri nets originated as mathematical representations of concurrent systems. They were introduced by C. A. Petri and since then they went through theoretical and applied developments that saw them significantly advancing the fields of parallel and distributed systems. Moreover, they helped define the modern studies of complex systems (including biological systems) and workflow management. The most attractive feature of Petri nets is their simple definition capturing many of the basic notions and issues of concurrent systems.

In this chapter we consider two models of Petri nets: elementary net systems and place/transition systems. We study these systems as computing devices and we link some components present in them to their computational power. We prove that the combination of different components leads to a hierarchy of computational processes.

The results presented in this chapter facilitate the study of the computational power of abstract devices. In the following chapters we see how these results can be applied to some of the models of membrane systems we consider.

4.1 Basic definitions

Definition 4.1 *An* elementary net system *(or* EN system*) is a tuple* $N = (P, T, F, C_{\text{in}})$, *where:*

(i) (P, T, F) *is a* net, *that is:*

    1. *P and T are sets with* $P \cap T = \emptyset$;

    2. $F \subseteq (P \times T) \cup (T \times P)$;

    3. *for every* $t \in T$ *there exist* $p, q \in P$ *such that* $(p, t), (t, q) \in F$;

    4. *for every* $t \in T$ *and* $p, q, \in P$, *if* $(p, t), (t, q) \in F$, *then* $p \neq q$;

(ii) $C_{\text{in}} \subseteq P$ *is the* initial configuration *(or* initial marking*).*

Elements of P are called *places* (graphically represented with circles), elements of T are called *transitions* (graphically represented with full rectangles), and elements of $X = P \cup T$ are called *elements (of N)*. We allow P and T to have an infinite number of elements. In EN systems the presence of an unbounded number of tokens or an infinite number of transitions is explicitly mentioned, while the presence of a finite number of tokens or transitions is implicitly assumed.

F is called the *flow relation* and its elements are graphically represented by directed arrows connecting places and transitions.

We consider *connected* nets except when otherwise stated.

Definition 4.2 *A net (P, T, F) is* connected *if, when any element of the flow relation is replaced by a bi-directional connection, for each $x_1, x_2 \in X$ there is a path from x_1 to x_2.*

A set $C \subseteq P$ is a *configuration*; graphically, a configuration is represented by placing a 'token' (that is, a bullet •) in every circle corresponding to a place in C. In the literature of EN systems what we indicated as being a configuration is normally called *marking*. Anyhow, in this monograph we prefer to use the term *configuration* for this concept. Given an EN system $N = (P, T, F, C_{in})$, (P, T, F) is the *underlying net* of N. A net is depicted in Fig. 4.1.

For each $x \in X$, $^\bullet x = \{y \in X \mid (y, x) \in F\}$ is the *input set* of x, while $x^\bullet = \{y \in X \mid (x, y) \in F\}$ is the *output set* of x. It is possible that in a net (P, T, F) if either $x_1, x_2 \in P$ or $x_1, x_2 \in T$, $x_1 \neq x_2$, then $^\bullet x_1 = {}^\bullet x_2$ or $x_1^\bullet = x_2^\bullet$.

Let $N = (P, T, F, C_{in})$ be an EN system, $t \in T$ and $C, D \subseteq P$. The transition t can be *fired* in C if $^\bullet t \subseteq C$ and $t^\bullet \cap C = \emptyset$. The transition t fires from C to D if t fires in C and $D = (C \backslash {}^\bullet t) \cup t^\bullet$, this is denoted by $C[t\rangle D$. If for $C_1, C_{n+1} \subseteq P$, $t_i \in T$, $1 \leq i \leq n$, $n \in \mathbb{N}_+$, $C_j[t_j\rangle C_{j+1}$ with $1 \leq j \leq n$, $x = t_1 \ldots t_n$, then we say that x fires in C_1 and we write $C_1[x\rangle C_{n+1}$.

A configuration in which no transition can fire is called *final*.

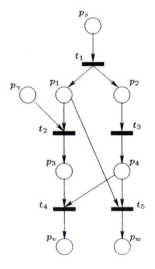

Fig. 4.1 A net. © With kind permission of Springer Science and Business Media [81]

Given an EN system $N = (P, T, F, C_{\text{in}})$, $x \in T^*$ is a *firing sequence of N* if x fires in C_{in}. The set of all firing sequences of N is denoted by $FS(N)$. A configuration $C \subseteq P$ is said to be a *reachable configuration of N* if there exist an $x \in FS(N)$ with $C_{\text{in}}[x\rangle C$. The set of all reachable configurations of N is denoted by $\mathbb{C}_N$.

Given an EN system $N = (P, T, F, C_{\text{in}})$ we define the *sequential configuration graph of N*, $SCG(N) = (\mathbb{C}_N, E, label)$, an edge-labelled directed multigraph, such that:

$E = \{(C, D) \mid C, D \in \mathbb{C}_N, \exists t \in T, C[t\rangle D\};$

$label : \mathbb{C}_N \times \mathbb{C}_N \rightarrow T$ such that if $(C, D) \in E$, $\exists t \in T$ and $C[t\rangle D$, then
  $label(C, D) = t$.

Let $N = (P, T, F, C_{\text{in}})$ be an EN system, $C, D \subseteq P$, $U \subseteq T$, $U \neq \emptyset$, $|U| \in \mathbb{N}_+$. The set U is a *step from C to D*, denoted by $C[U\rangle D$, if:

(i) for each two distinct $t_1, t_2 \in U, ({}^\bullet t_1 \cup t_1^\bullet) \cap ({}^\bullet t_2 \cup t_2^\bullet) = \emptyset$;

(ii) ${}^\bullet U \subseteq C$ and $U^\bullet \cap C = \emptyset$ (where ${}^\bullet U = \bigcup_{t \in U} {}^\bullet t$ and $U^\bullet = \bigcup_{t \in U} t^\bullet$);

(iii) $D = (C \backslash {}^\bullet U) \cup U^\bullet$.

If a step $U \subseteq T$ is such that $|U| \geq 2$, then U is called *concurrent step*. It is important to notice that even if in an EN system the number of transitions can be infinite, as steps have finite cardinality, then the number of tokens in the places is always finite.

Given an EN system $N = (P, T, F, C_{\text{in}})$ we define the *configuration graph of N*, $CG(N) = (\mathbb{C}_N, E, label)$, an edge-labelled directed multigraph, such that:

$E = \{(C, D) \mid C, D \in \mathbb{C}_N, \exists U \subseteq T,\ U \text{ a step },\ C[U\rangle D\};$

$label : \mathbb{C}_N \times \mathbb{C}_N \rightarrow \mathcal{P}(T) \setminus \{\emptyset\}$ such that if $(C, D) \in E$, $\exists U \subseteq T$ and $C[U\rangle D$,
  then $label(C, D) = U$.

Let $N = (P, T, F, C_{\text{in}})$ be an EN system and $U \subseteq T$ a step. The step U is a *maximal strategy step* if there are $C, D, D' \in \mathbb{C}_N$, $C[U\rangle D$ and there is no $U' \subseteq T$ such that $U \subset U'$ with $C[U'\rangle D'$. We define the *maximal strategy configuration graph of N*, $MSCG(N) = (\mathbb{C}_N, E, label)$, an edge-labelled directed multigraph, such that:

$E = \{(C, D) \mid \exists U \subseteq T,\ C[U\rangle D \text{ and } U \text{ a maximal strategy step}\};$

$label : \mathbb{C}_N \times \mathbb{C}_N \rightarrow \mathcal{P}(T) \setminus \{\emptyset\}$ such that if $(C, D) \in E$, $\exists U \subseteq T$, $C[U\rangle T$ with
  U a maximal strategy step, then $label(C, D) = U$.

The initial configurations of the configuration graphs (of any kind) considered in this monograph are pointed to by a wavy arrow $\leadsto$.

In Fig. 4.2 a *SCG*, a *CG* and a *MSCG* for the net depicted in Fig. 4.1 are depicted.

We are also going to consider *EN systems with priorities*. In this case the definition of the EN system includes the priority function $Pri : T \to \mathbb{N}_+$. Let $N = (P, T, F, C_{\text{in}}, Pri)$ be an EN system with priorities and let $C, D, D' \in \mathbb{C}_N, U \subseteq T$. The step U can occur in N only if U is a maximal strategy step and there is no maximal strategy step U', $U' \neq U$, $C[U'\rangle D'$ such that there exists $t' \in U'$, $Pri(t') > Pri(t)$ for each $t \in U$. In this case the step U is a called *priority step*. The *priority configuration graph of N* is only defined for EN systems with priorities. Given such a system $N = (P, T, F, C_{\text{in}}, Pri)$ we define the *priority configuration graph of N*, $PrCG(N) = (\mathbb{C}_N, E, label)$, an edge-labelled directed multigraph, such that:

$E = \{(C, D) \mid C, D \in \mathbb{C}_N, \exists U \subseteq T, \text{ and } U \text{ is a priority step}\};$

$label : \mathbb{C}_N \times \mathbb{C}_N \to \mathcal{P}(T) \setminus \{\emptyset\}$ such that if $(C, D) \in E$, $\exists U \subseteq T$, $C[U\rangle T$ and U is a priority step, then $label(C, D) = U$.

If we consider the net depicted in Fig. 4.1 underlying an EN system with priorities having $C_{\text{in}} = \{p_s, p_\gamma\}$, $Pri(t_2) = 2$ and $Pri(t_1) = Pri(t_3) = Pri(t_4) = Pri(t_5) = 1$, then Fig. 4.2(c) depicts a priority configuration graph for this net.

EN systems with inhibitor arcs $N = (P, T, F, I, C_{\text{in}})$ add to Definition 4.1 the following:

     5. $I \subseteq P \times T$.

An inhibitor arc is depicted with a line from a place to a transition ending with an empty circle (see Fig. 4.3) and it affects the dynamics of a net such that if $(p, t) \in I$, $C \subseteq P$ and $p \in C$, then t cannot fire in C.

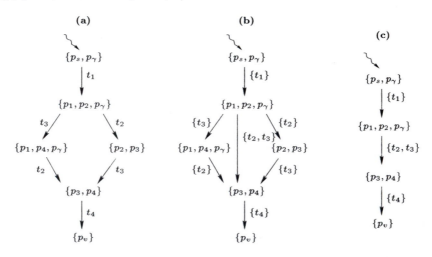

Fig. 4.2 (a) a SCG, (b) a CG and (c) a $MSCG = PriCG$ for the net depicted in Fig. 4.1

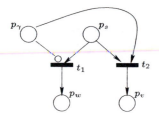

Fig. 4.3 Net with inhibitor arc

Formally, let $N = (P, T, F, I, C_{in})$ be an EN system with inhibitor arcs, $t \in T$ and $C, D \subseteq P$. The transition t can be *fired* in C if

$^\bullet t \subseteq C$;

$(p, t) \notin I$ for each $p \in C$;

$t^\bullet \cap C = \emptyset$.

We assume that the firing of transitions is instantaneous, that is, if transition t fires in C, then all the tokens in $^\bullet t$ are removed at the same time and then all the the tokens in $t^\bullet$ are added at the same time. In this way the definitions of *firing sequence*, *reachable configuration* and *SCG* can be extended to EN systems with inhibitor arcs. The definition of *step* requires the addition of

(iv)   there is no $p \in C$ and $t \in U$ such that $(p, t) \in I$.

In this way the definition of *concurrent step* and *configuration graph* can also be extended to EN systems with inhibitor arcs.

Definition 4.3 *A place/transition system (P/T system) is a tuple $N = (P, T, F, W, C_{in})$, where:*

(i) (P, T, F) is a net:

    1. *P and T are sets with $P \cap T = \emptyset$;*

    2. *$F \subseteq (P \times T) \cup (T \times P)$;*

    3. *for every $t \in T$ there exist $p, q \in P$ such that $(p, t), (t, q) \in F$;*

(ii) $W : F \to \mathbb{N}_+$ is a weight function;

(iii) $K : P \to \mathbb{N}_+ \cup \{+\infty\}$ is a capacity function;

(iv) $C_{in} : P \to \mathbb{N}$ is the initial configuration.

Notice that the definition of a net for a P/T system differs from that given for an EN system in the lack of item 4. We allow P and T to have an infinite number of elements. In P/T systems the presence of an unbounded number of tokens or an infinite number of transitions is explicitly mentioned, while the presence of a finite number of tokens or transitions is implicitly assumed. We only consider P/T systems such that the capacity function always returns $+\infty$. For this reason we omit the indication of K from the definitions of P/T systems.

A *configuration* of a P/T system is a function from P to $\mathbb{N}$. The dynamic behaviour of a P/T system is analogous to that of an EN system, but considering the weight function (and the capacity function).

When a net underlying a P/T system is depicted the value of the weight function is indicated by a number placed close to a directed arrow; if no number is indicated it is assumed that the weight function returns 1. With only a few exceptions in this monograph we only consider P/T systems such that $W(f) = 1$ for $f \notin F$.

Let $N = (P, T, F, W, C_{in})$ be a P/T system, $t \in T$ and let C be a configuration of N. The transition t can be *fired* in C if $W(p, t) \leq C(p)$ for each $p \in {}^\bullet t$. The transition t fires from C to D if t fires in C and for all $p \in P$:

$D(p) = C(p) - W(p, t)$ if $p \in {}^\bullet t$;

$D(p) = C(p) + W(t, p)$ if $p \in t^\bullet$;

$D(p) = C(p)$ otherwise.

This is denoted by $C[t\rangle D$.

The definitions of *firing sequence, reachable conguration,* and *sequential configuration graph* for P/T systems follow from those given for EN systems considering that the vertices of the graphs are vectors.

Let $N = (P, T, F, W, C_{in})$ be a P/T system, $C, D : P \to \mathbb{N}$ be two configurations of N, and $U \subset T, U \neq \emptyset$. The set U is a *step* (for a P/T system) from C to D, denoted by $C[U\rangle D$, if:

(i) $\sum_{t \in U} W(p, t) \leq C(p)$;

(ii) $D(p) = C(p) - \sum_{t \in U} W(p, t) + \sum_{t \in U} W(t, p)$.

If a P/T step $U \subset T$ is such that $|U| \geq 2$, then U is called a *P/T concurrent step*.

The definitions of *configuration graph* and *maximal strategy configuration graph* for a P/T system follow from that given for EN systems when concurrent steps for P/T systems are considered instead of concurrent steps.

In this monograph the vertices in the configuration graphs associated with P/T systems are depicted as vectors. For the sake of clarity some abuse of notation is present in the indication of these configurations. If, for instance $P = \{p_1, p_2, p_3\}$ and a configuration C is such that $C(p_1) = 1, C(p_2) = 5$ and $C(p_3) = 0$, then the vector indicating C is $(p_1, p_2 = 5)$. That is, we only indicate the places having at least one token. If the token is just one, then we indicate the place, otherwise the place together with the number of tokens. The definitions of *P/T system with priorities* and *P/T system with inhibitor arcs* follow from those of EN systems with priorities and EN systems with inhibitor arcs, respectively.

Let $N = (P, T, F, W, C_{in})$ be a P/T system, C and D configurations of N and $U \in \mathbb{N}^{|T|}$, $U = (u_1, \ldots, u_{|T|})$, $u_i \neq 0$, $1 \leq i \leq |T|$. The vector U is a *maximally parallel concurrent step from C to D*, denoted by $C[U\rangle^{mp} D$, if:

(i) for each $p \in P$ $\sum_{i=1}^{|T|} u_i W(p, t_i) \leq C(p)$;

(ii) there is no $U' = (u'_1, \ldots, u'_{|T|})$ such that $u'_i \geq u_i$ for $1 \leq i \leq |T|$ with at least one $j \in \{1, \ldots, |T|\}$, $u'_j > u_j$ with $\sum_{i=1}^{|T|} u'_i W(p, t_i) \leq C(p)$ for each $p \in P$.

If this holds, then we say that U *mp-fires* from C to D with $D(p) = C(p) - \sum_{i=1}^{|T|} u_i W(p, t_i) + \sum_{i=1}^{|T|} u_i W(t_i, p)$ for each $p \in P$.

Given a P/T system $N = (P, T, F, W, C_{in})$, $x = (x_1, \ldots, x_n)$ with $x_1, \ldots, x_n \in \mathbb{N}^{|T|}$, is a *maximally parallel firing sequence of N* if there are $C_1, \ldots, C_n$ configurations of N such that $C_{in}[x_1\rangle^{mp} C_1, \ldots, C_{n-1}[x_n\rangle^{mp} C_n$. In this case we say that x is a *firing sequence*, that x *mp-fires* in C_{in} and we write $C_{in}[x\rangle^{mp} C_n$. The set of all mp-firing sequences of N is denoted by $MPFS(N)$. A configuration C of N is said a *maximally parallel reachable configuration of N* if there is a maximally parallel firing sequence $x \in MPFS(N)$ with $C_{in}[x\rangle^{mp} C$. The set of maximally parallel reachable configurations of N is denoted by $\mathbb{C}_N^{mp}$.

Given a P/T system $N = (P, T, F, W, C_{in})$ we define the *maximally parallel configuration graph of N*, $MPCG(N) = (\mathbb{C}_N^{mp}, E, label)$, as an edge-labelled directed multigraph, with:

$E = \{(C, D) \mid C, D \in \mathbb{C}_N^{mp}, \exists x \in \mathbb{N}^{|T|}, C[x\rangle^{mp} D\}$;

$label : \mathbb{C}_N^{mp} \times \mathbb{C}_N^{mp} \to \mathbb{N}^{|T|}$ such that if $(C, D) \in E$, $\exists x \in \mathbb{N}^{|T|}$ and $C[x\rangle^{mp} D$, then $label(C, D) = x$.

If we consider the P/T system depicted in Fig. 4.4(a), then the CG, $MPCG$ and $MSCG$ are depicted in Fig. 4.4(b), (c) and (d), respectively.

EN systems can be regarded as a specific kind of P/T systems having $W : F \to \{1\}$, $K : P \to \{1\}$, and such that item 4 of Definition 4.1 holds.

Lemma 4.1 *Let N be a P/T system. It is possible to define a P/T system N' and two bijections α and β such that $MPCG(N')$ strongly bisimulates $MSCG(N)$ according to α and β.*

Proof Let us consider $N = (P, T, F, W, C_{in})$ and $N' = (P', T, F', W', C'_{in})$ with:

$P' = P \cup \{p_t \mid t \in T\}$;

$F' = F \cup \{(p_t, t), (t, p_t) \mid t \in T\}$;

$$W'(x, y) = \begin{cases} W(x, y) & \text{if } (x, y) \in F; \\ 1 & \text{otherwise;} \end{cases}$$

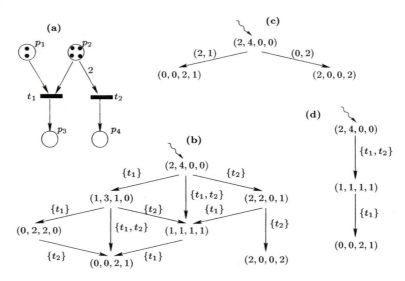

Fig. 4.4 **(a)** A P/T system and its **(b)** CG, **(c)** $MPCG$ and **(d)** $MSCG$

$$C'_{\text{in}}(p) = \begin{cases} C_{\text{in}}(p) & \text{if } (x,y) \in F; \\ 1 & \text{otherwise.} \end{cases}$$

The P/T system N' is such that for each configuration any transition can fire at most once. This should be clear from the definition of N'.

Let $MSCG(N) = (Q, E, label)$, $MPCG(N') = (Q', E', label')$, and $B = \{p_t \mid t \in T\}$. If $C \in Q$, then $C \cup B \in Q'$. If $(C, D) \in E$, then $(C \cup B, D \cup B) \in E'$. If $label(C, D) = \{t_{i_1}, \ldots, t_{i_n}\}$, $1 \leq n \leq |T|$, then $label'(C \cup B, D \cup B) = (f(t_1), \ldots, f(t_{|T|}))$ where $f : T \to \{0, 1\}$ and

$$f(t) = \begin{cases} 0 & \text{if } t \notin \{t_{i_1}, \ldots, t_{i_n}\}; \\ 1 & \text{if } t \in \{t_{i_1}, \ldots, t_{i_n}\}. \end{cases}$$

The bijection α is such that $(C, C \cup B) \in \alpha$ for all $C \in Q$ while the bijection β is such that $(\{t_{i_1}, \ldots, t_{i_n}\}, (f(t_1), \ldots, f(t_{|T|}))) \in \beta$ for $\{t_{i_1}, \ldots, t_{i_n}\} \subseteq T$. Nothing else is in α and β. The statement easily follows. $\qquad\square$

If we denote by N the P/T system depicted in Fig. 4.4(a), then $MSCG(N)$ is depicted in Fig. 4.4(d). Following the proof of the previous lemma, it is possible to define N', another P/T system (depicted in Fig. 4.5(a)), such that $MPCG(N')$ (depicted in Fig. 4.5(c)) strongly simulates $MSCG(N)$ according to the relations defined in Lemma 4.1.

Lemma 4.2 *Let N be a P/T system. It is possible to define a P/T system N' with an infinite number of transitions, a bijection α and a surjection β such that $MPCG(N)$ strongly simulates $MSCG(N')$ according to α and β.*

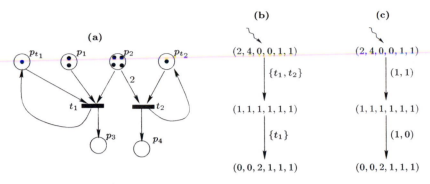

Fig. 4.5 (a) A P/T system, its (b) $MSCG$ and (c) its $MPCG$

Proof Let us consider $N = (P, T, F, W, C_{in})$, $T = \{t_1, \ldots, t_{|T|}\}$, and $N' = (P, T', F', W', C_{in})$ such that:

for each $t_i \in T$, $1 \leq i \leq |T|$, there are sets $T_i = \{t_{i,1}, t_{i,2}, \ldots\}$, $T = \bigcup_{i=1}^{|T|} T_i$ (an infinite set of transitions) with ${}^{\bullet}t_{i,j} = {}^{\bullet}t_i$ and $t_{i,j}^{\bullet} = t_i^{\bullet}$, $j \in \{1, 2, \ldots\}$, $1 \leq i \leq |T|$;

the set F' contains $(p, t_{i,j})$ for each $p \in {}^{\bullet}t_i$ and $(t_{i,j}, p)$ for each $p \in t_i^{\bullet}$, $j \in \{1, 2, \ldots\}$, $1 \leq i \leq |T|$;

the function W' is such that $W'(p, t_{i,j}) = W(p, t_i)$ and $W'(t_{i,j}, p) = W(t_i, p)$ for $p \in P$, $t \in T$, $j \in \{1, 2, \ldots\}$, $1 \leq i \leq |T|$.

For each configuration in N in which a transitions t_i fires n times there are $t_{i,k_1}, \ldots, t_{i,k_n}$ transitions in N' each firing once, $1 \leq i \leq |T|$, $k_1, \ldots, k_n \in \{1, 2, \ldots\}$ and such that all the k symbols are distinct. Formally: for each $C, D \in \mathbb{C}_N^{mp}$ such that $C[U\rangle^{mp}D$ for $U = (u_1, \ldots, u_{|T|})$ there is $U' \subset T'$, $U' = \{t_{i,k_1}, \ldots, t_{i,k_{u_1}}, \ldots, t_{|T|,k_1}, \ldots, t_{|T|,k_{u_{|T|}}}\}$ such that $C[U'\rangle D$ in N'.

Let $MPCG(N) = (Q, E, label)$ an edge-labelled directed multigraph and $MSCG(N') = (Q, E', label')$ an edge-labelled directed multigraph. If $C, D \in Q$ and $(C, D) \in E$, then there are an infinite number of edges from C to D in E'. If $label(C, D) = (n_{t_1}, \ldots, n_{t_{|T|}})$, then $label'(C, D) = \cup_{i=1}^{|T|} A_{n_{t_i}, t_i}$ where $A_{n_{t_i}, t_i} = \{B \mid B \subset \mathcal{P}(T_i), |B| = n_{t_i}\}$, $1 \leq i \leq |T|$. The bijection α is such that $(C, C) \in \alpha$ for each $C \in Q$, while the surjection β is such that $((n_{t_1}, \ldots, n_{t_{|T|}}), \cup_{i=1}^{|T|} A_{n_{t_i}}) \in \beta$ for $(n_{t_1}, \ldots, n_{t_{|T|}}) \in \mathbb{N}^{|T|}$. Nothing else is in α and β. $\square$

If we denote by N the P/T system depicted in Fig. 4.4(a), then $MPCG(N)$ is depicted in Fig. 4.4(c). Following the proof of the previous lemma, it is possible to define N', another P/T system (depicted in Fig. 4.6(a), where the dotted lines suggests the presence of an infinite number of transitions), such that $MSCG(N')$ (depicted in Fig. 4.6(b)) simulates $MPCG(N)$. In Fig. 4.6(b) $A_{2,t_1} = \{B \mid B \subset$

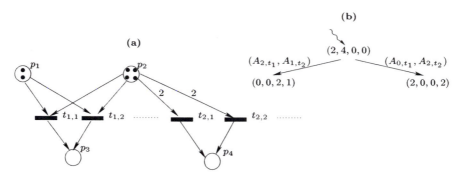

Fig. 4.6 (a) A P/T system and its (b) *MSCG*

$\{t_{1,1}, t_{1,2}, \ldots\}, |B| = 2\}$, $A_{1,t_2} = \{B \mid B \subset \{t_{2,1}, t_{2,2}, \ldots\}, |B| = 1\}$, $A_{0,t_1} = \{B \mid B \subset \{t_{1,1}, t_{1,2}, \ldots\}, |B| = 0\} = \emptyset$, $A_{2,t_2} = \{B \mid B \subset \{t_{2,1}, t_{2,2}, \ldots\}, |B| = 2\}$.

In this monograph we consider (EN systems and) P/T systems as either accepting or generating computing devices. The definition of accepting P/T systems includes the indication of a set $P_{\text{in}} \subset P$ of *input places* and one *final place* $p_f \in P \setminus P_{\text{in}}$. The places in $P \setminus P_{\text{in}}$ are called *work places*.

An *accepting P/T system* N with input C_{in} is denoted by $N(C_{\text{in}}) = (P, T, F, W, P_{\text{in}}, p_f)$ where C_{in} is the initial configuration of the input places. All the remaining places are empty in the initial configuration. A priority function is indicated if priorities are used. A configuration $C_{\text{fin}} \in \mathbb{C}_N$ is said to be *final* if no firing is possible from C_{fin}. In the following we use DG to refer, in general, to sequential configuration graphs, configuration graphs, maximal strategy configuration graphs and priority configuration graphs.

We say that a P/T system $N(C_{\text{in}}) = (P, T, F, W, P_{\text{in}}, p_f)$ with $P_{\text{in}} = \{p_{\text{in},1}, \ldots, p_{\text{in},k}\}$, $k \in \mathbb{N}_+$, *accepts* the vector $(C_{\text{in}}(p_{\text{in},1}), \ldots, C_{\text{in}}(p_{\text{in},k}))$ if in DG there is a final configuration C_{fin} such that:

$C_{\text{fin}}(p_f) > 0$;

there is at least one path from C_{in} to C_{fin};

no other configuration D in the paths from C_{in} to C_{fin} is such that $D(p_f) > 0$.

The *set of vectors accepted* by N is denoted by $\mathbb{N}^k(N)$ and it is composed of the vectors $(C_{\text{in}}(p_{\text{in},1}), \ldots, C_{\text{in}}(p_{\text{in},k}))$ accepted by N.

The definition of generating P/T systems includes the indication of a set $P_{\text{out}} \subset P$ of *output places* and one *final place* $p_f \in P$. The places in $P \setminus P_{\text{in}}$ are called *work places*.

A *generating P/T system* is denoted by $N = (P, T, F, W, C_{\text{in}}, P_{\text{out}}, p_f)$ and it is such that $C_{\text{in}}(p_f) = 0$. A priority function is indicated if priorities are used. A configuration $C_{\text{fin}} \in \mathbb{C}_N$ is said to be *final* if no firing is possible from

C_{fin}. In the following we use DG to refer, in general, to sequential configuration graphs, configuration graphs, maximal strategy configuration graphs and priority configuration graphs.

We say that a generating P/T system $N = (P, T, F, W, C_{\text{in}}, P_{\text{out}}, p_f)$ with $P_{\text{out}} = \{p_{\text{out},1}, \ldots, p_{\text{out,k}}\}$, $k \in \mathbb{N}_+$, *generates* the vector $(C_{\text{fin}}(p_{\text{out},1}), \ldots, C_{\text{fin}}(p_{\text{out,k}}))$ if in DG there is a final configuration C_{fin} such that:

$C_{\text{fin}}(p_f) > 0$;

there is at least one path from C_{in} to C_{fin};

no other configuration D in the paths from C_{in} to C_{fin} is such that $D(p_f) > 0$.

In this case N *halts* with output P_{out}. If, from the initial configuration, the system N reaches a final configuration C_{fin} with $C_{\text{fin}}(p_f) = 0$, then we say that N *stops*.

The *set of vectors generated* by N is denoted by $\mathbb{N}^k(N)$ and it is composed of the vectors $(C_{\text{fin}}(p_{\text{out},1}), \ldots, C_{\text{fin}}(p_{\text{out,k}}))$ generated by N.

The definitions of *accepting* and *generating* EN systems follow from the above.

4.2 Four equivalent features

In this section we study how different kinds of configuration graphs related to EN systems and P/T systems can (weakly or strongly) simulate instructions of the kind (s, γ^-, v, w) present in register machines. We start with the net (underlying a P/T system) depicted in Fig. 4.1. In this net the work places p_s, p_v and p_w are associated with the states s, v and w, respectively; the input place p_γ is associated with the register γ. The remaining work places and transitions are specific to the simulated instruction.

One $MSCG$ for this net, having (p_s) as initial configuration, is depicted in Fig. 4.7(a). Here if a token is present in p_s and no token is present in any of the other places in the net in Fig. 4.1, then a token will be present in p_w. Another $MSCG$ for the same net, having one token in p_s and $n \in \mathbb{N}_+$ tokens in p_γ as initial configuration, is depicted in Fig. 4.7(c). Here if a token is present in p_s, $n \in \mathbb{N}_+$ tokens are present in p_γ and no token is present in any of the other places, then a token will be present in p_v and $n - 1$ tokens will be present in p_γ.

It should be clear that this $MSCG$ weakly simulates instructions of the kind (s, γ^-, v, w). This is because the P/T system needs to fire more than one transition to pass from a configuration associated with a state of the register machine to another configuration associated with another state. It is essential for us to note that, as the work places p_1, p_2, p_3 and p_4 and the transitions are specific to the simulated instruction, then none of these configurations is used in

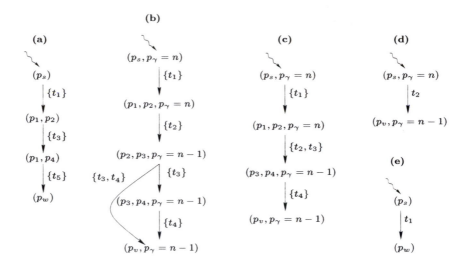

Fig. 4.7 (a), *PrCG* and *MSCG*, **(b)** *PrCG*, **(c)** *MSCG* for the net in Fig. 4.1; **(d)**, **(e)** *SCGs* for the net in Fig. 4.3 and Fig. 4.8, $n \in \mathbb{N}_+$. © With kind permission of Springer Science and Business Media [81]

the simulation of other instructions. Moreover, as the value of the register γ is unbounded, the number of tokens present in this P/T system is unbounded, too.

Let us now consider the net depicted in Fig. 4.1 as underlying a P/T systems with priorities such that transition t_2 has higher priority than the other transitions (all having the same priorities). The *PrCG* associated with this net for the initial configuration (p_s) is depicted in Fig. 4.7(a), while that for the initial configuration $(p_s, p_\gamma = n)$, $n \in \mathbb{N}_+$, is depicted in Fig. 4.7(b). Also in this case there is a weak simulation between these *PsCGs* and instructions of the kind (s, γ^-, v, w).

It is possible to define a simpler P/T system with priorities such that its *PrCG* simulates instructions of the kind (s, γ^-, v, w). The net underlying this system is depicted in Fig. 4.8 where t_1 has priority 2 and t_2 has priority 1.

In Fig. 4.7(d) and 4.7(e) two *PrCGs* associated with this net, one having $(p_s, p_\gamma = n)$, $n \in \mathbb{N}_+$, and the other having (p_s) as initial configuration, are depicted. In this case the *PrCGs* strongly simulates instructions of the kind (s, γ^-, v, w).

It is important to notice that in the P/T system just considered there are as many such nets as there are instructions of the kind (s, γ^-, v, w) present in the simulated register machine. Moreover, as the register machine can be in only one state per time, then at most one place p_s can have a token.

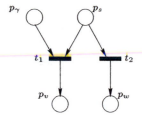

Fig. 4.8 Net underlying a P/T system with priorities simulating (s, γ^-, v, w)

Let us consider again the net depicted in Fig. 4.1 with one token in p_s, $n \in \mathbb{N}_+$ tokens in p_γ and no tokens in the remaining places. It is possible to define an EN system having the same dynamics of the P/T system just considered. The net underlying such EN system is depicted in Fig. 4.9 (without considering the dashed place and arrow). In this net the dotted lines suggest the presence of an infinite number of places, transitions and elements of the flow relation. Moreover, if we consider the presence of priorities, all the t_2^i, $1 \leq i \leq n$, have higher priority than the other transitions (all having the same priority). If the place p_γ in the P/T system has $i \in \{1, \ldots\}$ tokens, then in the EN system the place $p_{\gamma=i}$ has one token and $p_{\gamma=j}$, $j \neq i$ have no token (we discuss the case $i = 0$ in a while).

The *MSCG* and *PrCG* for this EN system, weakly simulating instructions of the kind (s, γ^-, v, w), can easily be derived by those depicted in Fig. 4.2(c) and Fig. 4.7(a).

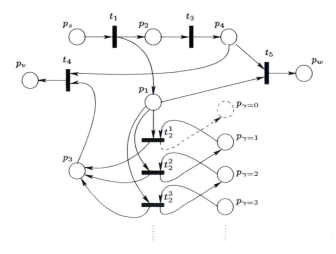

Fig. 4.9 Net underlying an EN system. © With kind permission of Springer Science and Business Media [81]

It should be clear that none of the different kinds of configuration graphs related to these P/T systems and EN system can simulate the instruction (s, γ^-, v, w) if maximal strategies or priorities are not considered. If in the net depicted in Fig. 4.1, for instance, p_s and p_γ have one token and the other places have no tokens, then the firing of t_1, t_3, t_5 would result in a token in p_w (not a simulation of (s, γ^-, v, w)).

As in the net underlying the P/T system in Fig. 4.1, the place p_γ can contain an unbounded number of tokens, then in the net underlying the EN system in Fig. 4.9 the number of transitions t_2^i, places $p_{\gamma=i}, i \in \mathbb{N}_+$ and, consequently, the number of elements in the flow relation is infinite.

The simulation of instructions of the kind (s, γ^-, v, w) can also be performed by the configuration graph related to the P/T system with inhibitor arcs depicted in Fig. 4.3. Two sequential configuration graphs of this P/T system with inhibitor arcs, one having $(p_s, p_\gamma = n)$, $n \in \mathbb{N}_+$, and the other having p_s as initial configuration, are depicted in Fig. 4.7(d) and 4.7(e). It should be clear that also here the lack of the inhibitor arc would not allow that net to simulate the 0-test. These sequential configuration graphs strongly simulate instructions of the kind (s, γ^-, v, w).

The graphs depicted in Fig. 4.7 are such that from the configuration at the top the system can only evolve toward the configuration at the bottom. So we can write 'reduced' graphs. This is done in Fig. 4.10(a) and Fig. 4.10(c).

If in the net depicted in Fig. 4.9 we consider the dashed place $p_{\gamma=0}$ and the dashed arrow coming into this place, then the dynamics of this net are not changed; it is actually extended to model a register machine also when a register γ is empty. The resulting maximal strategy and priority configuration graphs depicted in Fig. 4.7(a) are changed, adding the place $p_{\gamma=0}$ to every configuration. This means that its 'reduced' graph changes into that depicted in Fig. 4.10(b).

If we now consider the 'reduced' graphs depicted in Fig. 4.10(b) and Fig. 4.10(c) (and we change $p_\gamma = n$ into $p_{\gamma=n}$ and $p_\gamma = n - 1$ into $p_{\gamma=n-1}$), then we can define another net underlying an EN system whose SCG simulates the instruction (s, γ^-, v, w). This net is depicted in Fig. 4.11, where the dotted lines

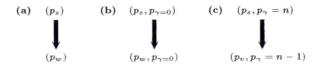

Fig. 4.10 'Reduced' graphs, $n \in \mathbb{N}_+$

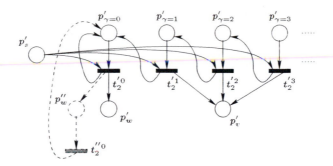

Fig. 4.11 Net underlying the 'reduced' EN system simulating (s, γ^-, v, w). © With kind permission of Springer Science and Business Media [81]

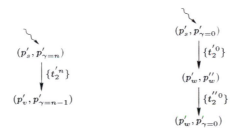

Fig. 4.12 SCGs for the net in Fig. 4.11, $n \in \mathbb{N}_+$

suggest the presence of an infinite number of places, transitions and elements of the flow relation.

As $(p'_{\gamma=0}, t_2'^0), (t_2'^0, p'_{\gamma=0}) \in F$ what is depicted in Fig. 4.11 is not a net underlying an EN system as defined in Definition 4.1 (as item 4 is not satisfied). We can overcome this by simply removing $(t_2'^0, p'_{\gamma=0})$ and adding the dashed place, transition and arrow depicted in Fig. 4.11.

The net underlying the EN system depicted in Fig. 4.11 can be regarded as a rewriting of that depicted in Fig. 4.9. Two SCGs for this EN system, one having $(p'_s, p'_{\gamma=n})$, $n \in \mathbb{N}_+$, and another having $(p'_s, p'_{\gamma=0})$ as initial configuration, are depicted in Fig. 4.12. These SCGs can weakly simulate instructions of the kind (s, γ^-, v, w).

Recalling (from Section 4.1) that an EN system can be regarded as a specific kind of a P/T system we can say that:

Theorem 4.1 *For each instruction of the kind (s, γ^-, v, w) there is:*

(i) a P/T system N such that $MSCG(N)$ weakly simulates (s, γ^-, v, w);

(ii) a P/T system N with priorities such that PrCG(N) strongly simulates
      (s, γ^-, v, w);
(iii) a P/T system N with inhibitor arcs such that SCG(N) strongly simulates
      (s, γ^-, v, w);
(iv) a P/T system N with an infinite number of places and transitions such that
      SCG(N) weakly simulates (s, γ^-, v, w).

In Section 4.4 we see that, considering the way the nets are built, item *(iv)* of the previous theorem can be rewritten so as to have places as the only infinite element.

The title of the present section refers to the fact that maximal strategy, priorities, inhibitor arcs and an infinite number of places are equivalent features in the context of the previous theorem.

4.3   About gambling

The simulation of instructions of the kind (s, γ^-, v, w) or of pairs of instructions $(s, \gamma^-, v), (s, \gamma^{=0}, w)$ can be performed differently from what we saw in the previous section.

Let us consider the net depicted in Fig. 4.13 as underlying a P/T system. In this system the work places and transitions are specific to the simulated instruction. In Fig. 4.14 two *MSCG*s for this system having either (p_s) or $(p_s, p_\gamma = n), n \in \mathbb{N}_+$, as initial configurations are depicted.

We can explain this dynamics in the following way. When a token is in p_s the system can non-deterministically 'gamble' if there is at least a token in p_γ. The system reaches a configuration with a token in either p_v or p_w if and only if the gamble is correct.

This dynamics will become clearer when we go into the details of a proof using it (Theorem 5.3, for instance).

For the moment we note a few things. The *MSCG*s depicted in Fig. 4.14, one having (p_s) and the other having $(p_s, p_\gamma = n)$ as initial configuration, are also the *PrCG* of the net depicted in Fig. 4.13 if it underlies a P/T system with priorities such that transition t_4 has priority 2 while all the remaining transitions have priority 1. The graphs depicted in Fig. 4.14 weakly simulate instructions of the kind (s, γ^-, v, w) (or of pairs of instructions $(s, \gamma^-, v), (s, \gamma^{=0}, w)$) present in register machines. As in the previous section, as the place and the transitions present in the P/T system are specific to the simulated instruction, then none of these configurations is used in the simulation of other instructions of the kind (s, γ^-, v, w).

The presence of the place $p_{\bar{\gamma}}$ is of interest for our discussion. We name this place a *conflicting place* and $\bar{\gamma}$ a *conflicting register*. This is because the presence

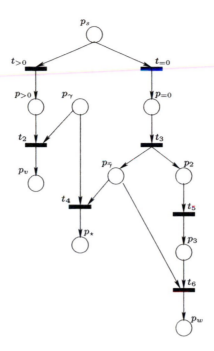

Fig. 4.13 Another net underlying a P/T system simulating (s, γ^-, v, w). © With kind permission of Springer Science and Business Media [81]

of a token in it 'conflicts' with the presence of at least one token in p_γ forcing the system to fire t_4.

If we consider what we said in Section 4.2, then we can define an EN system N with an infinite number of places and transitions such that $SCG(N)$ weakly simulates instructions of the kind (s, γ^-, v, w) (or of pairs of instructions $(s, \gamma^-, v), (s, \gamma^{=0}, w)$). The net underlying N does not have conflicting places but a place for each different value of the simulated register (see Fig. 4.11). We will consider such an EN system in Section 8.5.

4.4 On the computing power of building blocks

Let us introduce the nets depicted in Fig. 4.15 and call them *building blocks*, *join* and *fork* in particular, as depicted in that figure. The places present in each building block are distinct.

Definition 4.4 *Let* $x, y \in \{\text{join}, \text{fork}\}$ *be building blocks and let* $\bar{t}_x$ *and* $\hat{t}_y$ *be the transitions present in* x *and* y *respectively.*

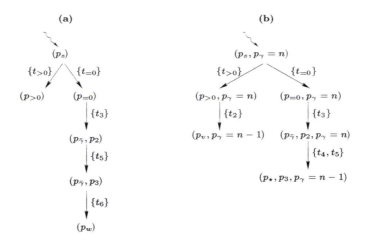

Fig. 4.14 $MSCG$s and $PrCG$s for the net in Fig. 4.13, $n \in \mathbb{N}_+$

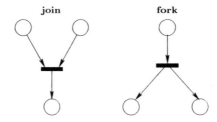

Fig. 4.15 Building blocks: *join* and *fork*. © With kind permission of Springer Science and Business Media [81]

We say that y comes after x (or x is followed by y, or x comes before y or x and y are in sequence) if $\bar{t}_x^{\bullet} \cap {}^{\bullet}\hat{t}_y \neq \emptyset$. We say that x and y are in parallel if ${}^{\bullet}\bar{t}_x \cap {}^{\bullet}\hat{t}_y \neq \emptyset$.

We say that a net is composed of building blocks (it is composed of x) if it can be defined by building blocks (it is defined by x) sharing places but not transitions.

If Fig. 4.18(a), Fig. 4.18(c) and Fig. 4.16 building blocks in sequence and in parallel are depicted.

4.4.1   A hierarchy of processes

In this section we study the computational power of accepting EN systems and P/T systems whose underlying net is composed of building blocks.

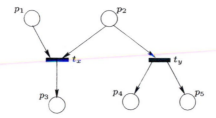

Fig. 4.16 A *join* and a *fork* in parallel

Theorem 4.2 *For each accepting (generating) register machines M there is an accepting (generating) P/T system N whose underlying net is composed of building blocks such that MSCG(N) weakly simulates M. Moreover, for each accepting (generating) P/T system N whose underlying net is composed of building blocks there is an accepting (generating) register machine weakly simulating MSCG(N).*

Proof We prove the statement for accepting devices. The proof for the generating devices is similar.

Part I: (Such MSCGs weakly simulate accepting register machines) Let $M = (S, R, s_0, f)$ be an accepting register machine with n registers. We define $N(C_{\text{in}}) = (P, T, F, W, P_{\text{in}}, p_f)$ an accepting P/T system as in the statement of the theorem. The composition of *join* and *fork* depicted in Fig. 4.17(a) allows us to define a net underlying the P/T system simulating instructions of the kind $(s, \gamma^-, v, w) \in R$. In this net the places and transitions are specific to the simulated instruction. In Fig. 4.17(b) and Fig. 4.17(c) the *MSCGs* for that net are depicted. In this net the transitions t_1 and t_3 belong to a *fork* while the remaining transitions belong to a *join*.

The instructions of the kind $(s, \gamma^+, v) \in R$ can be simulated by a *fork* with $^\bullet t_+ = \{p_s\}$ and $t_+^\bullet = \{p_\gamma, p_v\}$ (where t_+ is the transition present in the *fork* and it is specific to the simulated instruction).

As it is possible that the register machine is non-deterministic (that is, once in a state more than one instruction can be applied), then some building blocks are in parallel. Considering the net depicted in Fig. 4.17(a) and that simulating (s, γ^+, v), then it may be that $^\bullet t_1 = {}^\bullet t_+$.

The input places P_{in} are those associated with the registers of M and p_{s_0}, the place associated with the initial state of M. The initial configuration of the P/T system N has as many tokens in the input places associated with the registers of the register machine as the values associated with these registers; there is also one token in the place associated with s_0, the initial state of the register machine.

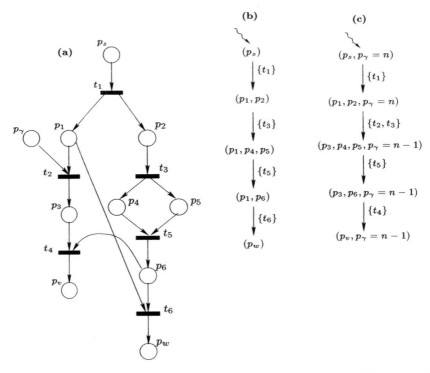

Fig. 4.17 (a) Net composed of building blocks underlying a P/T system simulating (s, γ^-, v, w), **(b)** and **(c)** $MSCG$s for this net, $j \in \mathbb{N}, n \in \mathbb{N}_+$

The P/T system simulates one instruction of the register machine per time. This is due to the fact that during the simulation only one place (any one) associated with the state of the register machine has a token (this should be clear by the way the instructions are simulated). The computation of the P/T system halts (eventually) when p_f, the place associated with the final state of the register machine, has a token in it.

The P/T system accepts its input only if the register machine accepts it, too. It should be clear that the simulation is a weak one.

Part II: (Accepting register machines weakly simulate such $MSCG$s) Each place of the accepting P/T system N has two different registers associated with it in the register machine $M = (S, I, s_1, s_f)$. We call these registers *original* and *copy* and we assume that the input registers of M are the original registers associated with the places in P_{in}. Moreover, each transition of N has a different register associated with it.

Given a place $p \in P$ in the initial configuration of N, the content of the original and copy register associated with p is equal to $C_{\text{in}}(p)$, that is, the number of tokens present in p in the initial configuration of N, the content of the input register being the number of tokens in the input places. The registers associated with the transitions are initially empty.

The register machine tries to simulate in a fixed order the firing of the transitions of the P/T systems. For each transition t this simulation is divided into two stages.

The register machine tries to subtract $W((p,t))$ from each copy register associated with p, $p \in {}^\bullet t$. The state of the register machine records the finite number of steps required to do so. If it is not possible to subtract $W((p,t))$ for at least one place $p \in {}^\bullet t$, then the content of the copy registers is restored. This can be done using the states of the register machine. If instead it is possible to subtract $W((p,t))$ for all the places $p \in {}^\bullet t$, then M adds 1 to the register associated with t and then it tries to simulate the next transition in the order.

After M tried to simulate the last transition in the order, for each register associated with transition t whose content is bigger than 1:

M decreases this register by 1;

M subtracts $W((p,t))$ from each original register associated with $p \in {}^\bullet t$;

M adds $W((t,p'))$ to each copy and original register associated with $p' \in t^\bullet$.

At the end of this process M changes its state into s_f only if the content of the original register associated with p_f is bigger than 1. If no simulation of the firing of any transition was performed, that is, if none of the registers associated with the transitions was 1, then M halts in a state different from s_f, otherwise M tries to simulate again the firing of the transitions in N. It should be clear that the simulation is a weak one. □

Because of Theorem 4.1, Theorem 4.2 is valid also for: *PrCGs* of P/T system with priorities, *CGs* of P/T systems with inhibitor arcs, and *SCGs* of P/T systems having an infinite number of places and transitions. Because of Theorem 4.1 and Lemma 4.1, Theorem 4.2 is valid also for *MPCGs* of P/T systems.

Proposition 4.1 *A connected net composed of building blocks is such that if the number of places is infinite, then the number of transitions is infinite, too. Conversely, it may be that if the number of transitions is infinite the number of places is finite.*

The first sentence of the previous proposition is clearly true. For the second sentence: imagine, for instance, a net underlying a P/T system having an infinite number of transitions belonging to a *join*, all these transitions having the same input and output sets but different weights associated with the elements of the flow relation.

Framingham State College
Framingham, MA

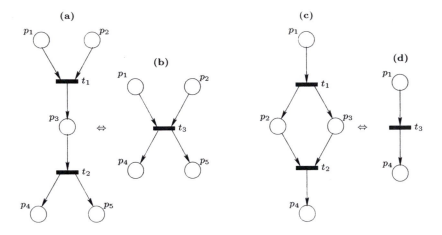

Fig. 4.18 (a) *Join* and *fork* in sequence and (b) rewritten in a 'compressed' form; (c) *fork* and *join* in sequence and (d) rewritten in a 'compressed' form

Sequences of building blocks can be *rewritten* in a 'compressed' form as depicted in Fig. 4.18.

Without changing its dynamics the net depicted in Fig. 4.1 can be rewritten as composed of building blocks. The resulting net is depicted in Fig. 4.17(a). In a similar way the nets depicted in Fig. 4.9 and Fig. 4.11 can be rewritten as composed of building blocks. This allows us to change item (iv) of Theorem 4.1 into:

(iv) a P/T system with an infinite number of places such that SCG(N) weakly simulates (s, γ^-, v, w).

Notice that also the net depicted in Fig. 4.13 can be obtained as composed of building blocks. This can be done, for instance, by considering the rewritings depicted in Fig. 4.18 and Fig. 4.19.

Theorem 4.3 *For each accepting partially blind register machines M there is an accepting P/T system N whose underlying net is composed of building blocks such*

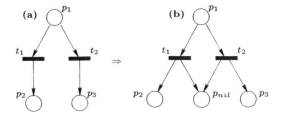

Fig. 4.19 A non-determinism building block simulated by two *forks* in parallel

that $SCG(N)$ strongly simulates M. Moreover, for each accepting P/T system N whose underlying net is composed of building blocks there is a partially blind register machine weakly simulating $SCG(N)$.

Proof *Part I: (Such SCGs strongly simulate partially blind register machines)* This proof follows from the proof of Theorem 4.2, Part I, considering that instructions of the kind (s, γ^-, v) can be simulated by a *join* with ${}^\bullet t_- = \{p_s, p_\gamma\}$ and $t_-^\bullet = \{p_v\}$ (where t_- is the transition present in the *join* and it is specific to the simulated instruction) and that instructions of the kind (s, γ^+, v) can be simulated by a *fork* with ${}^\bullet t_+ = \{p_s\}$ and $t_+^\bullet = \{p_\gamma, p_v\}$ (where t_+ is the transition present in the *fork* and it is specific to the simulated instruction).

If the simulated register machine tries to subtract 1 from an empty register, then the P/T system never reaches a configuration with a token in its final place. As the value of the register γ is unbounded, the number of tokens present in this P/T system is unbounded, too. It should be clear that the simulation is a strong one.

Part II: (Partially blind register machines weakly simulate such SCGs) We recall that we consider P/T systems such that every place can store an unbounded number of tokens. Each place of the accepting P/T system N has one different register associated with it in the register machine $M = (S, R, s_0, f)$. In the initial configuration the values of the registers are equal to the number of tokens present in the associated place in the initial configuration of the P/T system.

The machine M simulates non-deterministically the firing of transitions of N. For each transition $t \in T$ the register machine tries to subtract $W(p, t)$ from each register associated with $p \in {}^\bullet t$. This is performed by a sequence of instructions each trying to subtract one unit from the register associated with $p \in {}^\bullet t$. If this cannot be done, then M stops. Otherwise $W(t, p)$ is added to each register associated with $p \in t^\bullet$. When 1 is added to p_f, the register associated with the final place of N, then M goes into its final state and halts.

The register machine accepts the input only if the P/T system accepts it, too. It should be clear that the simulation is a weak one. $\qquad\square$

Theorem 4.4 *For each restricted register machines M there is an accepting P/T system N whose underlying net is composed of a join and fork in sequence such that $MSCG(N)$ weakly simulates M. Moreover, for each accepting P/T system N whose underlying net is composed of a join and fork in sequence there is a restricted register machine weakly simulating $MSCG(N)$.*

Proof *Part I: (Such MSCGs weakly simulate restricted register machines)* We said that the *join* and *fork* sequence depicted in Fig. 4.18(a) can be 'rewritten' as depicted in Fig. 4.18(b).

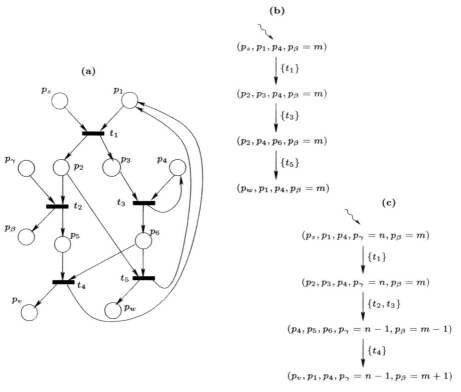

Fig. 4.20 **(a)** Net underlying a P/T system simulating $(s, \gamma^-, \beta^+, v, w)$ with **(b)** and **(c)** *MSCG*s associated with it, $n, m \in \mathbb{N}_+$

If we consider this 'rewriting', then the net depicted in Fig. 4.20(a) can underlie a P/T system whose *MSCG* simulates the instruction $(s, \gamma^-, \beta^+, v, w)$. The places and transitions present in this net are specific to the simulated instruction. We can see this from the *MSCG*s depicted in Fig. 4.20(b) and 4.20(c) associated with this net (notice that places p_1 and p_4 have a token in the initial and final configurations).

The rest of the proof follows from that of Theorem 4.2, Part I. It should be clear that the simulation is a weak one.

Part II: (Restricted register machines weakly simulate such MSCGs) The proof follows from that of Theorem 4.2, Part II, considering that the register machine has one more register called a *buffer*. This register stores no value in the initial configuration. Every time a *join* block is simulated the value of the buffer is increased by one, when a *fork* block is simulated the value of the buffer is decreased by one. The register machine simulates one *join* and *fork* in sequence at one time. As only *join* and *fork* in sequence are present in the net underlying the P/T system, then only one buffer is needed. It should be clear that the simulation is a weak one. □

As from Theorem 4.1 and Proposition 4.1 we know that a maximal strategy can be regarded as the presence of an infinite number of places, then the previous three theorems can be rewritten in the following way:

Theorem 4.5

[Theorem 4.2] *For each register machine M there is an accepting P/T systems N whose underlying net is composed of building blocks having an infinite number of places (and using an unbounded number of tokens) such that $SCG(N)$ can weakly simulate M. Moreover, for each accepting P/T system N whose underlying net is composed of building blocks having an infinite number of places (and using an unbounded number of tokens) there is a register machine weakly simulating $SCG(N)$.*

[Theorem 4.3] *For each partially blind register machine M there is an accepting P/T systems N whose underlying net is composed of building blocks (and using an unbounded number of tokens) such that $SCG(N)$ can strongly simulate M. Moreover, for each accepting P/T systems N whose underlying net is composed of building blocks (and using an unbounded number of tokens) there is a partially blind register machine weakly simulating $SCG(N)$.*

[Theorem 4.4] *For each restricted register machine M there is an accepting P/T systems N whose underlying net is composed of building blocks and having an infinite number of places (and using a finite number of tokens) such that $SCG(N)$ can weakly simulate M. Moreover, for each accepting P/T system N whose underlying net is composed of building blocks and having an infinite number of places (and using a finite number of tokens) there is a restricted register machine weakly simulating $SCG(N)$.*

Theorem 4.6 *EN systems whose underlying net is composed of joins can accept only finite sets of vectors.*

Proof *Part I: (Such EN systems can accept finite sets of vectors)* Now we are dealing with EN systems (at most one token in a place) and not with P/T systems as in the previous theorems in this section. Let $W = \{w_1, \ldots, w_{|W|}\}$ be a finite set of vectors. Without loss of generality we assume that all the vectors of W have the same dimension d (if this is not the case, then 0's are added to the vectors of lower dimension so as to reach dimension d) and that $w_i = (w_{i,1}, \ldots, w_{i,d}), 1 \leq i \leq |W|$. Moreover, let m be the maximum number in $w_i, 1 \leq i \leq |W|$. We define an accepting EN system $N = (P, T, F, V, s_f)$ with $P = V + S$ and

$$V = \{p_{j,k}^{(i)} \mid 1 \le i \le |W|, 1 \le j \le d, 0 \le k \le m\} \cup \{p_\star\};$$
$$S = \{s_j^{(i)} \mid 1 \le i \le |W|, 1 \le j \le d\} \cup \{s_f, s_\star\}.$$

The places in V are input places, while the places in S are work places.

The transitions are introduced in the following:

initial: $t_1^{(i)}$ are transitions with ${}^\bullet t_1^{(i)} = \{s_1^{(1)}, p_{1,w_{i,1}}^{(i)}\}$ and $t_1^{(i)\bullet} = \{s_2^{(i)}\}$ with $1 \le i \le |W|$.

    Informally: if $w_{i,1}$ is the first number of vector i, then checking the second number of vector i is allowed.

general: $t_j^{(i)}$ are transitions with ${}^\bullet t_j^{(i)} = \{s_j^{(i)}, p_{j,w_{i,j}}^{(i)}\}$ and $t_j^{(i)\bullet} = \{s_{j+1}^{(i)}\}$ with $1 \le i \le |W|$ and $2 \le j \le d-1$.

    Informally: if $w_{i,j}$ is the jth number in vector i, then checking the $j+1^{th}$ number in vector i is allowed.

final: $t_d^{(i)}$ are transitions with ${}^\bullet t_d^{(i)} = \{s_d^{(i)}, p_{d,w_{i,d}}^{(i)}\}$ and $t_d^{(i)\bullet} = \{s_f\}$ with $1 \le i \le |W|$.

    Informally: if the last number of the ith vector is there, then put a token in the final place s_f.

extra: $t_\star^{(q)}$ are transitions with ${}^\bullet t_\star^{(q)} = \{s_f, p\}$ and $t_\star^{(q)\bullet} = \{s_\star\}$ with $1 \le q \le |V|$ and $p \in V$.

    Informally: if the input did not encode a vector in W, then remove the token from the final place.

All these transitions belong to a *join* block. If $w = (w_1, \ldots, w_g)$ is a vector, then the initial configuration of the EN systems is $s_1^{(1)}$ and $p_{j,w_j}^{(i)}$ for $1 \le j \le d$ and $i \in \{1, \ldots, |W|\}$. If $g < d$, then also $p_{j,0}^{(i)}$ have a token. If $g > d$, then $p_\star$ also has a token in the initial configuration.

The EN system N is such that if $w \in W$, then N can accept w; if $w \notin W$, then N does not accept w. The EN system works in the following way: if possible an *initial* transition fires. This can happen only if the first number of the vector w is also the first number of (at least) one vector in W. If the first number of w is not the first number of a vector in W, then no firing occurs (and N does nor reach a configuration with a token in its final place). The following sequence of firing, mainly involving *general* transitions, tries to match the vector w with one of the vectors in W. Notice that if $w = w_{i'} \in W$ but in the initial configuration the places $p_{j,w_j}^{(i)}, i \ne i'$, got a token, then the EN system does not reach a configuration with a token in its final place.

As long as the numbers in w match those of the chosen vector i in W, the EN system keeps firing. If at a certain number a mismatch is present (which includes the input vector having dimension smaller than d), then the EN system stops firing with no token in the final place. If the vector w is at least of dimension d, then it is possible that the firing matching the last symbol of the chosen vector in W lets a token be in s_f (this occurs because of the firing of a *final* transition).

An *extra* transition fires if the vector w has a dimension bigger than d or if in the initial configuration the input places of N did not encode a vector in W. This last situation is present, for instance, if input places with different superscripts have a token in the initial configuration. If an *extra* transition fires, then the EN system does not accept the input.

It should be clear that the EN system N accepts only (encodings of) vectors in W.

Example 4.1 describes such an EN system accepting the set of vectors $W = \{(1, 0, 1)\}$.

Part II: (The sets of vectors accepted by such EN systems are finite) This fact derives from the following two observations. As we are dealing with EN systems the number of tokens present in them is at most equal to the number of places; moreover as they are composed of *joins*, during firing, the number of tokens can only decrease (by one for each firing).

So, if such an EN system has p places and $t \leq p$ tokens in its initial configuration, then it is able to accept a set of vectors that has at most t elements, that is, a finite set of vectors. □

Example 4.1 An EN system whose underlying net is composed of *joins* accepting the set $W = \{(1, 0, 1)\}$.

This example is related to the proof of Theorem 4.6, Part I. As the EN system resulting from this proof are very complex, we consider a small set of vectors W. The EN system related to this example is depicted in Fig. 4.21.

In this net the input places are $V = \{p_{1,0}^{(1)}, p_{1,1}^{(1)}, p_{2,0}^{(1)}, p_{2,1}^{(1)}, p_{3,0}^{(1)}, p_{3,1}^{(1)}, p_\star\}$ while the work places are $S = \{s_1^{(1)}, s_2^{(1)}, s_3^{(1)}, s_f, s_\star\}$. The only *initial* transition is $t_1^{(1)}$, the only *general* transition is $t_2^{(1)}$, the only final transition is $t_3^{(1)}$, while the transitions $t_\star^{(i)}, 1 \leq i \leq 7$, are *extra*.

This EN system halts only if the initial configuration is $\{s_1^{(1)}, p_{1,1}^{(1)}, p_{2,0}^{(1)}, p_{3,1}^{(1)}\}$ In this case the firing sequence is $t_1^{(1)}, t_2^{(1)}, t_3^{(1)}$.

Any other initial configuration lets the EN system reach a final configuration without s_f in it. ◇

For the following theorem we need to introduce a restricted form of P/T systems such that the firing of any transition allows the number of tokens present in the system either to remain unchanged or to decrease. We denote by *P/T′ systems* this kind of P/T systems. Formally: if T is the set of transitions and P is the set of places of a P/T′ system, then $\forall t \in T, \sum_{p \in {}^\bullet t} W((p, t)) \geq \sum_{p \in t^\bullet} W((t, p))$.

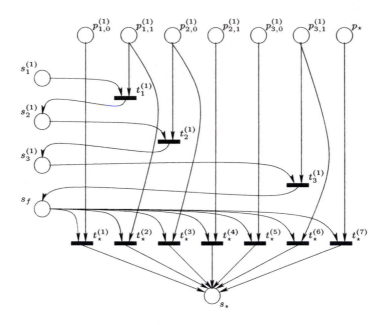

Fig. 4.21 Net underlying an EN system accepting the set of vectors $W = \{(1, 0, 1)\}$

Theorem 4.7 *For each accepting (generating) register machine M there is an accepting (generating) P/T ' system N having an infinite number of places and whose underlying net is composed of joins such that $MSCG(N)$ weakly simulates M.*

Proof We prove the statement for accepting devices. The proof for generating devices is similar.

We describe a non-deterministic P/T′ system that has as many register places p_γ as there are registers present in the register machine $M = (S, I, s_1, s_f)$; for each state $s \in S$ of the register machine the P/T′ system has state place p_s; for each instruction of the kind $(s, \gamma^-, v, w) \in I$ the P/T′ system has places p'_v and p''_v. Moreover, there is an infinite number of places p_{r_i}, $i \in \{1, 2, \ldots\}$, and places p_{rep} and p'.

In the initial configuration the p_{r_i} places have one token each and p' has three tokens. Moreover, there are as many tokens in the register places as the initial value of the registers in the register machine. The remaining places, being the work places, do not have any token.

The net depicted in Fig. 4.22, having an infinite number of places (suggested by the dotted line present in it), allows us to have a random even number of

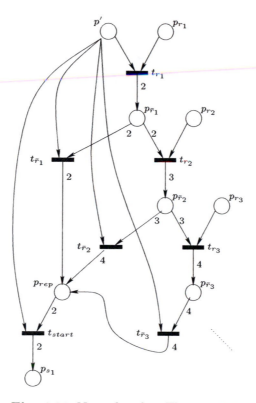

Fig. 4.22 Net related to Theorem 4.7

tokens in p_{rep} and two tokens in p_{s_1}, the state place associated with the initial state of M.

The first transition to file is t_{r_1} so that two tokens are present in $p_{\bar{r}_1}$. When this happens either transition $t_{\bar{r}_1}$ or t_{r_2} can fire. In general, every time transition t_{r_i}, $i \in \{1, 2, \ldots\}$, fires, then the place $p_{\bar{r}_i}$ gets $i+1$ tokens. In this way either transition $t_{\bar{r}_i}$ or $t_{r_{i+1}}$ can fire. If transition $t_{r_{i+1}}$ fires, then the resulting configuration is similar to the one just discussed. If instead transition $t_{\bar{r}_i}$ fires, then p_{rep} gets either $i+1$ tokens, if i is odd, or $i+2$ tokens is i is even. In any case p_{rep} gets an even number of tokens. When this happens t_{start} fires so that p_{s_1} gets two tokens and the simulation of instructions of M can start.

The function of the repository place p_{rep} is to provide tokens. In case p_{rep} runs out of tokens before a token is present in p_f, the final place of the P/T′ system, then the system stops.

The simulation of instructions of the kind (s, γ^+, v) is performed by the net underlying a P/T′ system depicted in Fig. 4.23.

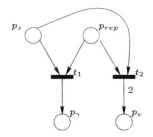

Fig. 4.23 Net underlying a P/T′ system simulating (s, γ^+, v)

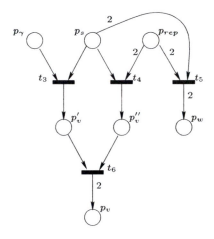

Fig. 4.24 Net underlying the P/T′ system simulating (s, γ^-, v, w)

If two tokens are present in p_s and at least two tokens are present in p_{rep}, then maximal strategy forces t_1 and t_2 to fire in parallel so that one token is added to p_γ and two tokens are added to p_v.

The simulation of instructions of the kind (s, γ^-, v, w) is performed by the net underlying a P/T′ system depicted in Fig. 4.24.

If at least one token is present in p_γ and two tokens are present in p_s, then the maximal strategy forces t_3 and t_4 to fire in parallel. In this way one token is subtracted from p_γ, two tokens are subtracted from p_s and from p_{rep} and a token is added to p_v' and p_v''. When this happens transition t_6 fires removing the tokens from p_v' and p_v'' and adding two tokens to p_v.

If instead no token is present in p_γ when p_s contains two tokens, then two things can happen:

Transition t_5 fires. Two tokens are subtracted from p_s and p_{rep} and two tokens are added to p_w.

Transition t_4 fires twice, one after the other. The resulting presence of two tokens in p_v'' allows the P/T$'$ system to stop.

From the above it should be clear that p_{rep} always has an even number of tokens or no token; the simulation of the instructions in I allows the state places to have either no token or two tokens.

All these transitions belong to *join* and the number of tokens removed by a firing is greater than or equal to the number of tokens put by the same firing.

The P/T$'$ system can simulate the register machine in a faithful way. If, starting from an input (tokens put in the register places), the state place p_f associated with the final state of the register machine gets a token, then the register machine accepts that input. It should be clear that the simulation is a weak one. □

4.5 Vector addition systems

Formal systems very closely related to P/T systems are *vector addition systems*.

Definition 4.5 *A* vector addition system *(VAS) of dimension k is a pair $W = (v_0, Y)$, where $v_0 \in \mathbb{N}^k$ is a vector called the* initial vector *and Y is a finite set of vectors over $\mathbb{Z}^k$.*

A VAS $W = (v_0, Y)$ operates starting from the initial vector v_0 and sequentially adding vectors in Y. Each addition is called a *transition*. The obtained vectors have to have elements in $\mathbb{N}$. If no vector in Y can be added to the obtained vector such that the result of the sum is a vector with elements in $\mathbb{N}$, then the VAS *halts*.

The *reachability set* of a VAS $W = (v_0, Y)$ is the set $\mathbb{C}_W$ containing the initial vector v_0 and all vectors in $\mathbb{N}^k$ that can be obtained adding elements of Y to v_0. The *halting reachability set* of a VAS $W = (v_0, Y)$ is the set $\mathbb{C}_W^h = \{z \mid z \in \mathbb{C}_W,\ z + w \not\geq 0\ \forall w \in Y\}$.

Definition 4.6 *A* vector addition system with states *(VASS) of dimension k is a tuple $W = (v_0, S, H, s_0)$ where:*

$v_0 \in \mathbb{N}^k$ *is a vector called the* initial vector;

S *is a set of* states;

$H : S \to \mathbb{Z}^k \times S$ *is the* transition function;

$s_0 \in S$ *is the* initial state.

If $H(s) = (w, s')$, then the tuple (s, w, s') is called a *transition*. A VASS $W = (v_0, S, H, s_0)$ being in state s' and having current vector v' operates by adding $w \in \mathbb{Z}^k$ to v' and changing state into s'' only if $H(s') = (w, s'')$ and

$v'' = v' + w \geq 0$. The vector v'' then becomes the current vector and s'' the state of W. Initially the state of W is s_0 and the current vector is the initial vector v_0. If when in state s and having v as current vector no element in the transition function H can be applied (either because $H(s)$ is not defined or because $H(s) = (w''', s''')$ and $v + w'''$ is not greater than or equal zero), then W *halts*.

Definition 4.7 *A VAS or VASS $W = (v_0, S, H, s_0)$ is communication-free if for each transition $(s, w, s') = (s, (w_1, \ldots, w_k), s')$ at most one w_i is negative and, if such, equal to -1.*

The *reachability set* $\mathbb{C}_W$ and *halting reachability set* $\mathbb{C}_W^h$ for a VASS W are defined similarly as for a VAS.

The *reachability problem* for a VAS (VASS) W is to determine whether $v \in \mathbb{C}_W$, given v. The *equivalence problem* for two VASs (VASSs) W' and W'' is to determine whether $\mathbb{C}_{W'} = \mathbb{C}_{W''}$. The *reachability problem for halting configurations* and the *equivalence problem for halting configurations* are defined in a similar way.

It is known that:

Theorem 4.8

(i) *If W is an n-dimensional VASS, then it is possible to define an $(n + 3)$-dimensional VAS W' simulating W.*

(ii) *If W is a two-dimensional VASS, then $\mathbb{C}_W \in \mathsf{SLS}_2$.*

(iii) *There are three-dimensional VASS V such that $\mathbb{C}_W \notin \mathsf{SLS}_3$.*

(iv) *If W is a five-dimensional VAS, then $\mathbb{C}_W \in \mathsf{SLS}_5$.*

(v) *There are six-dimensional VAS W such that $\mathbb{C}_W \notin \mathsf{SLS}_6$.*

(vi) *The (halting) reachability problem for VAS and for VASS is decidable.*

(vii) *The equivalence problem for VAS and for VASS is undecidable.*

4.6   P/T systems and membrane systems

As already said the results presented in this chapter are applicable to several models of membrane systems considered in this monograph.

The reader could then wonder why do we study membrane systems (or any other model that can be simulated by a P/T system) instead of solely studying the computational power of P/T systems in terms of building blocks, their combinations and other factors. The results obtained by such a study are indeed general and applicable to a vast number of abstract computing devices. However they do not say much about the details, of how, for instance, a specific computing device simulates instructions of the kind (s, γ^-, v, w) and they do not consider the features defining the way a computer device operates. In some models of membrane systems, for instance, the topological structure of the compartments is a cell-tree and symbols can pass from one compartment to another only following the topological structure. A simulation could translate this feature into a particular labelling of the places in the P/T system; this labelling would then put limits in the flow relation of the P/T system. This implies that the simulation relations α and β could be quite complex. This would limit further studies of the specific membrane system or result in very unnatural (and inelegant) definitions (see the beginning of Section 6.7).

So, if the aim is to understand how a specific computing device works, then its direct study is the best path to follow. As we will see in the following chapters, some of the results considered in the present chapter help the study of the computational power of models of membrane systems. In the majority of the chapters presenting models of membrane systems we hint at how EN systems and P/T systems can simulate models of membrane systems.

Some models of membrane systems allow us to have an infinite number of symbols in the initial configuration but these systems are such that only a finite number of these symbols is used for each computation. This is comparable to models of register machines having registers with initially an unbounded value. How can P/T system simulate such systems or machines? One way to do this considers that membrane systems with an unbounded number of symbols use only a finite number of them in each computation. The P/T systems performing the simulation can then randomly fix the number of available symbols at the beginning of each computation. This number is independent of the initial configuration of the P/T system. In Theorem 4.7 we already saw a net with an infinite number of places allowing a place to have a random even number of tokens. Other nets are those composed of *join* and depicted in Fig. 4.25(a) and Fig. 4.25(b) (the notation used in this figure is consistent with that used in the present chapter). The subsequent simulation can then halt with a token in p_{acc} only if the random amount initially fixed was sufficient for that computation.

4.7 Final remarks and research topics

We think that the line of research presented in this chapter is important and worth continuing. Looking at the dynamics of a system it is possible to unify (that is, regard as similar) elements that otherwise would be regarded as different (see, for instance, Theorem 4.1). This, together with the study of the dynamical properties of a system, allows us to unify different computing models,

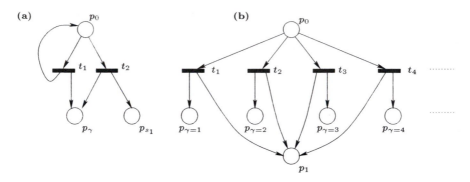

Fig. 4.25 Two nets to fix random quantities

facilitates the study of their computational power and introduces new measures of complexity.

If it is not strictly needed to know the details of the system accepting or generating certain sets of vectors, then this kind of study is very useful as it avoids tedious proofs.

In the following chapters we will see how this can be applied to the models of membrane systems we consider. It should be clear that these applications are not limited to membrane systems but they could be extended to Brane Calculi, Diophantine equations, etc.

Theorem 4.7 is not satisfactory as it is not only based on the dynamics (that is, the presence of only *joins*) but considers systems with limitations in the weight function. For this reason we have:

Suggestion for research 4.1 *Determine the computational power of P/T systems whose underlying net is composed of* joins. *Is there a simulation between the MPCGs of these systems and register machines?*

Only in Theorem 4.4 have we considered limitations in the arrangements of the building blocks. What about other limitations in the arrangement?

Suggestion for research 4.2 *Study further the computational power of P/T systems whose relative arrangement of pairs of* fork *and* join *is limited.*

In particular one could focus on the limitations needed to generate semilinear sets.

The overall aim of this research is to:

Suggestion for research 4.3 *Create full hierarchies of accepting and generating computational processes in terms of* join *and* fork, *their combinations and the functions W and K.*

4.8 Bibliographical notes

Petri nets were introduced in [197] and since then they have been extensively studied. Readers interested in knowing more about Petri nets can refer to [58, 227, 228].

P/T systems having an underlying net that depicted in Fig. 4.1 were considered also in [35]. There the authors considered maximum concurrency (there called *maximal strategy*) as a way to operate.

The term *conflicting register* was introduced in [87]. A concept similar to conflicting register was also used in [69]. Even if explicitly defined in the above papers, the concept of conflicting register was already implicitly present in all the proofs of membrane systems using symbols generating $\mathbb{N}$ RE and using 'gamble' ([203], for instance). Some of these proofs were so made that once in p_w the computation could still enter the p_{wrong} place (for instance, Theorem 1 in [79]).

Studies of EN systems as a computing device and of the dynamical properties of specific computing devices through EN systems are reported, for instance, in [35, 111, 134, 145–147].

Some parts presented in this chapter were published in [81]. The present chapter is mainly based on [83].

The results summarised in Theorem 4.8 are from [24, 104, 112, 178, 240]. It is relevant to say that communication-free VAS are equivalent to communication-free EN systems and commutative context-free grammars [61, 114].

4.8.1 About the notation

Readers not familiar with Petri nets may wonder why we used the symbol • to denote both the input and output sets of elements in X and the tokens present in places. Here we followed the very well established notation present in this field of research.

5 Symport/antiport

In this chapter we present one of the most studied models of membrane systems: *P systems with symport/antiport*. Their simple and elegant way of operating has caught the interest of many researchers and several papers have been written on this model or inspired by the operations used by it. Some issues studied relate to descriptional complexity, others considered restricted ways in which operations can be performed, others used these operations on other platforms, etc.

All in all, it would be very well possible to write a monograph dedicated only to P systems with symport/antiport. In this chapter we give an overview and provide the proofs of the latest results on this model.

5.1 Biological motivations

One of the functions of the plasma membrane is to maintain the internal composition of the cell. This membrane forms a barrier that blocks the free exchange of most biological molecules between the cytoplasm and the environment.

A small number of molecules can diffuse in and out of a cell depending on their relative concentration inside and outside it. This process is called *passive diffusion* and it always concerns the passage from a compartment with high concentration to one with a lower concentration of the molecule.

Facilitated diffusion is performed by specific transport proteins, *channel* and *carrier* proteins, that mediate the selective passage of molecules across the membrane.

Channel proteins form open pores in the membrane, allowing small molecules of the appropriate size and charge to freely pass through the lipid bilayer. Examples of channel proteins are *porins*, which permit the free passage of ions and small polar molecules; *aquaporins*, which permit water molecules to be able to cross the membrane much more rapidly than without such proteins; and *ion channels*, which mediate the passage of ions across the plasma membrane.

Some of these channels, ions channels in particular, are not permanently open. Instead their activity is regulated by 'gates' that transiently open in response to specific stimuli. *Ligand-gated channels*, for instance, open in response to the binding of neurotransmitters or other signalling molecules; *voltage-gated channels* open in response to changes in electric potential across the plasma membrane (see Section 7.1).

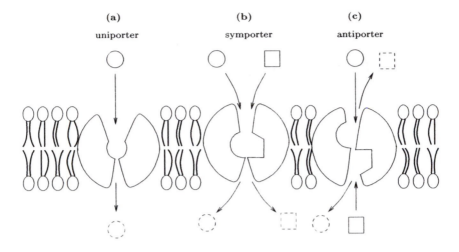

Fig. 5.1 **(a)** A uniporter, **(b)** a symporter and **(c)** an antiporter

Carrier proteins (also called *carriers, permeases* or *transporters*) bind specific molecules to be transported and undergo conformational changes that allow the molecules to pass across the membrane. Carrier proteins perform either *passive transport* or *active transport*. In passive transport the transported molecules are moved according to the concentration gradient; in active transport the molecules are moved against the concentration gradient.

Uniporters transport a single type of molecule and perform passive transport. The activity of a uniporter is depicted in Fig. 5.1.

In *active transport* the activity of the carriers (also called *pumps*) is tightly coupled with a source of metabolic energy. This kind of transport is performed by *symporters* and *antiporters*.

Symporters can transport two molecules in the same direction, while *antiporters* can transport two molecules in opposite directions. The activity of symporters and antiporters is depicted in Fig. 5.1.

5.2   Basic definitions

Definition 5.1 *A* (generating) P system with symport/antiport *of degree m is a construct*

$$\Pi = (V, \mu, L_0, L_1, \ldots, L_m, R_1, \ldots, R_m, comp)$$

where:

V is an alphabet;

$\mu = (Q, E)$ is a cell-tree underlying Π where:

$Q \subset \mathbb{N}$ *contains* vertices. *For simplicity we define* $Q = \{0, 1, \dots, m\}$. *Each*
vertex in Q *defines a* compartment *of* Π;

$E \subseteq Q \times Q$ *defines* arcs *between vertices, denoted by* (i, j), $i, j \in Q, i \neq j$;

L_i, $0 \leq i \leq m$, *are multisets over* V. *All the elements in* L_0 *have infinite*
multiplicity *while the elements in* $L_1, \dots, L_m$ *have finite multiplicity*;

R_i, $1 \leq i \leq m$, *are finite sets of rules of the kind:* $(x; in), (x; out)$, *called* symport
rules, *or* $(y; out/x; in)$, *called* antiport *rules, with both* x *and* y *multisets over*
V *with a finite and not-empty support. This means that* $(\phi; in), (\phi; out)$ *and*
$(\phi; out/\phi; in)$ *are not allowed*;

$comp \in Q$ *defines the* output compartment *which has to be elementary.*

Vertex 0 identifies the environment, while vertex 1 the skin compartment of
μ. The rules R_i associated with membrane i defining compartment i, $1 \leq i \leq m$,
can change the multisets M_i and M_j over V, associated with vertices i and j,
respectively, in μ where i is the parent of j in the following way:

if $(x; in) \in R_j$ and $x \subseteq M_i$, then M_i changes into $M'_i = M_i \setminus \{x\}$ while M_j
changes into $M'_j = M_j \cup \{x\}$;

if $(x; out) \in R_j$ and $x \subseteq M_j$, then M_j changes into $M'_j = M_j \setminus \{x\}$ and M_i
changes into $M'_i = M_i \cup \{x\}$;

if $(y; out/x; in) \in R_j$, $x \subseteq M_i$ and $y \subseteq M_j$, then M_i changes into $M'_i =
M_i \setminus \{x\} \cup \{y\}$ and M_j changes into $M'_j = M_j \cup \{x\} \setminus \{y\}$.

In general, if a multiset x is subtracted from M_i and united to M_j we say
that x *passes* from compartment (vertex) i to compartment (vertex) j.

A *configuration* of a P system Π with symport/antiport of degree m is given
by the $(m+1)$-tuple $(M_0 \setminus L_0, M_1, \dots, M_m)$ of multisets over V associated with
the compartments of Π. Note that the configuration does not record the symbols
in the environment occurring with infinite multiplicity as they are invariant to
any configuration. The $(m+1)$-tuple $(\phi, L_1, \dots, L_m)$ is called the *initial configu-*
ration. For two configurations $(M_0 \setminus L_0, M_1, \dots, M_m)$ and $(M'_0 \setminus L_0, M'_1, \dots, M'_m)$
of Π we write $(M_0 \setminus L_0, M_1, \dots, M_m) \Rightarrow (M'_0 \setminus L_0, M'_1, \dots, M'_m)$ to indicate a
transition from $(M_0 \setminus L_0, M_1, \dots, M_m)$ to $(M'_0 \setminus L_0, M'_1, \dots, M'_m)$, that is, the
application of a multiset of rules associated with each compartment under the
requirement of maximal parallelism. The reflexive and transitive closure of $\Rightarrow$ is
denoted by $\Rightarrow^*$.

A *computation* is a sequence of transitions between configurations of a system Π starting from the initial configuration $(\phi, L_1, \dots, L_m)$. If a computation
is finite, then the last configuration is called *final* and we say that the system *halts*. The result of a finite computation is the cardinality of the multiset of symbols present in the final configuration in the output compartment
comp.

The *weight* of a rule is given by $|v|$ (that is, the cardinality of the multiset v) in case of a symport $(v; \text{in})$ or $(v; \text{out})$ and by $max(\{|v|, |w|\})$ in case of an antiport $(v; \text{out}/w; \text{in})$.

The set $N(\Pi)$ denotes the set of numbers generated by a P system Π with symport/antiport. The family of sets $N(\Pi)$ generated by P systems with symport/antiport of degree at most m, using symports of weight at most p and antiports of weight at most q is denoted by $N\,OP_m(\text{sym}_p, \text{anti}_q)$.

P systems with symport/antiport have been also considered as *accepting* devices. In this case a distinguished compartment, *comp* in Definition 5.1, here called an *input compartment*, contains, in the initial configuration, the input. If such a system halts, then it is said to accept the input. The family of sets accepted by such P systems with symport/antiport are indicated as the families generated by such systems but with an 'a' after the N. So, for instance, the family of sets accepted by P systems with symport/antiport of degree at most m, using symports of weight at most p and antiports of weight at most q, is denoted by $N\,aOP_m(\text{sym}_p, \text{anti}_q)$.

Other models of P systems with symport/antiport considered in this monograph are introduced in the following sections.

5.3   Examples

The P systems with symport/antiport considered in this section generate very similar or the same sets of numbers. The presented systems differ in the kind of rules they use and aim to show how similar sets of numbers can be generated by different kinds of P systems with symport/antiport.

Example 5.1 A P system with symport/antiport of degree 3 using only symports of weight 2 generating any non-negative even number.

The formal definition of such a P system is:

$$\Pi_1 = (\{a, b\}, (\{0, 1, 2, 3\}, \{(0, 1), (1, 2), (1, 3)\}), \{a\}, \{b\}, \phi, \phi, R_1, R_2, R_3, 3)$$

with
$R_1 = \{(b; \text{out}), (ba; \text{in})\};$
$R_2 = \{(aa; \text{in}), (b; \text{in})\};$
$R_3 = \{(aa; \text{in})\}.$

In the initial configuration there is an infinite number of occurrences of a in the environment, the skin compartment contains b, while the two remaining compartments are empty. Notice that, as required by Definition 5.1, compartment 3, the final compartment, is elementary.

The system works as follows: occurrences of a's can pass from the environment to the skin compartment. This is performed with the help of the symbol b: it can pass from the skin compartment to the environment by itself and pass back together with an occurrence of a. When pairs of a's are present in the skin compartment, then they can pass into compartment 2 and from here to compartment 3. At any moment the symbol b can pass into compartment 2, in this way halting the computation.

The number of a's present in compartment 3 represents the output of the system. It is possible that when the system halts one occurrence of a is present in the skin compartment. This occurrence does not contribute to the set of numbers generated by Π_1. We can say that Π_1 generates $\{2n \mid n \in \mathbb{N}\}$. $\diamond$

What about if one wants a P system with symport/antiport able to generate any multiple of a number n? A possibility is to modify Example 5.1 so that the rules in compartments 2 and 3 are $(a^n; \text{in})$ instead of $(a^2; \text{in})$, that is, multisets of n a's. The resulting systems would have a weight equal to n. Another possibility is to define a P system with symport/antiport having a fixed weight independently of n. Here are two examples, one using only symports of weight 3 and the other using symports of weight 1 and antiports of weight 2.

Example 5.2 A P system with symport/antiport of degree 2 using only symports of weight 3 generating any multiple of n, $n \in \mathbb{N}_+$.

The formal definition of such a P system is:

$$\Pi_2 = (V, \mu, L_0, L_1, \phi, R_1, R_2, 2)$$

with:

$$
\begin{aligned}
V &= \{a, b_1, \dots, b_{n+1}, c_1, \dots, c_{n+1}\}; \\
\mu &= (\{0, 1, 2\}, \{(0, 1), (1, 2)\}); \\
L_0 &= \{a, c_2, \dots, c_{n+1}\}; \\
L_1 &= \{c_1, b_1, \dots, b_{n+1}\}; \\
R_1 &= \{(b_i c_i; \text{out}) \mid 1 \le i \le n+1\} \cup \{(b_i c_{i+1} a; \text{in}) \mid 1 \le i \le n\} \cup \\
&\quad\ \{(b_{n+1} c_1; \text{in}), (c_1; \text{out})\}; \\
R_2 &= \{(a; \text{in})\}.
\end{aligned}
$$

As required by Definition 5.1, compartment 2, the final compartment, is elementary. The system works as follows: b_i and c_i pass from the skin compartment to the environment (initially b_1 and c_1 pass). When this happens b_i, c_{i+1} and a can pass from the environment to the skin compartment. Because of $(a; \text{in}) \in R_2$ one occurrence of a can pass from the skin compartment to compartment 2. At the same time b_{i+1} and c_{i+1} can pass to the environment. This 'chain' goes on until $(b_n c_{n+1} a; \text{in})$ is applied. When this happens n occurrences of a passed in the skin compartment, and from here to compartment 2. When c_{n+1} is in the skin compartment $(b_{n+1} c_{n+1}; \text{out})$ is applied. This is followed by the application of $(b_{n+1} c_1; \text{in})$ so that the 'chain' can be repeated. When c_1 is in the skin compartment, then $(c_1; \text{out})$ can be applied halting Π_2.

The symbols a can enter the skin compartment and then compartment 2 in 'chains' of n. We can say that Π_2 generates $\{jn \mid j \in \mathbb{N}\}$ for a given $n \in \mathbb{N}_+$. $\diamond$

Example 5.3 A P system with symport/antiport of degree 2 using symports of weight 1 and antiports of weight 2 generating any multiple of n, $n \in \mathbb{N}_+$.

The formal definition of such a P system is:

$$\Pi_3 = (\{a, b_1, \ldots, b_{n+1}\}, (\{0, 1, 2\}, \{(0, 1), (1, 2)\}), L_0, L_1, \phi, R_1, R_2, 2)$$

with

$L_0 = \{b_2, \ldots, b_{n+1}\};$
$L_1 = \{b_1\};$
$R_1 = \{(b_i; \text{out}/ab_{i+1}; \text{in}) \mid 1 \leq i \leq n\} \cup \{(b_{n+1}; \text{out}/b_1; \text{in}), (b_1; \text{out})\};$
$R_2 = \{(a; \text{in})\}.$

As required by Definition 5.1, compartment 2, the final compartment, is elementary. The system works as follows: b_i passes from the skin compartment to the environment at the same moment when b_{i+1} and a pass from the environment to the skin compartment. Initially the applied rule is $(b_1; \text{out}/b_2a; \text{in})$. The presence of b_i in the skin compartment allows the application of $(b_i; \text{out}/b_{i+1}a; \text{in})$. The occurrences of a present in the skin compartment pass into compartment 2 because of $(a; \text{in}) \in R_2$.

This 'chain' goes on until b_{n+1} is brought in the skin compartment. During this process n occurrences of a passed from the environment to the skin compartment. When b_{n+1} passes to the skin compartment, then $(b_1; \text{out}/b_{n+1}; \text{in})$ can be applied. In this way the 'chain' can be repeated. When b_1 is in the skin compartment, then $(b_1; \text{out})$ can be applied halting Π_3.

We can say that Π_3 generates $\{jn \mid j \in \mathbb{N}\}$ for a given $n \in \mathbb{N}_+$. $\diamond$

5.4   P systems with symport/antiport and P/T systems

Let $\Pi = (V, \mu, L_0, L_1, \ldots, L_m, R_1, \ldots, R_m, comp)$, $\mu = (Q, E)$, be a P system with symport/antiport. The presence of k occurrences of a symbol $x \in V$ in a compartment $i \in Q$ can be indicated in a P/T system by a place $x_{[i]}$ with k tokens associated with it. If k is an unbounded quantity, then this can be simulated by a P/T system as described in Section 4.6. Symport rules of the kind $(x; \text{in})$ and $(x; \text{out})$, x a multiset over V, can be simulated by P/T systems having an underlying net with transitions having as input and output sets a number of places equal to the different elements of V present in x. The weight functions associated with the flow relation depend on the cardinality of the elements on V present in x. Let us assume that in a P system with symport/antiport vertex j is the parent of vertex i. The nets underlying the P/T system simulating

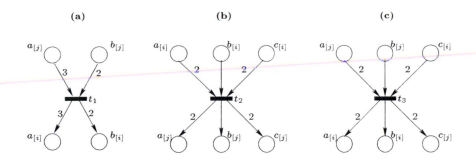

Fig. 5.2 Nets underlying P/T systems simulating **(a)** $(a^3b^2; \text{in}) \in R_i$, **(b)** $(a^2bc^2; \text{out}) \in R_i$ and **(c)** $(c^2; \text{out}/a^2b; \text{in}) \in R_i$

$(a^3b^2; \text{in}) \in R_i$ and $(a^2bc^2; \text{out}) \in R_i$, $a, b, c \in V$, are depicted in Fig. 5.2(a) and Fig. 5.2(b), respectively.

Antiport rules of the kind $(y; \text{out}/x; \text{in})$, x and y multisets over V, can be simulated by transitions having as input and output sets a number of places equal to the different elements of V present in $x \cup y$. The weight functions associated with the flow relation depend on the cardinality of the elements on V present in x and y. Let us assume again that in a P system with symport/antiport vertex j is the parent of vertex i. The net underlying the P/T system simulating $(c^2; \text{out}/a^2b; \text{in}) \in R_i$, $a, b, c, \in V$, is depicted in Fig. 5.2(c).

We know from Fig. 4.18 how the net depicted in Fig. 5.2(a) can be rewritten as a composition of building blocks. Now we show how nets such as those depicted in Fig. 5.2(b) and Fig. 5.2(c) can be rewritten in a similar fashion. We consider the net depicted in Fig. 5.2(c). Figure 5.3 shows how to rewrite it as a composition of building blocks. The net depicted in Fig. 5.3 underlies a P/T system such that its $MPCG$ can weakly simulate $(c^2; \text{out}/a^2b; \text{in}) \in R_i$. The place d (for *dummy*) has one token in the initial configuration. In order to fire $t_3^{(1)}$ needs the presence of two tokens in $a_{[j]}$ and one token in $b_{[j]}$; in order to fire transition $t_3^{(2)}$ needs the presence of two tokens in $c_{[i]}$ and one token in d.

Let us assume that in a configuration C the places $a_{[j]}$ and $b_{[j]}$ have at least 2 and 1 tokens, respectively, while $c_{[i]}$ has less than 2 tokens. Transition $t_3^{(1)}$ and $t_3^{(3)}$ fire in sequence letting the initial configuration C to be restored. Similarly if transition $t_3^{(1)}$ cannot fire while transition $t_3^{(2)}$ can fire. The subsequent firing of transition $t_3^{(4)}$ allows the original configuration to be restored.

Transition $t_3^{(5)}$ can then fire only if transitions $t_3^{(1)}$ and $t_3^{(2)}$ fired in the same configuration. This is followed by the firing of transition $t_3^{(6)}$ and then $t_3^{(7)}$ and $t_3^{(8)}$ in parallel. At the end of this firing sequence some tokens have been removed from $a_{[j]}$, $b_{[j]}$, $c_{[i]}$ and d and the same number of tokens has been put into $a_{[i]}$, $b_{[i]}$, $c_{[j]}$ and d, respectively.

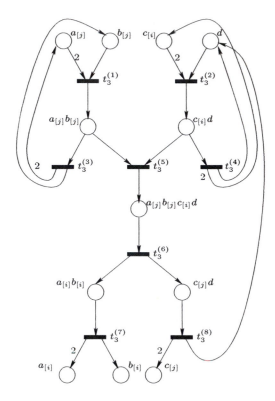

Fig. 5.3 Figure 5.2(c) rewritten as a composition of building blocks

The simulation of symports or antiports involving more than three different symbols with nets composed of building blocks can be performed in a similar way. The dummy place is only present if in these rules there is an odd number of different symbols.

In the next section we show how P/T systems can simulate instructions of the kind $(s, \gamma^{=0}, v)$ when their underlying net is composed of nets such as those depicted in Fig. 5.2.

5.5   Minimal degree and weight

Several researchers challenged themselves by studying the computational power of P systems with symport/antiport having minimal degree and weight. It was very surprising to find that relatively small numbers for these measures allowed these systems to be computationally complete.

Theorem 5.1 $\mathbb{N} OP_1(sym_1, anti_1) \subseteq \mathbb{N} FIN$.

Proof The allowed antiports do not change the number of symbols present in the skin compartment. This number can only change by the application of symports. If a symport of the kind $(a; \text{in})$ is associated with the skin membrane and the symbol a is initially present in the environment, then the P system never halts. If instead a was not initially present in the environment, then it was initially present (with a finite number of occurrences) in the skin compartment. Such symbols can pass to the environment either by symports of the kind $(a; \text{out})$ or by antiports of the kind $(a; \text{out}/b; \text{in})$, $a \neq b$. In the former case the P system never halts, in the latter case the number of symbols present in the skin compartment does not change.

The empty set can be generated by a P system with symport/antiport with no symbols in the skin compartment in the initial configuration. $\square$

Theorem 5.2 $\mathbb{N} \, OP_1(sym_2, anti_0) \subseteq \mathbb{N} \, \text{FIN}$.

Proof Let $\Pi = (V, (\{0,1\}, \{(0,1)\}), L_0, L_1, R_1, 1)$ be a P system with symport/antiport, symports of weight 2 and no antiports. We prove that Π cannot increase the number of symbols initially present in the skin compartment without going on forever.

Let $L_0 = \{a_1, \ldots, a_p\}$ and $L_1 = \{b_1, \ldots, b_q\}$, $p, q \in \mathbb{N}_+$.

One occurrence of a symbol initially present in the environment, let us say a_1, can pass into the skin compartment only with another symbol b_j initially present in the skin compartment by $(a_1 b_j; \text{in})$. Without loss of generality, let $j = 1$.

As the symbol b_1 is not present in the environment in the initial configuration, then first it has to pass there so that $(a_1 b_1; \text{in})$ can be applied. This can happen by $(b_1; \text{out})$ or by $(b_1 c; \text{out})$. In the first case the application of rules $(a_1 b_1; \text{in})$ and $(b_1; \text{out})$ allows Π to never halt. In the second case there are several possibilities. The symbol c can be initially present in the skin compartment. If $c = b_1$, then the application of rules $(a_1 b_1; \text{in})$ and $(b_1 c; \text{out})$ allows Π to never halt. If $c \neq b_1$ then the computation is finite. In this case the occurrences of c present in compartment 1 are substituted with those of a_1 not increasing the number of symbols present in this compartment. If c is not initially present in the environment, then $c \neq a_1$. This is because the application of the rules $(a_1 b_1; \text{in})$ and $(b_1 a_1; \text{out})$ allows Π to never halt. Let us say then that $c = a_2$.

The symbol a_2 is not initially present in the skin compartment, so some rule lets it pass into this compartment. The application of rules $(a_2; \text{in})$ and $(a_2 a_j; \text{in})$, $1 \leq j \leq p$ allows Π to never halt. The only possible way to let a_2 pass into the skin compartment is by a rule of the kind $(a_2 b_j; \text{in})$, with $1 \leq j \leq q$. If $j = 1$ the application of rules $(a_1 b_1; \text{in})$, $(b_1 a_2; \text{out})$ and $(a_2 b_1; \text{in})$ allows Π to never halt. If $j \neq 1$ let, without loss of generality, $j = 2$, then the rules $(a_1 b_1; \text{in})$, $(b_1 a_2; \text{out})$ and $(a_2 b_2; \text{in})$ are in R_1.

Now we can say of b_2 what we said before for b_1: as it is not initially present in the environment, then it has to pass there from the skin compartment. In this way the rule $(a_2b_2; \text{in})$ can be applied. This can only happen if a rule of the kind $(b_2c'; \text{out})$ is in R_1. If the symbol c' is initially present in the environment and $c' = b_1$, then the rules $(a_1b_1; \text{in})$, $(b_1a_2; \text{out})$, $(a_2b_2; \text{in})$ and $(b_2b_1; \text{out})$ allows Π to never halt. Similarly if $c' = b_2$. Let $c' = b_j$, $3 \leq j \leq q$ and, without loss of generality, $j = 3$, then the application of rules $(a_1b_1; \text{in})$, $(b_1a_2; \text{out})$, $(a_2b_2; \text{in})$ and $(b_2b_3; \text{out})$ allows Π to never halt. If c' is initially in the environment, then either for $c' = a_1$ or $c' = a_2$ no computation can be performed. If $c' = a_3$, then we can say of a_3 what we said for a_2.

At this point it should be clear that if a rule of the kind $(a_{i_1} b_{j_1}; \text{in})$ is present in the system, then also the rule $(b_{j_1} a_{i_2}; \text{out})$, with $i_2 \neq i_1$, has to be present. This implies that also the rule $(a_{i_2} b_{j_2}; \text{in})$, with $j_2 \neq j_1$, has to be present. This implies that $(b_{j_2} a_{i_3}; \text{out})$, with $i_3 \neq i_1, i_2$ has to be present. This implies that the rule $(a_{i_3} b_{j_3}; \text{in})$, with $j_3 \neq j_1, j_2$, has to be present, and so on.

In general, we can say that if $(b_{j_{h'}} a_{i_{k'}}; \text{out})$ is present, then $i_{k'} \neq i_1, \ldots, i_{k'-1}$, and that if $(a_{i_{k'}} b_{j_{h''}}; \text{in})$ is present, then $j_{h''} \neq j_1, \cdots, j_{h''-1}$. If we consider that q and p are finite, then the only operations that can be performed by such systems are either to maintain the initial configuration or to pass symbols from the skin compartment to the environment. $\square$

The following proofs consider register machines with instructions of the kind: $(s, \gamma^-, v), (s, \gamma^{=0}, w)$ and (s, γ^+, v).

Theorem 5.3 $\mathbb{N} \, OP_1(sym_1, anti_2) = \mathbb{N} \, RE$.

Proof Let us consider a register machine $M = (S, I, s_1, s_f)$ with n registers generating numbers. We define $\Pi = (V, \mu, L_0, L_1, R_1, 1)$ to be a P system with symport/antiport simulating M. The alphabet of Π contains the state-symbols s_i, s_i' and s_i'', for each state $s_i \in S$, it contains register-symbols γ_i and conflicting register-symbols $\bar{\gamma}_i, 1 \leq i \leq n$, for each of the registers of M, it also contains the symbol $\star$. The cell-tree underlying Π is composed of the environment and the skin compartment, this last being the final compartment. The multiset L_0 contains all the symbols in V, while L_1 contains s_1, the state-symbol associated with the initial state of M, and $val(\gamma_i)$ occurrences of the register-symbols $\gamma_i, 1 \leq i \leq n$ (where $val(\gamma_i)$ is the content of register γ_i).

If by γ we indicate a generic register in M, then the rules in R_1 associated with the instructions in M are:

instructions	rules
(s, γ^+, v)	$(s; \text{out}/\gamma v; \text{in})$
(s, γ^-, v)	$(s\gamma; \text{out}/v; \text{in})$
$(s, \gamma^{=0}, v)$	$(s; \text{out}/\bar{\gamma}v'; \text{in})$
	$(v'; \text{out}/v''; \text{in})$
	$(v''\bar{\gamma}; \text{out}/v; \text{in})$

Notice that three rules are associated with instructions of the kind $(s, \gamma^{=0}, v)$. The rules $(\gamma\bar{\gamma}; \text{out}/\star; \text{in}), (\star; \text{out}/\star; \text{in}), (s; \text{out}/s; \text{in}), s \in S \backslash \{s_f\}$ and $(s_f; \text{out})$ are also in R_1.

The presence of a state-symbol s in the skin compartment indicates that Π is simulating M being in state s. The repeated application of $(s; \text{out}/s; \text{in})$ allows Π to never halt.

The simulation of instructions of the kind (s, γ^+, v) and (s, γ^-, v) is straight-forward: the state-symbol s passes to the environment while the state-symbol v passes into the skin compartment. At the same time an occurrence of a register-symbol also passes from one compartment to another. If Π tries to simulate an instruction of the kind (s, γ^-, v) when no occurrences of γ are present in the skin compartment, then $(s\gamma; \text{out}/v; \text{in})$ cannot be performed while $(s; \text{out}/s; \text{in})$ is performed so that Π never halts.

The simulation of instructions of the kind $(s, \gamma^{=0}, v)$ uses conflicting register-symbols. Initially the rule $(s; \text{out}/\bar{\gamma}v'; \text{in})$ is applied. If a register-symbol γ is present in the skin compartment, then first $(\gamma\bar{\gamma}; \text{out}/\star; \text{in})$ and then $(\star; \text{out}/\star; \text{in})$ are applied. The subsequent continuous application of $(\star; \text{out}/\star; \text{in})$ allows Π to never halt. If instead no occurrence of γ is present in the skin compartment when $\bar{\gamma}$ passes into this compartment, then $(v'; \text{out}/v''; \text{in})$ and $(v''\bar{\gamma}; \text{out}/v; \text{in})$ are applied. The state-symbol v is present in the skin compartment only if no occurrence of γ is present in this compartment.

When the state-symbol s_f, associated with the final state of M, is present in the skin compartment, then $(s_f; \text{out})$ is applied and the only symbols present in the skin compartment are the register-symbols γ indicating the value of register γ in a final configuration of M. $\square$

Let us assume that in the P systems with symport/antiport related to the previous theorem the only symport used $(s_f; \text{out})$, where s_f is the final state of the register machine M, is removed. The resulting P system uses only antiports and when it halts it always has at least one symbol, s_f, in the skin compartment. So we have:

Corollary 5.1 $\mathbb{N}_1 \, OP_1(sym_0, anti_2) = \mathbb{N}_1 \, RE$.

Before going on introducing more results on the computational power of P systems with symport/antiport considered in this chapter we want to show using P/T systems how the simulation of instructions of the kind $(s, \gamma^{=0}, v)$ has been performed in the proof of Theorem 5.3. We do this following the labelling of the vertices as indicated in Section 5.4. The net underlying this P/T system is depicted in Fig. 5.4 where, for the moment, we consider all the dashed places, transitions and edges as they had full lines.

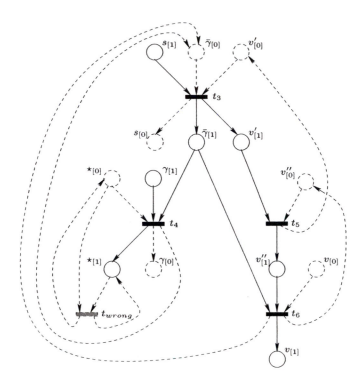

Fig. 5.4 The net underlying a P/T system simulating $(s, \gamma^{=0}, v)$ related to Theorem 5.3

Transition t_3 in this figure refers to $(s; \text{out}/\bar{\gamma}v'; \text{in})$, transition t_5 refers to $(v'; \text{out}/v''; \text{in})$, transition t_6 refers to $(v''\gamma; \text{out}/v; \text{in})$, transition t_4 refers to $(\gamma\bar{\gamma}; \text{out}/\star; \text{in})$ and transition t_{wrong} refers to $(\star; \text{out}/\star; \text{in})$. If we now look at the net depicted in Fig. 5.4 disregarding the dashed places, transitions and edges, then we see that the remaining net is similar to part of the net depicted in Fig. 4.13. The similarity is about transitions t_3, t_4, t_5 and t_6, their input and output sets and related flow relations. So, in Theorem 5.3 the simulation of instructions of the kind $(s, \gamma^{=0}, v)$ follows (part of) what was described in Section 4.3. In Fig. 5.4 it is important to notice that the simulation of $(s, \gamma^{=0}, v)$, that is, the firing of t_3, t_5 and t_6, allows the initial configuration to be different from the final one only for the s and v places. The remaining places involved in this firing, $\bar{\gamma}, v'$ and v'', have a token in the final configurations only if they had one in the initial configuration.

Nets similar to that in Fig. 5.4 can be depicted for the remaining theorems in this section. We do not do this as such nets are quite complex.

If one considers only symports, then:

Theorem 5.4 $\mathbb{N}_7 \, OP_1(sym_3, antio) = \mathbb{N}_7 \, RE.$

Proof Let $M = (S, I, s_1, s_f)$ be a register machine with n registers generating numbers. We define $\Pi = (V, \mu, L_0, L_1, R_1, 1)$ to be a P system with symport/antiport simulating M. We consider $I' = I \cup \{(s, \bar{\gamma}^+, s_v), (s_v, \bar{\gamma}^-, v) \mid (s, \gamma^{=0}, v) \in I, \gamma \text{ register in } M\}$. The instructions in I' have unique labels $l \in \{1, \ldots, |I'|\}$. In the following we indicate the components of Π.

$$
\begin{aligned}
V &= L_0 \cup L_1; \\
\mu &= (\{0,1\}, \{(0,1)\}); \\
L_0 &= S \cup \{s_v \mid (s, \gamma_i^{=0}, v) \in I, 1 \le i \le n\} \cup \\
&\quad \{\gamma_i, \gamma_i', \bar{\gamma}_i, \bar{\gamma}_i' \mid 1 \le i \le n\} \cup \{x_2, x_3, \star\} \cup \\
&\quad \{l_j \mid l \in \{1, \ldots, |I'|\}, j \in \{2,4,5,6\}\}; \\
L_1 &= \{s_1, z, x_1, x_4, x_5, x_6, u\} \cup \{l_1, l_3 \mid l \in \{1, \ldots, |I'|\}\}; \\
R_1 &= \{1 : (z\gamma_i\bar{\gamma}_i; \text{out}) \mid 1 \le i \le n\} \cup \\
&\quad \{2 : (z\star; \text{out}), 3 : (z\star; \text{in}), 4 : (x_1 x_2 x_3; \text{in}), 5 : (x_2 x_4 x_5; \text{out}), \\
&\quad \ 6 : (x_3 x_6; \text{out}), 7 : (s_f u; \text{in})\} \cup \\
&\quad \{8 : (s_f u\alpha; \text{out}) \mid \alpha \in \{l_1, l_3 \mid l \in \{1, \ldots, |I'|\}\} \cup \{\gamma_i' \mid 1 \le i \le n\} \cup \\
&\quad \{9 : (sl_1 x_1; \text{out}), 10 : (l_1 l_2 x_4; \text{in}), 11 : (l_2 l_3 \gamma_i'; \text{out}), \\
&\quad \ 12 : (l_3 v x_5; \text{in}), 13 : (\gamma_i' x_6 \gamma_i; \text{in}) \mid l : (s, \gamma_i^+, v) \in I, \ 1 \le i \le n\} \cup \\
&\quad \{15 : (l_2 l_3 z; \text{out}), 16 : (sl_1 x_1; \text{out}), 17 : (l_1 l_2 x_4; \text{in}), \\
&\quad \ 18 : (l_2 l_3 \gamma_i; \text{out}), 19 : (l_3 l_4 x_5; \text{in}), 20 : (l_4 l_5; \text{out}), 21 : (l_5 x_6 v; \text{in}) \mid \\
&\quad \hspace{7cm} l : (s, \gamma_i^-, v) \in I, \ 1 \le i \le n\}.
\end{aligned}
$$

In order to facilitate the explanation, rules have been numbered. The presence of a symbol s in the skin compartment indicates that Π can simulate M being in state s. The content of a register γ in M is indicated by the number of occurrences of the symbol γ present in the skin compartment.

The simulation of instructions of the kind (s, γ_i^+, v), $1 \le i \le n$, is performed by the sequence of rules: 9, 4, (5, 6), 10, 11, (12, 13). Rules 5 and 6 and 12 and 13 are applied in parallel. If the symbol s is in the skin compartment, then rule 9 can be applied. In this way s, l_1 and x_1 pass to the environment. After this some combinations of the x and l symbols in the skin compartment and in the environment define configurations in which only a few rules can be applied. As a result of this, one occurrence of γ_i passes into the skin compartment, simulating the addition of 1 to register γ_i.

The simulation of instructions of the kind (s, γ_i^-, v), $1 \le i \le n$, in case there is an occurrence of γ_i in the skin compartment is performed by the sequence of rules: 16, 4, (5, 6), 17, 18, 19, 20 and 21. Rules 5 and 6 are applied in parallel. Similarly to what was indicated before, the application of rule 16 primes a sequence of configurations. As a result of this, one occurrence of γ_i passes from the skin compartment to the environment. This simulates the subtraction of 1 from register γ_i.

In case there is no occurrence of γ_i in the skin compartment, then the simulation of instructions of the kind (s, γ_i^-, v), $1 \leq i \leq n$, leads to the application of the sequence of rules: 16, 4, (5, 6), 17, 15. The last rule allows the symbol z to pass to the environment. If this happens, then the continuous application of rules 2 and 3 forces the P system to never halt.

The simulation of instructions of the kind $(s, \gamma_i^{=0}, v)$ is performed simulating first $(s, \bar{\gamma}^+, s_v)$ and then $(s_v, \bar{\gamma}^-, v)$. If $\bar{\gamma}_i$ and γ_i are present in the skin compartment, then maximal parallelism forces the application of rule 1 letting z pass to the environment. If this happens, then the continuous application of rules 2 and 3 forces the P system to never halt.

When the symbol s_f passes into the skin compartment, then the P system simulates the register machine being in the final state. Rules of the kind 8 and 7 are continuously applied to allow several symbols to pass from the skin compartment to the environment. The only symbols different from γ_1, the output register of M, which are left in the skin compartment are: s_f, z, u, x_1, x_4, x_5 and x_6, that is, seven symbols.   □

What if one allows only symports and antiports of weight 1? This is called *minimal cooperation* as it refers to the minimum weight that the rules can have.

Theorem 5.5 $\mathbb{N}_1 \, OP_2(sym_1, anti_1) = \mathbb{N}_1 \, RE$.

Proof Let $M = (S, I, s_1, s_f)$ be a generating register machine with n registers. Moreover, γ_1, the register rendering the output, can only be incremented.

Without loss of generality we assume that $(s, \gamma^{=0}, s_f)$ is the only instruction that lets M change state into s_f and we define $I' = I \setminus \{(s, \gamma^{=0}, s_f)\}$. The set of instructions is such that $I = I^+ \cup I^- \cup I^{=0}$ with I^+ containing only instructions of the kind (s, γ^+, v), I^- containing only instructions of the kind (s, γ^-, v) and $I^{=0}$ containing only instructions of the kind $(s, \gamma^{=0}, v)$. Moreover, $|I^+| = l^+$, $|I^-| = l^-$, $|I^{=0}| = l^{=0}$ so that $|I| = l^+ + l^- + l^{=0}$. The instructions in I have unique labels. The labels of the instructions in I^+ are $\{1, \ldots, l^+\}$, the labels of the instructions in I^- are $\{l^+ + 1, \ldots, l^+ + l^-\}$, the labels of the instructions in $I^{=0}$ are $\{l^+ + l^- + 1, \ldots, |I|\}$ with $(s, \gamma^{=0}, s_f)$ having label $|I|$.

We define $\Pi = (V, \mu, L_0, L_1, L_2, R_1, R_2, 2)$, a P system with symport/antiport simulating M, with:

$$V = S \cup \{a_i, a_i'', b_i, d_i, e_i \mid 1 \leq i \leq |I|\} \cup \{a_i' \mid 1 \leq i \leq |I| - 1\} \cup$$
$$\{\gamma_i \mid 1 \leq i \leq n\} \cup \{J_\gamma, h_0, h_1, h_2, f, g, r_0, r_1, r_2, r_3, r_4, r_5, r_6, t_1, t_2, t_3\};$$
$$\mu = (\{0, 1, 2\}, \{(0, 1), (1, 2)\});$$

$$L_0 = S \cup \{a_i, a_i'', e_i \mid 1 \le i \le |I|\} \cup \{a_i' \mid 1 \le i \le |I| - 1\} \cup$$
$$\{\gamma_i \mid 1 \le i \le n\} \cup \{h_0, r_0, r_1, r_2, r_3, r_4, r_5, r_6\};$$
$$L_1 = \{J_\gamma, h_1, h_2\};$$
$$L_2 = \{b_i, d_i \mid 1 \le i \le |I|\} \cup \{t_1, t_2, t_3, f, g\};$$
$$R_i = R_{i,1} \cup R_{i,2} \cup R_{i,3}, \ i = 1, 2.$$

The content of $R_{i,1}$, $R_{i,2}$ and $R_{i,3}$, $i = 1, 2$, is defined in the following.

During the simulation the content of the registers γ_i, $1 \le i \le n$, in M is indicated by the number of occurrences of symbols γ_i in compartment 2. The presence of a symbol $s_j \in S$ in compartment 1 indicates that the simulation of M being in state s_j can begin.

The computation performed by Π can be logically divided into three phases: *start, simulate* and *finish*. The rules in $R_{1,1}$ and $R_{2,1}$ belong to the *start* phase, the rules in $R_{1,2}$ and $R_{2,2}$ belong to the *simulate* phase, and the rules in $R_{1,3}$ and $R_{2,3}$ belong to the *finish* phase. Only a correct simulation of M results in Π halting.

Following the original proof of this theorem we use a pictorial representation of the compartments in Π together with the only symbols relevant for the considered explanation. The skin compartment and the compartment present in it are depicted by rectangles; the environment is not depicted. In order to have a clearer explanation in the following we focus on explaining one phase at a time, not indicating the symbols that are not subjects of rules in that phase. The transitions indicate the labels of the applied rules.

Start phase: arbitrary numbers $k_1, \ldots, k_n \in \mathbb{N}$ of occurrences of symbols γ_i, $1 \le i \le n$, pass from the environment to the skin compartment. The system Π can simulate M only if a sufficient number of occurrences of γ_i is present in the skin compartment. During every computation of Π the symbol g passes between the skin compartment and compartment 2. This can be ended only if the tree phases have been correctly completed by Π.

$$R_{1,1} = \{r_{1,1} : (J_\gamma; \text{out}/\gamma_i; \text{in}) \mid 1 \le i \le n\} \cup$$
$$\{r_{1,2} : (J_\gamma; \text{in}), r_{1,3} : (f; \text{out}/s_1; \text{in})\};$$
$$R_{2,1} = \{r_{2,1} : (g; \text{out}), r_{2,2} : (g; \text{in}), r_{2,3} : (f; \text{out}/J_\gamma; \text{in})\}.$$

The symbol J_γ is used to allow occurrences of γ_i, $1 \le i \le n$, symbols to pass from the environment (where they are present with infinite multiplicity) to the skin compartment:

$\gamma_1 \ldots \gamma_n s_1 \boxed{J\gamma \boxed{fg}} \Rightarrow_{r_{1,1}, r_{2,1}} J_\gamma \gamma_1 \ldots \gamma_n s_1 \boxed{\gamma_1 g \boxed{f}} \Rightarrow_{r_{1,2}, r_{2,2}}$

$\gamma_1 \ldots \gamma_n s_1 \boxed{J\gamma\gamma_1 \boxed{fg}} \Rightarrow_{r_{1,1}, r_{2,1}} \ldots \Rightarrow_{r_{1,1}, r_{2,1}}$

$J_\gamma \gamma_1 \ldots \gamma_n s_1 \boxed{\gamma_1^{k_1} \ldots \gamma_n^{k_n} g \boxed{f}} \Rightarrow_{r_{1,2}, r_{2,2}}$

$$\gamma_1 \ldots \gamma_n s_1 \boxed{J_\gamma \gamma_1^{k_1} \cdots \gamma_n^{k_n} \boxed{fg}} \Rightarrow_{r_{2,1}, r_{2,3}} \gamma_1 \ldots \gamma_n s_1 \boxed{fg\gamma_1^{k_1} \cdots \gamma_n^{k_n} \boxed{J_\gamma}} \Rightarrow_{r_{1,3}, r_{2,2}}$$

$$\gamma_1 \ldots \gamma_n f \boxed{s_1 \gamma_1^{k_1} \cdots \gamma_n^{k_n} \boxed{gJ_\gamma}}$$

Eventually J_γ passes into compartment 2 while f passes from this compartment to the skin compartment. When this happens, then $(f; \mathrm{out}/s_1; \mathrm{in}) \in R_{1,1}$ can be applied and the *simulate* phase starts.

Simulate phase: the instructions of M are simulated. During this phase several rules, with label written in **bold**, can start to be applied forever. If this happens, then Π never halts. We call *scenario 1* the transitions in which h_2 passes to the environment. The transitions in which $\mathbf{r_{2,9}}$ instead of rule $r_{2,8}$ is applied, letting h_1 pass into compartment 2, are called *scenario 2*. Informally, if during this phase h_2 passes to the environment, (*scenario 1*), then it remains there. During the next simulation rule $\mathbf{r_{2,9}}$ instead or rule $r_{2,8}$ is applied. This means that h_1 passes into compartment 2 and the computation never halts (see *scenario 2* below).

We now explain the synchronisation of a_j passing to the environment and b_j, $1 \leq i \leq |I|$. The symbol a_j allows h_0 to pass into the skin compartment while b_j lets h_1 pass to the environment. When this happens the application of rule $r_{1,15}$ lets h_0 and h_1 pass back to the environment and to the skin compartment, respectively. If a_j passes to the environment while b_j remains there, h_1 remains in the skin compartment (or compartment 2) and h_0 passes to the skin compartment (*scenario 2*), then rule $\mathbf{r_{2,19}}$ is applied (as rules $r_{1,1}$ and $r_{1,2}$ can always be applied). If b_j passes into the skin compartment while a_j remains there, h_0 remains in the environment and h_1 passes there (*scenario 3*), then rule $\mathbf{r_{1,16}}$ is applied letting h_2 pass to the environment. The application of rule $\mathbf{r_{1,16}}$ is immediate if an instruction of the kind (s, γ^+, v) was simulated while it happens after a few transitions if an instruction of the kind $(s, \gamma^{=0}, v)$ was simulated.

The application of the rule $\mathbf{r_{1,13}}$ causes *scenario 1*. The application of rule $\mathbf{r_{2,4}}$ leads to the application of rule $\mathbf{r_{2,11}}$. The application of either rule $\mathbf{r_{2,11}}$ or rule $\mathbf{r_{2,16}}$ causes h_1 to remain forever in compartment 2 and leads, eventually, to *scenario 2*. The application of rule $\mathbf{r_{2,14}}$ causes a continuous application of rules $r_{1,1}$ and $r_{1,2}$. So, in Π, in order to have a finite number of transitions no rule with a label written in bold should be applied.

$$\begin{aligned}
R_{1,2} = \ &\{r_{1,4} : (s_i; \mathrm{out}/a_j; \mathrm{in}) \mid l_j : (s_i, \gamma, v) \in I\} \cup \\
&\{r_{1,5} : (e_j; \mathrm{out}/s_i; \mathrm{in}) \mid l_j : (s, \gamma, s_i) \in I\} \cup \\
&\{r_{1,6} : (b_i; \mathrm{out}/a_i'; \mathrm{in}) \mid i \in I'\} \cup \\
&\{r_{1,7} : (a_i; \mathrm{out}/h_0; \mathrm{in}), r_{1,8} : (h_1; \mathrm{out}/b_i; \mathrm{in}), r_{1,9} : (d_i; \mathrm{out}/e_i; \mathrm{in}) \mid \\
&\hspace{7cm} i \in \{1, \ldots, |I|\} \cup
\end{aligned}$$

$$\{r_{1,10} : (a_i'; \text{out}/d_i; \text{in}) \mid i \in \{1,\dots,l^+, l^+ + l^- + 1,\dots,|I|\}\} \cup$$
$$\{r_{1,11} : (a_i'; \text{out}/a_i''; \text{in}), r_{1,12} : (a_i''; \text{out}/d_i; \text{in}) \mid$$
$$i \in \{l^+ + 1,\dots,l^+ + l^-\}\} \cup$$
$$\{\mathbf{r_{1,13}} : (h_2; \text{out}/d_i; \text{in}) \mid i \in \{1,\dots,l^+\}\} \cup$$
$$\{r_{1,14} : (\gamma_f; \text{out}/a_{|I|}; \text{in}), r_{1,15} : (h_0; \text{out}/h_1; \text{in}),$$
$$\mathbf{r_{1,16}} : (h_2; \text{out}/h_1; \text{in}), r_{1,17} : (e_{|I|}; \text{out}/h_0; \text{in}), r_{1,18} : (b_{|I|}; \text{out})\};$$
$$R_{2_2} = \{\mathbf{r_{2,4}} : (\gamma_i; \text{out}/a_j'; \text{in}) \mid l_j : (s, \gamma_i^{=0}, v) \in I\} \cup$$
$$\{r_{2,5} : (\gamma_i; \text{out}/a_j''; \text{in}) \mid l_j : (s, \gamma_j^-, v) \in I\} \cup$$
$$\{r_{2,6} : (a_j'; \text{out}/\gamma_i; \text{in}) \mid l_j : (s, \gamma_i^+, v) \in I\} \cup$$
$$\{r_{2,7} : (b_i; \text{out}/a_i; \text{in}), r_{2,8} : (a_i; \text{out}/h_2; \text{in}), \mathbf{r_{2,9}} : (a_i; \text{out}/h_1; \text{in}),$$
$$r_{2,10} : (d_i; \text{out}/b_i; \text{in}) \mid i \in \{1,\dots,|I|\}\} \cup$$
$$\{r_{2,11} : (a_i'; \text{out}/h_1; \text{in}) \mid i \in \{1,\dots,l^+, l^+ + l^- + 1,\dots,|I|\}\} \cup$$
$$\{r_{2,12} : (h_2; \text{out}/d_i; \text{in}) \mid i \in \{l^+ + 1,\dots,|I|\}\} \cup$$
$$\{r_{2,13} : (a_i''; \text{out}), \mathbf{r_{2,14}} : (J_\gamma; \text{out}/a_i''; \text{in}) \mid i \in \{l^+ + 1,\dots,l^+ + l^-\}\} \cup$$
$$\{r_{2,15} : (e_i; \text{out}/d_i; \text{in}), \mathbf{r_{2,16}} : (e_i; \text{out}/h_1; \text{in}), r_{2,17} : (e_i; \text{in}),$$
$$r_{2,18} : (h_2; \text{out}/a_i'; \text{in}) \mid i \in \{1,\dots,l^+\}\} \cup$$
$$\{\mathbf{r_{2,19}} : (J_\gamma; \text{out}/h_0; \text{in})\}.$$

Simulation of instructions of the kind (s_i, γ_k^+, s_p) *with some* γ_k *in the skin compartment.*

$$s_p e_j a_j a_j' h_0 \boxed{s_i \gamma_k h_1 h_2 \boxed{b_j d_j}} \Rightarrow_{r_{1,4}} s_i s_p e_j a_j' h_0 \boxed{a_j \gamma_k h_1 h_2 \boxed{b_j d_j}} \Rightarrow_{r_{2,7}}$$

$$s_i s_p e_j a_j' h_0 \boxed{b_j \gamma_k h_1 h_2 \boxed{a_j d_j}} \Rightarrow_{r_{1,6}, r_{2,8}} s_i s_p e_j b_j h_0 \boxed{a_j' \gamma_k h_1 a_j \boxed{h_2 d_j}} \Rightarrow_{r_{1,7}, r_{1,8}, r_{2,18}}$$

$$s_i s_p e_j a_j h_1 \boxed{\gamma_k h_0 b_j h_2 \boxed{a_j' d_j}} \Rightarrow_{r_{1,15}, r_{2,6}, r_{2,10}} s_i s_p e_j a_j h_0 \boxed{a_j' h_1 d_j h_2 \boxed{\gamma_k b_j}} \Rightarrow_{r_{1,9}}$$

$$s_i s_p d_j a_j h_0 \boxed{a_j' h_1 e_j h_2 \boxed{\gamma_k b_j}} \Rightarrow_{r_{1,10}, r_{2,17}} s_i s_p a_j a_j' h_0 \boxed{d_j h_1 h_2 \boxed{e_j \gamma_k b_j}} \Rightarrow_{r_{2,15}}$$

$$s_i s_p a_j a_j' h_0 \boxed{e_j h_1 h_2 \boxed{d_j \gamma_k b_j}} \Rightarrow_{r_{1,5}} s_i e_j a_j a_j' h_0 \boxed{s_p h_1 h_2 \boxed{d_j \gamma_k b_j}}$$

Except for s_i being replaced by s_p and γ_k passing from the skin compartment to compartment 2, all the remaining symbols are returned to the compartment in which they originally were.

Simulation of instructions of the kind (s_i, γ_k^+, s_p) *with no* γ_k *in the skin compartment.* The first four transitions are as in the previous case. Then:

$$s_i s_p e_j a_j h_1 \boxed{h_0 b_j h_2 \boxed{a_j' d_j}} \Rightarrow_{r_{1,15}, r_{2,10}} s_i s_p e_j a_j h_0 \boxed{h_1 d_j h_2 \boxed{a_j' b_j}}$$

From this configuration rule $\mathbf{r_{2,11}}$ can be applied forever.

Simulation of instructions of the kind (s_i, γ_k^-, s_p) *with some* γ_k *in compartment 2.*

$$s_p e_j a_q a_j a_j' a_j'' h_0 \boxed{s_i h_1 h_2 \boxed{b_j \gamma_k d_j}} \Rightarrow_{r_{1,4}}$$

$$s_i s_p e_j a_q a_j' a_j'' h_0 \boxed{a_j h_1 h_2 \boxed{b_j \gamma_k d_j}} \Rightarrow_{r_{2,7}}$$

$$s_i s_p e_j a_q a_j' a_j'' h_0 \boxed{b_j h_1 h_2 \boxed{a_j \gamma_k d_j}} \Rightarrow_{r_{1,6}, r_{2,8}}$$

$$s_i s_p e_j a_q a_j'' b_j h_0 \boxed{a_j' h_1 a_j \boxed{h_2 \gamma_k d_j}} \Rightarrow_{r_{1,7}, r_{1,8}, r_{1,11}}$$

$$s_i s_p e_j a_q a_j a_j' h_1 \boxed{h_0 b_j a_j'' \boxed{h_2 \gamma_k d_j}} \Rightarrow_{r_{1,15}, r_{2,5}, r_{2,10}}$$

$$s_i s_p e_j a_q a_j a_j' h_0 \boxed{h_1 d_j \gamma_k \boxed{a_j'' b_j h_2}} \Rightarrow_{r_{1,9}, r_{2,13}}$$

$$s_i s_p d_j a_q a_j a_j' h_0 \boxed{a_j'' h_1 e_j \gamma_k \boxed{b_j h_2}} \Rightarrow_{r_{1,5}, r_{1,12}}$$

$$s_i e_j a_q a_j a_j' a_j'' h_0 \boxed{s_p h_1 d_j \gamma_k \boxed{b_j h_2}}$$

Except for s_i being replaced by s_p and γ_k passing from compartment 2 to the skin compartment, all the remaining symbols are returned to the compartment in which they originally were. Symbol d_j returns to compartment 2 in the first transition related to the simulation of the next instruction:

$$s_i e_j a_q a_j a_j' a_j'' h_0 \boxed{s_p h_1 d_j \gamma_k \boxed{b_j h_2}} \Rightarrow_{r_{1,4}, r_{2,12}}$$

$$s_i s_p e_j a_j a_j' a_j'' h_0 \boxed{a_q h_1 h_2 \gamma_k \boxed{b_j d_j}}$$

Simulation of instructions of the kind (s_i, γ_k^-, s_p) *with no* γ_k *in compartment 2.* The first four transitions are as in the previous case. Then in configuration:

$$s_i s_p e_j a_q a_j a_j' h_1 \boxed{h_0 b_j a_j'' \boxed{h_2 \gamma_k d_j}}$$

rule $\mathbf{r_{2,14}}$ can be applied forever.

Simulation of instructions of the kind $(s_i, \gamma_k^{=0}, s_p)$ *with no* γ_k *in compartment 2.*

$$a_j s_p e_j a_q a_j' h_0 \boxed{s_i h_1 h_2 \boxed{b_j d_j}} \Rightarrow_{r_{1,4}} s_i s_p e_j a_q a_j' h_0 \boxed{a_j h_1 h_2 \boxed{b_j d_j}} \Rightarrow_{r_{2,7}}$$

$$s_i s_p e_j a_q a_j' h_0 \boxed{b_j h_1 h_2 \boxed{a_j d_j}} \Rightarrow_{r_{1,6}, r_{2,8}} s_i s_p e_j a_q b_j h_0 \boxed{a_j' h_1 a_j \boxed{h_2 d_j}} \Rightarrow_{r_{1,7}, r_{1,8}}$$

$$s_i s_p e_j a_q a_j h_1 \boxed{b_j a_j' h_0 \boxed{h_2 d_j}} \Rightarrow_{r_{1,15}, r_{2,10}} s_i s_p e_j a_q a_j h_0 \boxed{h_1 a_j' d_j \boxed{h_2 b_j}}$$

Except for s_i being replaced by s_p all the remaining symbols are returned to the compartment in which they originally were. Symbol d_j returns to compartment 2 in the first transition related to the simulation of the next instruction:

$$s_i s_p e_j a_q a_j h_0 \boxed{h_1 a'_j d_j \boxed{h_2 b_j}} \Rightarrow_{r_{1,4}, r_{2,12}} s_i s_p e_j a_j a'_j h_0 \boxed{a_q h_1 h_2 \boxed{b_j d_j}}$$

Simulation of instructions of the kind $(s_i, \gamma_k^{=0}, s_p)$ *with some* γ_k *in compartment 2.* The first three transitions are as in the previous case. Then in configuration:

$$s_i s_p e_j a_q b_j h_0 \boxed{a'_j h_1 a_j \boxed{h_2 d_j \gamma_k}}$$

rule $\mathbf{r_{2,4}}$ can be applied forever.

Let us consider b_j and d_j initially present in compartment 2 passing first to the environment and then back into compartment 2. We represent this by $2 \to 1 \to 0 \to 1 \to 2$ where the number refers to the compartment in which b_j and d_j are and $\to$ indicates the passage from one compartment to another. The symbols a_j, a'_j, a''_j and e_j are initially present in the environment and they pass first to compartment 2 and then back to the environment ($0 \to 1 \to 2 \to 1 \to 0$). We have to prove that if they *return* to the compartment in which they are originally present ($2 \to 1 \to 2$ or $0 \to 1 \to 0$) or *repeat* passing to the skin compartment before passing back to the compartment where they originally were ($2 \to 1 \to 0 \to 1 \to 0$ or $0 \to 1 \to 2 \to 1 \to 2$), then Π never halts. Here is the proof:

a_j, $j \in \{1, \dots, |I|\}$
    Return: see scenario 2.
    Repeat: impossible without b_j.
a'_j, $j \in \{1, \dots, l^+\}$
    Return: impossible without d_j.
    Repeat: impossible without h_2.
a''_j, $j \in \{l^+ + 1, \dots, l^+ + l^-\}$
    Return: impossible without d_j.
    Repeat: in the same transition s_p passes into the skin compartment. This requires h_2 in the skin compartment in three transitions. Since d_j remains in compartment 2 for at least two transitions h_2 is unavailable in the skin compartment for at least three transitions. So rule $\mathbf{r_{2,8}}$ is applied forever.
e_j, $j \in \{1, \dots, l^+\}$
    Return: since d_j does not pass into compartment 2, rule $\mathbf{r_{1,13}}$ is applied forever.
    Repeat: rule $\mathbf{r_{2,16}}$ is applied forever.
b_j, $j \in \{1, \dots, |I|\}$
    Return: if a_j passes to the environment, then scenario 2 occurs. Otherwise,

if a_j passes into compartment 2, then rule $\mathbf{r_{2,8}}$ is applied forever.

Repeat: The computation never halts because of the continuous application of $r_{2,1}$ and $r_{2,2}$.

d_j, $j \in \{1, \ldots, l^+\}$

Return: impossible without e_j.

Repeat: rule $\mathbf{r_{1,13}}$ is applied forever.

d_j, $j \in \{l^+ + l^- + 1, \ldots, |I|\}$

Return: e_j remains in the environment. The computation never halts because of the continuous application of $r_{2,1}$ and $r_{2,2}$.

Repeat: in the same transition a_q passes into compartment 2. This requires h_2 in compartment 1 in two transitions. As h_2 is unavailable in the skin compartment the continuous application of $\mathbf{r_{2,8}}$ allows Π to never halt.

Finish phase: When s_f, the symbol associated with the final state of the simulated register machines, passes into the skin compartment, then the finish phase starts. During this phase several symbols are removed from compartment 2. The only symbols remaining in this compartment are (several occurrences of) γ_1 and one occurrence of J_γ. The rules associated with this phase are:

$$R_{1,3} = \{r_{1,19} : (b_i; \mathsf{out}/t_3; \mathsf{in}), r_{1,20} : (d_i; \mathsf{out}/t_3; \mathsf{in}) \mid i \in \{1, \ldots, |I|\}\} \cup$$
$$\{r_{1,21} : (r_i; \mathsf{out}/r_{i+i}; \mathsf{in}) \mid 1 \leq i \leq 5\} \cup$$
$$\{r_{1,22} : (t_1; \mathsf{out}/r_1; \mathsf{in}), r_{1,23} : (t_2; \mathsf{out}), r_{1,24} : (g; \mathsf{out}/t_2; \mathsf{in}),$$
$$r_{1,25} : (h_2; \mathsf{out}/t_2; \mathsf{in}), r_{1,26} : (t_3; \mathsf{out})\};$$
$$R_{2,3} = \{r_{2,20} : (a_i; \mathsf{out}/r_6; \mathsf{in}), r_{2,21} : (b_i; \mathsf{out}/r_6; \mathsf{in}), r_{2,22} : (d_i; \mathsf{out}/r_6; \mathsf{in}) \mid$$
$$i \in \{1, \ldots, |I|\}\} \cup$$
$$\{r_{2,23} : (e_i; \mathsf{out}/r_6; \mathsf{in}) \mid i \in \{1, \ldots, l^+\}\} \cup$$
$$\{r_{2,24} : (t_i; \mathsf{out}/r_0; \mathsf{in}) \mid 1 \leq i \leq 3\} \cup$$
$$\{r_{2,25} : (h_2; \mathsf{out}/r_0; \mathsf{in}), r_{2,26} : (r_0; \mathsf{out}), r_{2,27} : (r_6; \mathsf{out})\}.$$

Informally, these rules perform the following: when s_f is in the skin compartment rule $r_{1,14}$ is applied so that s_f passes to the environment and a_n to the skin compartment. If in one of the previous transitions the symbols h_1 passed into compartment 2, then the application of $\mathbf{r_{2,19}}$ allows Π to never halt. If in one of the previous transitions the symbols h_2 passed into compartment 2, then the application of $\mathbf{r_{2,9}}$ allows Π to never halt. Otherwise the symbol h_2 is in the skin compartment, it passes into compartment 2 (rule $r_{2,8}$), then the symbol r_0 passes, in a few transitions, into the skin compartment (rule $r_{1,17}$).

The symbol r_0 is used to let the symbols t_1, t_2, t_3 and h_2 pass into the skin compartment. The symbol t_2 is used to bring the symbols h_2 and g to the environment. The symbol t_1 starts a series of rules whose net effect is to let r_6 pass into the skin compartment and to let the symbols a_j, b_j, d_j and e_j pass from compartment 2 to the skin compartment. The symbol t_3 is used to bring the symbols b_j and d_j to the environment. Here is an example of how this takes place:

$$r_1r_2r_3r_4r_5r_6 \mid r_0 \boxed{gt_1t_2t_3h_2b_jd_j} \Rightarrow_{r_{2,1},r_{2,24}}$$

$$r_1r_2r_3r_4r_5r_6 \mid t_1g \boxed{r_0t_2t_3h_2b_jd_j} \Rightarrow_{r_{1,22},r_{2,2},r_{2,26}}$$

$$r_2r_3r_4r_5r_6t_1 \mid r_1r_0 \boxed{gt_2t_3h_2b_jd_j} \Rightarrow_{r_{1,21},r_{2,1},r_{2,24}}$$

$$r_1r_3r_4r_5r_6t_1 \mid r_2t_3g \boxed{r_0t_2h_2b_jd_j} \Rightarrow_{r_{1,21},r_{1,26},r_{2,2},r_{2,26}}$$

$$r_1r_2r_4r_5r_6t_1t_3 \mid r_3r_0 \boxed{gt_2h_2b_jd_j} \Rightarrow_{r_{1,21},r_{2,1},r_{2,25}}$$

$$r_1r_2r_3r_5r_6t_1t_3 \mid r_4h_2g \boxed{r_0t_2b_jd_j} \Rightarrow_{r_{1,21},r_{2,2},r_{2,26}}$$

$$r_1r_2r_3r_4r_6t_1t_3 \mid r_5h_2r_0 \boxed{gt_2b_jd_j} \Rightarrow_{r_{1,21},r_{2,1},r_{2,24}}$$

$$r_1r_2r_3r_4r_5t_1t_3 \mid r_6h_2t_2g \boxed{r_0b_jd_j} \Rightarrow_{r_{1,23},r_{2,2},r_{2,21},r_{2,26}}$$

$$r_1r_2r_3r_4r_5t_1t_2t_3 \mid h_2b_jr_0 \boxed{gr_6d_j} \Rightarrow_{r_{1,19},r_{1,25},r_{2,1},r_{2,27}}$$

$$r_1r_2r_3r_4r_5b_jh_2t_1 \mid t_2t_3r_0gr_6 \boxed{d_j} \Rightarrow_{r_{1,23},r_{1,26},r_{2,2},r_{2,22}}$$

$$r_1r_2r_3r_4r_5b_jh_2t_1t_2t_3 \mid d_jr_0 \boxed{gr_6} \Rightarrow_{r_{1,20},r_{2,1},r_{2,27}}$$

$$r_1r_2r_3r_4r_5b_jd_jh_2t_1t_2 \mid t_3r_0gr_6 \boxed{} \Rightarrow_{r_{1,24},r_{1,26}}$$

$$r_1r_2r_3r_4r_5b_jd_jh_2gt_1t_3 \mid t_2r_0r_6 \boxed{} \Rightarrow_{r_{1,23}}$$

$$r_1r_2r_3r_4r_5b_jd_jh_2gt_1t_2t_3 \mid r_0r_6 \boxed{}$$

The computation goes on in this way until all symbols b_j and d_j (and possibly also the symbols a_j and e_j) have passed from compartment 2 to the environment. It is relevant to notice that none of the symbols γ_1 can move out of compartment 2 as in this phase the symbol h_2 is in the environment. So the symbols a'_j cannot let symbols γ_i pass into compartment 2. As the simulated register machine M can only increment the content of register γ_1, then we can say that Π can simulate M so that when Π halts the only symbols present in compartment 2 are (several occurrences of) γ_1 and one occurrence of J_γ. $\square$

Another case of minimal cooperation is when the symbols in a P system with symport/antiport can move in the same direction rather than in the opposite one, that is, only symports of weight 2 are allowed.

The result we consider here is:

Theorem 5.6 $\mathbb{N}_1 \, OP_2(sym_2, anti_0) = \mathbb{N}_1 \, RE$.

Proof As this proof follows the lines of that of Theorem 5.5 we only give the description of the P system $\Pi = (V, \mu, L_0, L_1, L_2, R_1, R_2, 2)$:

$$
\begin{aligned}
V \;=\;& S \cup \{\gamma_i \mid 1 \le i \le n\} \cup \{a_i, b_i, d_i, e_i, g_i, f_i, z_i \mid 1 \le i \le |I| - 1\}\cup \\
& \{s' \mid s \in S \setminus \{s_1\}\} \cup \{\bar{s} \mid s \in S \setminus \{s_f\}\}\cup \\
& \{J_\gamma, h, w_1, w_2, w_3, t_1, t_2, t_3, t_4, t_5, t_6, t_7, t_8, t_9, r, \star\}; \\
\mu \;=\;& (\{0, 1, 2\}, \{(0, 1), (1, 2)\}); \\
L_0 \;=\;& \{\gamma_i \mid 1 \le i \le n\} \cup \{s' \mid s \in S \setminus \{s_1\}\} \cup \{a_i, e_i, g_i \mid 1 \le i \le |I| - 1\}\cup \\
& \{\star, t_4, t_6, t_8, r\}; \\
L_1 \;=\;& \{f_i, z_i \mid 1 \le i \le |I| - 1\} \cup \{\bar{s} \mid s \in S \setminus \{s_f\}\}\cup \\
& \{s_1, J_\gamma, h, w_1, w_3, t_1, t_5, t_7, t_9\}; \\
L_2 \;=\;& \{b_i, d_i \mid 1 \le i \le |I| - 1\} \cup \{w_2, t_2, t_3\}; \\
R_i \;=\;& R_{i,1} \cup R_{i,2} \cup R_{i,3}, \;\; i = 1, 2 \text{ with} \\
R_{1,1} \;=\;& \{(J_\gamma; \text{out})\} \cup \{(J_\gamma \gamma_i; \text{in}) \mid 1 \le i \le n\}; \\
R_{2,1} \;=\;& \emptyset;
\end{aligned}
$$

$$
\begin{aligned}
R_{1,2} \;=\;& \{(s\bar{s}; \text{out}) \mid s \in S \setminus \{s_f\}\} \cup \{(a_j \bar{s}_i; \text{in}) \mid l_j : (s_i, \gamma, v) \in I\}\cup \\
& \{(a_j b_j; \text{out}), (b_j g_j; \text{in}) \mid j \in \{1, \ldots, |I| - 1\}\}\cup \\
& \{(w_1 w_2; \text{out}), (\star w_2; \text{in})\} \cup \{(g_j d_j; \text{out}) \mid j \in \{1, \ldots, l^+ + l^-\}\}\cup \\
& \{(d_j s_i; \text{in}) \mid l_j : (s, \gamma^-, s_i) \in I\} \cup \{(z_j s_i; \text{in}) \mid l_j : (s, \gamma^{=0}, s_i) \in I\}\cup \\
& \{(z_j g_j; \text{out}) \mid j \in \{l^+ + l^- + 1, \ldots, |I|\}\}\cup \\
& \{(d_j e_j; \text{in}), (e_j f_j; \text{out}), (f_j s_i; \text{in}) \mid l_j : (s, \gamma^+, s_i) \in I\};
\end{aligned}
$$

$$
\begin{aligned}
R_{2,2} \;=\;& \{(z_j; \text{out}), (b_j h; \text{in}), (d_j e_j; \text{in}), (d_j w_3; \text{in}), (e_j; \text{out}), (g_i w_3; \text{in}) \mid \\
& \hspace{6cm} j \in \{1, \ldots, l^+\}\}\cup \\
& \{(z_j w_2; \text{out}), (a_j w_2; \text{out}), (d_j h; \text{in}) \mid j \in \{l^+ + 1, \ldots, l^+ + l^-\}\}\cup \\
& \{(b_j w_3; \text{in}), (z_j g_j; \text{out}), (g_j w_2; \text{out}) \mid j \in \{l^+ + l^- + 1, \ldots, |I|\}\}\cup \\
& \{(a_j z_j; \text{in}), (a_j b_j; \text{out}) \mid j \in \{1, \ldots, |I| - 1\}\}\cup \\
& \{(b_j g_j; \text{in}) \mid j \in \{l^+ + l^- + 1, \ldots, |I|\}\}\cup \\
& \{(g_j d_j; \text{out}) \mid j \in \{1, \ldots, l^- + l^-\}\}\cup \\
& \{(z_j \gamma_k; \text{out}) \mid l_j : (s, \gamma_k^-, v) \in I\} \cup \{(g_j \gamma_k; \text{in}) \mid l_j : (s, \gamma_k^+, v) \in I\}\cup \\
& \{(z_j \gamma_k; \text{out}) \mid l_j : (s, \gamma_k^{=0}, v) \in I\}\cup \\
& \{(\star; \text{in}), (\star; \text{out}), (w_2 w_3; \text{out}), (h; \text{out})\}; \\
R_{1,3} \;=\;& \{(s_f t_3; \text{out}), (w_1 t_3; \text{out}), (w_3 t_3; \text{out}), (h t_3; \text{out}), (t_3; \text{in}), \\
& (t_1 t_2; \text{out}), (t_2 t_4; \text{in}), (t_4 t_5; \text{out}), (t_5 t_6; \text{in}), (t_6 t_7; \text{out}), (t_7 t_8; \text{in}), \\
& (t_8 t_9; \text{out}), (t_9 r; \text{in})\}; \\
R_{2,3} \;=\;& \{(J_\gamma b_j; \text{out}), (J_\gamma d_j; \text{out}) \mid j \in \{1, \ldots, |I| - 1\}\}\cup \\
& \{(s_f t_1; \text{in}), (s_f t_3; \text{out}), (t_1 t_2; \text{out}), (r; \text{out}), (r J_\gamma; \text{in}), (J_\gamma w_2; \text{out})\}.
\end{aligned}
$$

The constants l^+ and l^- are defined as in the proof of Theorem 5.5. $\square$

Considering what we said in Section 5.4 and Theorem 4.3:

Theorem 5.7 *The sets* $\mathbb{N}OP_1(sym_1, anti_2)$, $\mathbb{N}_1OP_1(sym_0, anti_2)$, $\mathbb{N}_7PP_1(sym_3,$ $anti_0)$, $\mathbb{N}_1\ OP_1(sym_2, anti_1)$ *and* $\mathbb{N}_1\ OP_2(sym_2, anti_0)$ *generated by systems operating in asynchronous mode are equal to the set generated by partially blind register machines.*

5.6 Following the traces

In the previous section we considered systems generating sets of numbers. Here we consider P systems with symport/antiport generating strings. This is done by introducing a distinguished symbol, informally called a *traveller*, and assigning elements of an alphabet W to the compartments of the P system. The language generated is the sequence of elements in W assigned to the compartments through which the traveller passed. That is, the generated strings result from the *trace* of the traveller and the resulting language is called the *trace language*.

The definition of such P systems with symport/antiport is:

$$\Pi = (V, \mu, L_0, L_1, \dots, L_m, R_1, \dots, R_m, t, \mathcal{L})$$

with $\mu = (Q, E)$ and where one occurrence of the traveller $t \in V$ is initially present only in the skin compartment. The function $\mathcal{L} : Q \to W \cup \{\epsilon\}$ is a labelling function over the *set of labels* W where $\mathcal{L}(0) = \epsilon$ and 0 is the root, that is, the environment, of the cell-tree μ. The rest of the elements in Π are defined as in Section 5.2 including the definitions of the initial configuration, configuration and computation. In order to define L(Π), the language generated by such a system, we have to introduce some notions.

Let $c = (M_0 \backslash L_0, M_1, \dots, M_m)$ be a configuration of Π. We define the function $\mathcal{T}_t : (V \to \mathbb{N})^{(m+1)} \to W \cup \{\epsilon\}$ such that $\mathcal{T}_t(M_0 \setminus L_0, M_1, \dots, M_m) = \mathcal{L}(j)$, where $t \in M_j$ for $0 \le j \le m$. So, given a configuration of a P system with symport/antiport the function $\mathcal{T}_t$ returns the label of the compartment j (or the environment) containing t (under the assumption that there is only one occurrence of the symbol t in Π).

Let then $C_\Pi = c_0, \dots, c_l$, where $c_0 \Rightarrow c_1 \Rightarrow \cdots \Rightarrow c_l$, be a finite sequence of configurations defining a computation of Π. Given a computation C_Π of a P system Π with symport/antiport we define the *trace string* as $trace(t, C_\Pi) = \mathcal{T}_t(c_0)\mathcal{T}_t(c_1)\dots\mathcal{T}_t(c_l)$.

The language generated by Π, called the *trace language*, is defined as:

$$L(\Pi) = \{trace(t, C_\Pi) \mid C_\Pi \text{ is a computation of } \Pi\}.$$

It should be clear that the degree of Π has to be at least equal to the cardinality of W. We denote by ℓ RE the family of RE languages over alphabets of size ℓ (where $\mathbb{N}$ RE is equal to 1 RE).

The family of trace languages over alphabets of cardinality ℓ, defined by P systems with symport/antiport with degree m, symports of weight at most p and antiports of weight at most q, is denoted by $\ell \, \text{LP}_m(\text{sym}_p, \text{anti}_q)$.

If we consider the results obtained in the previous section, then we have

Theorem 5.8 *For each $\ell \geq 1$*

$\ell \, \text{LP}_{\ell+1}(sym_0, anti_2) = \ell \, \text{RE};$

$\ell \, \text{LP}_{\ell+1}(sym_3, anti_0) = \ell \, \text{RE};$

$\ell \, \text{LP}_{\ell+2}(sym_2, anti_0) = \ell \, \text{RE}.$

Proof This proof is an elementary extension of some of the proofs in the previous section. We only prove the inclusion $\ell \, \text{LP}_{\ell+1}(\text{sym}_3, \text{anti}_0) \supset \ell \, \text{RE}$ and we sketch how to prove similar inclusions for the other equalities.

Let $M = (W, S, R, s_1, s_f)$ be a register machine with input tape with $W = \{w_1, \ldots, w_\ell\}$. We define $\Pi = (V, \mu, L_0, L_1, L_{w_1}, \ldots, L_{w_\ell}, R_1, R_{w_1}, \ldots, R_{w_\ell}, t, \mathcal{L})$, a P system with symport/antiport, with $N = \{0, 1\} \cup W$. The labelling function $\mathcal{L} : N \to W \cup \{\epsilon\}$ is such that $\mathcal{L}(w) = w$ if $w \in W$, $\mathcal{L}(w) = \epsilon$ otherwise. The cell-tree $\mu = (Q, E)$ underlying Π has vertex 0 the parent of vertex 1 and this last parent of all the remaining vertices in W.

In the initial configuration compartment 1 of Π has the same multisets of symbols as compartment 1 of the P system described in the proof of Theorem 5.4 united to one occurrence of the symbol t. All the inner compartments of Π are initially empty.

The simulation of instructions of the kind (s, γ^+, v), (s, γ^-, v) and $(s, \gamma^{=0}, v)$ is similar to that performed by the P systems described in the proof of Theorem 5.4. Instructions of the kind (s, a, v), $a \in W$, are simulated as instructions of the kind (s, γ^+, v). This means that a symbol a passes from the environment to the skin compartment and in this way it causes the traveller t to pass from the skin compartment into the compartment having label a. This is performed by the rules (a, t, in), $(t, \text{out}) \in R_a$. The symbol a remains in compartment a.

In a computation C_Π, the traveller follows the letters that are successively passed into the skin compartment 1. As this sequence equals the letters read by the register machine, $trace(t, C_\Pi) = x$ the string present on the input tape of the simulated register machine.

Note that this extension to our previous construction needs symport rules of weight two, and needs a compartment for each element of W. The result $\mathbb{N}_7 \, \text{OP}_1(\text{sym}_3, \text{anti}_0) = \mathbb{N}_7 \, \text{RE}$ from Theorem 5.4 generalises to $\ell \, \text{RE} = \ell \, \text{LP}_{\ell+1}(\text{sym}_3, \text{anti}_0)$ as the seven remaining symbols are not visible in the defined language by the P system Π.

Similarly the result $\mathbb{N}_6 \, \text{OP}_2(\text{sym}_2, \text{anti}_0) = \mathbb{N}_6 \, \text{RE}$ from Theorem 5.6 generalises to $\ell \, \text{RE} = \ell \, \text{LP}_{\ell+2}(\text{sym}_2, \text{anti}_0)$ (again, the remaining symbol is irrelevant to the trace language).

The result $\mathbb{N}_1 \, \mathsf{OP}_1(\mathrm{sym}_0, \mathrm{anti}_2) = \mathbb{N}_1 \, \mathsf{RE}$ from Corollary 5.1 generalises to $\ell \, \mathsf{RE} = \ell \, \mathsf{LP}_{\ell+1}(\mathrm{sym}_0, \mathrm{anti}_2)$ where the added compartments have $L_w = \{k\}$ and $R_w = \{(t; \mathrm{out}/k; \mathrm{in}), (k; \mathrm{out}/wt; \mathrm{in})\}$ for $w \in W$. $\qquad\qquad\square$

In the case of one-letter alphabets, $\ell = 1$, we obtain:

Corollary 5.2 $1 \, \mathsf{LP}_2(sym_0, anti_2) = 1 \, \mathsf{LP}_2(sym_3, anti_0) = 1 \, \mathsf{LP}_3(sym_2, anti_0) = 1 \, \mathsf{RE}$.

The trace of the traveller in Theorem 5.8 is the sequence of symbols in W passing from the environment to the skin compartment (that is, the string x present on the input tape of the register machine).

Recalling what we defined earlier in this section and considering that at most one element of W is present in the skin compartment of the P system, we define the function $\mathcal{S}_W : (V \to \mathbb{N})^2 \to W$ as $\mathcal{S}_W(M_1, M_1') = w$ if $(M_0 \setminus L_0, M_1, \cdots, M_m) \Rightarrow (M_0' \setminus L_0, M_1', \cdots, M_m')$ is a transition of the system and if there exists a unique $w \in W$ such that $M_1'(w) > M_1(w)$, otherwise $\mathcal{S}_W(M_1, M_1') = \epsilon$. Informally, the function $\mathcal{S}_W$ returns w if this is the symbol of W introduced in the skin compartment by a transition, otherwise it returns the empty string.

Given a computation $C_\Pi = c_0, \cdots, c_l$ of Π, a P system Π with symport/antiport, we define the *incoming string* as $incoming(W, C_\Pi) = \mathcal{S}_W(M_1^{(0)}, M_1^{(1)}) \, \mathcal{S}_W(M_1^{(1)}, M_1^{(2)}) \cdots \mathcal{S}_W(M_1^{(l-1)}, M_1^{(l)})$.

The language generated by Π, denoted as the *incoming language*, is defined as:

$$\mathsf{L}(\Pi) = \{incoming(W, C_\Pi) \mid C_\Pi \text{ is a computation of } \Pi\}.$$

The family of languages over an alphabet W, defined by the symbols of W passing from the environment into the skin compartment of P systems with symport/antiport having at most m compartments, symport of weight at most x, and antiport of weight at most y, is denoted by $\mathsf{rLP}_m(\mathrm{sym}_x, \mathrm{anti}_y)$.

Theorem 5.9 $\mathsf{rLP}_1(sym_3, anti_0) = \mathsf{rLP}_2(sym_2, anti_0) = \mathsf{RE}$.

Proof The inclusion $\mathsf{RE} \subseteq \mathsf{rLP}_1(\mathrm{sym}_3, \mathrm{anti}_0)$ follows directly from Theorem 5.4 if every instruction of the kind (s, a, v) is simulated as the instruction (s, γ_a^+, v); see the proof of Theorem 5.8. Similarly, the inclusion $\mathsf{RE} \subseteq \mathsf{rLP}_2(\mathrm{sym}_2, \mathrm{anti}_0)$ follows from Theorem 5.6. $\qquad\qquad\square$

5.7 Counting transitions

The definition of P systems with symport/antiport has also been changed so that their output is given by the number of transitions needed to pass from one specific configuration to another. Such systems, called *timed P systems with symport/antiport*, are equipped with two regular expressions C_{start} and C_{stop} describing configurations. The number of transitions from a configuration satisfying C_{start} until a configuration satisfying C_{stop} is counted and this value is the output of the system.

Let us consider any of the P systems with symport/antiport Π generating N_x RE, $x \in \mathbb{N}$, in Section 5.5. We can then define a timed P system with symport/antiport Π' similar to Π such that C_{start} sees the presence of s_f in the skin compartment. This means that from now the transitions of Π' are counted. In the following transitions Π' allows the symbols associated with the output register of the simulated register machine to pass one by one from the compartment in which they are (let us say compartment y) to another one. The configuration of Π' having no such symbols in compartment y are in C_{stop}. In this way Π' generates $\mathbb{N}$ RE even if $x \geq 1$.

5.8 Small number of symbols

In this section we consider P systems with symport/antiport for which the number of compartments and symbols present in their alphabet is kept as small as possible. The family of sets generated by such P systems using s symbols, of degree at most m, using symports of weight at most p and antiports of weight at most q, is denoted by $\mathbb{N} \, O_s P_m(\text{sym}_p, \text{anti}_q)$. When p, q or m are not bounded then they are replaced by $*$.

Theorem 5.10 $\mathbb{N} \, O_1 P_1(sym_*, anti_*) = \mathbb{N} \, \text{FIN}$.

Proof *Part I: (These P systems can generate any finite set of numbers)* Let L be a finite set of numbers. We define the P system with symport/antiport

$$\Pi = (\{a\}, (\{0,1\}, \{(0,1)\}), \{a\}, \{a^m\}, \{(a^m; \text{out}/a^j; \text{in}) \mid j \in L\}, 1)$$

with $m = max(L) + 1$ generating L.

In the initial configuration any of the rules can be applied once. In this way any number $j \in L$ of occurrences of a can pass from the environment to the skin compartment. When this happens the system halts.

The empty set can be generated by a system with no symbols in the initial configuration.

Part II: (The set generated by these P systems is finite) This is a consequence of Theorem 5.1. □

Theorem 5.11 $\mathbb{N} \, O_1 P_2(sym_*, anti_*) \supseteq \mathbb{N} \, \text{REG}$.

Proof From the previous theorem we know that $\mathbb{N} \, O_1 P_1(\text{sym}_*, \text{anti}_*) = \mathbb{N} \, \text{FIN}$, so here we only concentrate on infinite sets of numbers. Let us consider K, K_0, K_1

and k from Proposition 3.1. Let $K = \mathbb{N}\,\mathrm{REG} \setminus \mathbb{N}\,\mathrm{FIN}$ and let m be the smallest element in K such that $m > max(K_0 \cup K_1 \cup \{2k\})$ and $m' = m+2k$ (so, $m' \in K$).

We define a P system with symport/antiport

$$\Pi = (\{a\}, (\{0,1,2\}, \{(0,1),(1,2)\}), \{a\}, \{a^{m'}\}, \phi, R_1, R_2, 2)$$

with

$$R_1 = \{1 : (a^{m'}; \mathrm{out}/a^i; \mathrm{in}) \mid i \in K_0\} \cup$$
$$\{2 : (a^{m'}; \mathrm{out}/a^m; \mathrm{in}), 3 : (a^m; \mathrm{out}/a^{m+k}; \mathrm{in})\} \cup$$
$$\{4 : (a^m; \mathrm{out}/a^i; \mathrm{in}) \mid i \in K_1\};$$
$$R_2 = \{5 : (a; \mathrm{in})\}.$$

Similarly to the proof of Theorem 5.5 a pictorial representation of the compartments present in Π is used in the present proof.

$\mathbb{N}\,O_1 P_2(\boldsymbol{sym_*}, \boldsymbol{anti_*}) \supseteq K$: The elements of K_0 are generated by one application of one of the rules 1. The remaining elements of K can be generated by

or by

where $\Rightarrow^i$ denotes i transitions.

$\mathbb{N}\,O_1 P_2(\boldsymbol{sym_*}, \boldsymbol{anti_*}) \subseteq K$: The other possible computations of Π are:

m' symbols pass into compartment 2;

m symbols pass into compartment 2 (possibly after k symbols passed there, too);

if in the initial configuration instead of the application of rule 1 rules 2 and 5 are applied, then the system generates $2k + (i + jk)$ or $2k + (m + jk)$.

Nothing else can happen because $m + k < m'$, $max(K_0 \cup K_1) < m$ and because all symbols in compartment 1 not involved in rules in R_1 pass into compartment 2. $\square$

Theorem 5.12 $\mathbb{N}\,O_2 P_1(sym_*, anti_*) \supseteq \mathbb{N}\,\mathrm{REG}$.

Proof Let us consider K, K_0, K_1 and k from Proposition 3.1. We define the P system with symport/antiport

$$\Pi = (\{a,b\}, (\{0,1\}, \{(0,1)\}), \{a,b\}, \{b^2\}, R_1, 1)$$

with
$$R_1 = \{(b^2; \text{out}/a^i; \text{in}) \mid i \in K_0\} \cup \{(b^2; \text{out}/ab; \text{in}), (ab; \text{out}/a^{k+1}b; \text{in})\} \cup \\ \{(ab; \text{out}/a^i; \text{in}) \mid i \in K_1\}.$$

In the initial configuration there are two occurrences of b in the skin compartment. The elements in K_0 can be generated by the application of one rule in $\{(b^2; \text{out}/a^i; \text{in}) \mid i \in K_0\}$. If instead $(b^2; \text{out}/ab; \text{in})$ is applied, then rule $(ba; \text{out}/a^{k+1}b; \text{in})$ can be repeatedly applied. This repeated application ends when any of the rules in $\{(ab; \text{out}/a^i; \text{in}) \mid i \in K_1\}$ is applied and the system halts. In this way $jk + i$ occurrences of a passed from the environment to the skin compartment.

This proves that $N(\Pi) = K_0 \cup \{i + jk \mid i \in K_i, j \in \mathbb{N}_+\} = K$. $\qquad\square$

Theorem 5.13 *Any partially blind register machine with n registers can be simulated by P systems with symport/antiport with 1 symbol and degree $n + 3$.*

Proof Let $M = (S, I, s_1, s_f)$ be a partially blind register machine with n registers and with $f = |I|$.

We define the P system with symport/antiport

$$\Pi = (\{a\}, \mu, \{a\}, L_1, \dots, L_{n+3}, R_1, \dots, R_{n+3}, 2)$$

simulating M with:

$$
\begin{aligned}
\mu &= (\{1, \dots, n+3\}, \{(1, i) \mid 2 \le i \le n+2\} \cup \{(n+2, n+3)\}); \\
L_1 &= c(1); \\
L_{j+1} &= b(j), \ 1 \le j \le n; \\
L_{n+2} &= \phi; \\
L_{n+3} &= \{a\}; \\
R_1 &= \{1 : (c(p); \text{out}/c^{(1)}(q)b(t)a; \text{in}), 2 : (c^{(1)}(q); \text{out}/c^{(2)}(q); \text{in}), \\
&\quad\ \ 3 : (b(t)c^{(2)}(q); \text{out}/c(q); \text{in}) \mid (s_p, \gamma_t^+, s_q) \in I\} \cup \\
&\quad\ \ \{4 : (c(p); \text{out}/b(t)c^{(3)}(q); \text{in}), 5 : (c^{(3)}(q); \text{out}/c^{(4)}(q); \text{in}), \\
&\quad\ \ 6 : (b(t)ac^{(4)}(q); \text{out}/c(q); \text{in}) \mid (s_p, \gamma_t^-, s_q) \in I\} \cup \\
&\quad\ \ \{7 : (c(|S|); \text{out})\}; \\
R_{j+1} &= \{8 : (b(t); \text{out}/b(t)a; \text{in}), 9 : (b(t)a; \text{out}/b(t); \text{in}) \mid 1 \le t \le m\} \\
&\qquad\qquad\qquad\qquad\qquad\qquad\qquad\qquad\qquad 1 \le j \le n; \\
R_{n+2} &= \{10 : (a; \text{in})\}; \\
R_{n+3} &= \{11 : (a; \text{out}/a; \text{in})\}.
\end{aligned}
$$

This proof requires the use of a unique-sum set $U_{n+5|S|+1}$ with $n + 5|S| + 1$ elements. Different multiplicities of a, where the multiplicities are elements in $U_{n+5|S|+1}$, are uniquely associated with each of the registers of M. This is performed by the function $b : \{1, \dots, n\} \to (\{a\} \to U_{n+5|S|+1})$. Other different multiplicities of a, where the multiplicities are elements in $U_{n+5|S|+1}$, are uniquely associated with the states in S by the five c functions c, $c^{(1)}$, $c^{(2)}$, $c^{(3)}$ and

$c^{(4)}$ all from $\{1, \ldots, |S|\}$ to $(\{a\} \to U_{n+5|S|+1})$. We indicate by *max* the biggest element of $U_{n+5|S|+1}$.

The exact definition of the six functions is irrelevant for the proof. Only one thing is essential: none of the six functions returns *max*.

The simulation performed by the P system Π is strongly based on the use of a unique-sum set and on the property that for such sets none of the elements can be obtained as a linear combination of the remaining elements in the set. During the computation of Π different occurrences of the symbol a are present in the skin compartment. Only unique (maximally parallel) sequences of multisets of applied rules allow Π to simulate instructions of M. Any other maximally parallel sequence of multisets of applied rules sees the presence of rule 10. If this rule is applied once, then the repeated application of rule 11 allows Π to never halt.

Each of the compartments $j + 1$, $1 \leq j \leq n$, contains $b(j)$ in its initial configuration. This number of occurrences of a represents 0 as the content of the registers in M. The addition of 1 to register γ_j in M is performed by adding one occurrence of a to compartment $j + 1$. Conversely for the subtraction. The presence of $c(s)$, $s \in S$, in the skin compartment indicates that Π simulates the register machine being in state s.

The simulation of instructions of the kind (s_p, γ_t^+, s_q) is performed by the sequential application of rules 1, (2, 8), 3, where rules 2 and 8 are applied in parallel.

The simulation of instructions of the kind (s_p, γ_t^-, s_q), if in compartment $t+1$ there are at least $b(t)+1$ occurrences of a, is performed by the sequential application of rules 4, (5, 9), 6, where rules 5 and 9 are applied in parallel. If in compartment $t+1$ there are less than $b(t)$ occurrences of a, then rules 4, (5, 10) are applied and the continuous application of rule 11 allows the P system to never halt.

When $c(|S|)$ occurrences of a are present in the skin compartment, then the application of rule 7 lets them pass into the environment, in this way halting the computation. In this configuration each compartment $1 + j$ has $b(j)$, $1 \leq j \leq n$, additional occurrences of a more than the value of register γ_j of the register machine in its final configuration.

It is important to notice that in the sets indicated by Proposition 3.2 only the biggest element (*max*) in each set is odd. This element is not returned by any of the six functions. $\qquad \square$

Knowing from Section 3.4 that partially blind register machines with one register can generate or accept $\mathbb{N}$ REG, we have:

Corollary 5.3 $\mathbb{N} \, O_1 P_4(sym_*, anti_*) \supseteq \mathbb{N}$ REG.

If we consider Theorem 4.1, what we said in Section 5.4, Fig. 5.4 and the discussion associated with it, then we have:

Theorem 5.14 *Any register machine with n registers can be simulated by P systems with symport/antiport with one symbol, degree $n+3$ and priorities between the rules.*

The next two theorems have a more general nature.

Theorem 5.15 *Any register machine with n registers can be simulated by a P system with symport/antiport with $2+p$ symbols and m compartments, $n = pm$.*

Proof Let $M = (S, I, s_1, s_f)$ be a register machine with n registers, $n = pm$, accepting numbers and let γ_1 be the input register. We define $\Pi = (V, \mu, L_0, L_1, \ldots, L_m, R_1, \ldots, R_m, 2)$, a P system with symport/antiport simulating M, with:

$V = \{a_j \mid 1 \le j \le p\} \cup \{b, \star\}$;

$\mu = (\{1, \ldots, m\}, \{(1, i) \mid 2 \le i \le m\})$;

$L_0 = V$;

$L_1 = c(s_1)$;

$L_i = \phi, \ 2 \le i \le m$;

$R_1 = \{1 : (c(s_k); \text{out}/c(s_q)a_j; \text{in}) \mid (s_k, \gamma_j^+, s_q) \in I, \ 1 \le j \le p\} \cup$
$\quad \{2 : (c(s_k); \text{out}/c(s_k)bh_+(j + (i - 1)p)a_j; \text{in}),$
$\quad \ \ 3 : (c(s_k)b; \text{out}/c(s_k)b^2; \text{in}), 4 : (c(s_k)b^2; \text{out}/c(s_k)b^3; \text{in}),$
$\quad \ \ 5 : (c(s_k)b^3 h_+(j + (i - 1)p); \text{out}/c(s_q); \text{in}) \mid$
$\quad \quad \quad \quad \quad \quad \quad (s_k, \gamma_{j+(i-1)p}^+, s_q) \in I, \ 1 \le j \le p, \ 2 \le i \le m\} \cup$
$\quad \{6 : (c(s_k)a_j; \text{out}/c(s_q); \text{in}), 7 : (c(s_k); \text{out}/c(s_k)bh_{=0}(j); \text{in}),$
$\quad \ \ 8 : (c(s_k)b; \text{out}/c(s_k)b^2; \text{in}), 9 : (h_{=0}(j)a_j; \text{out}/\star^{3g}; \text{in}),$
$\quad \ \ 10 : (c(s_k)b^2 h_{=0}(j); \text{out}/c(s_t); \text{in}) \mid (s_k, \gamma_j^-, s_q, s_t) \in I, \ 1 \le j \le p\} \cup$
$\quad \{11 : (c(s_k); \text{out}/c(s_k)bh_-(j + (i - 1)p); \text{in}),$
$\quad \ \ 12 : (c(s_k)b; \text{out}/c(s_k)b^2; \text{in}),$
$13 : (c(s_k)b^2; \text{out}/c(s_k)b^3; \text{in}),$
$14 : (c(s_k)b^3 h_-(j + (i - 1)p)a_j; \text{out}/c(s_q); \text{in}),$
$15 : (c(s_k); \text{out}/c(s_k)b^4 h_{=0}(j + (i - 1)p); \text{in}),$
$16 : (c(s_k)b^4; \text{out}/c(s_k)b^5; \text{in}),$
$17 : (c(s_k)b^5 h_0(j + (i - 1)p); \text{out}/c(s_t); \text{in}),$
$18 : (h_-(j + (i - 1)p); \text{out}/\star^{3g}; \text{in}) \mid$
$\quad \quad \quad (s_k, \gamma_{j+(i-1)p}^-, s_q, s_t) \in I, \ 1 \le j \le p, \ 2 \le i \le m\} \cup$
$\quad \{19 : (c(s_f); \text{out}), 20 : (b^r; \text{out}/\star^{3g}; \text{in}), 21 : (\star^g; \text{out}/\star^{3g}; \text{in})\}$;

$R_e = \{22 : (h_+(j + (i - 1)p)a_j; \text{in}), 23 : (h_+(j + (i - 1)p); \text{out}) \mid$
$\quad \quad \quad \quad \quad (s_k, \gamma_{j+(i-1)p}^+, s_q) \in I, \ 1 \le j \le p, \ 2 \le i \le m\} \cup$
$\quad \{24 : (a_j; \text{out}/h_-(j + (i - 1)p); \text{in}), 25 : (h_-(j + (i - 1)p); \text{out}),$
$\quad \ \ 26 : (a_j; \text{out}/h_{=0}(j + (i - 1)p); \text{in}) \mid$
$\quad \quad \quad (s_k, \gamma_{j+(i-1)p}^-, s_q, s_t) \in I, \ 1 \le j \le n, \ 2 \le i \le m\}$ for $2 \le e \le m$.

The functions c, h_+, h_-, $h_{=0}$ and the constants r and g are defined in the following.

The functions h_+, h_- and $h_{=0}$, all three from $\{\gamma_1, \ldots, \gamma_n\}$ to the multiset $\{b, \star\} \to \mathbb{N}_+$, encode the registers γ_i, $1 \leq i \leq n$, present in the instructions of M:

$$
\begin{aligned}
h_+(\gamma_i) &= \star^{(3n+i)} b^{r-(3n+i)} \quad \text{if } (s_k, \gamma_i^+, s_q) \in I; \\
h_-(\gamma_i) &= \star^{(4n+i)} b^{r-(4n+i)} \quad \text{if } (s_k, \gamma_i^-, s_q, s_t) \in I; \\
h_{=0}(\gamma_i) &= \star^{(5n+i)} b^{r-(5n+i)} \quad \text{if } (s_k, \gamma_i^-, s_q, s_t) \in I;
\end{aligned}
$$

where $r = 12n + 1$.

These encodings are such that the number of b symbols always exceeds that of $\star$ symbols and that the number of $\star$ symbols is unique for each encoding through the tree functions. The number of $\star$ symbols present in these encodings ranges from $3n + 1$ to $6n$ while the number of b ranges from $6n + 1$ to $9n + 2$.

The function $c : S \to (\{b\} \to \mathbb{N}_+)$, with $c(s_i) = b^{10n(2|S|-i)}$ encodes the instructions in S into unique multisets over $\{b\}$. This function is such that:

$|c(s_i)| + |c(s_j)| > |c(s_{|S|})|$ for $1 \leq i,\ j \leq |S|$, $i \neq j$. The sum of the codes of two instructions is larger than the largest code of an instruction in S.

$|c(s_{i+1})| - |c(s_i)| > 9n+2$, $1 \leq i \leq |S|-1$, where $9n+2$ is the largest multiplicity of b symbols defined by h_+, h_- and $h_{=0}$.

The c function we defined satisfies these two properties as:

$|c(s_i)| + |c(s_j)| > |c(s_{|S|})|$ holds as $10n(4|S| - i - j) > 10n|S|$;

$|c(s_{k+1})| - |c(s_k)| > 9n + 2$ holds as $10n(4|S| - 2k - 1) > 9n + 2$;

for $1 \leq i, j \leq |S|$, $1 \leq k \leq |S| - 1$.

The constant $g = |c(s_{|S|})| + 1$ where $s_{|S|} = s_f$ is the final state of the register machine M. As we see in the following description, this definition of g forces the P system Π to apply rule 21 forever, so that Π never halts.

The content of the registers $\gamma_{j+(i-1)p}$ is represented by the occurrences of the symbols a_j, $1 \leq j \leq p$, in compartment i, $2 \leq i \leq m$. Now we describe how the P system Π with symport/antiport can simulate M. The simulation of rules of the kind (s, γ^-, v, w), is performed by 'gambling' if γ can be decreased or not. In case of a wrong 'gamble', then the system applies rule 21 forever.

The P system Π works as follows:

the simulation of instructions of the kind (s_k, γ_j^+, s_q), $1 \leq j \leq p$, is performed by the application of rule 1;

the simulation of instructions of the kind $(s_k, \gamma_{j+(i-1)p}^+, s_q) \in I$, $1 \leq j \leq p$, $2 \leq i \leq m$, is performed by the application of rules 2, (3, 22), (4, 23) and 5, where rules in parentheses are applied in parallel;

the simulation of instructions of the kind $(s_k, \gamma_j^-, s_q, s_t) \in I$, $1 \leq j \leq p$, when the content of γ_j can be decreased, is performed by the application of rule 6;

the simulation of instructions of the kind $(s_k, \gamma_j^-, s_q, s_t) \in I$, $1 \leq j \leq p$, $2 \leq i \leq m$, when the content of γ_j can be decreased, is performed by the application of rules 11, (12, 24), (13, 25) and 14, where rules in parentheses are applied in parallel.

If the content of γ_j cannot be decreased, then instead of rule 14, rule 20 is applied. The following repeated application of rule 21 allows P to never halt:

the simulation of instructions of the kind $(s_k, \gamma_j^-, s_q, s_t) \in I$, $1 \leq j \leq p$, when the content of γ_j cannot be decreased, is performed by the application of rules 7, 8 and 10.

If, after the application of rule 7, rule 9 can be applied, then rule 8 is not applied. In this case rule 20 is applied. The following repeated application of rule 21 allows the P system to never halt;

the simulation of instructions of the kind $(s_k, \gamma_j^-, s_q, s_t) \in I$, $1 \leq j \leq p$, $2 \leq i \leq m$, when the content of γ_j cannot be decreased, is performed by the application of rules 15, 16 and 17.

If, after the application of rule 15, rule 26 can be applied, then rule 17 is not applied. In this case rule 20 is applied. The following repeated application of rule 21 allows the P system to never halt;

when $c(s_f)$, s_f being the final state of the register machine, is present in compartment 1, then rule 19 is applied. The P system halts in case rule 21 is not applied.

We can then say that Π halts only if M halts. When Π halts its final configuration represents the contents of the registers in M when it halts. $\square$

Considering the previous theorem and that register machines with three registers can generate $\mathbb{N}$ RE (see Section 3.4), we have:

Corollary 5.4 $\mathbb{N}\, O_5 P_1(sym_*, anti_*) = \mathbb{N}\, O_4 P_2(sym_*, anti_*) = \mathbb{N}$ RE.

A result similar to Theorem 5.15 is:

Theorem 5.16 *Any register machine with n registers can be simulated by a P system with symport/antiport with $p + 1$ symbols and $m + 1$ compartments, $n = (p + 1)(m + 1), n, m \geq 1$.*

Proof As this proof follows the lines of the proof of Theorem 5.15 we only give the description of the P system $\Pi = (V, \mu, L_0, L_1, \ldots, L_{m+1}, R_1, \ldots, R_{m+1}, 2)$:

$V \quad = \{a_j \mid 1 \leq j \leq p\} \cup \{b\};$

$\mu \quad = (\{1, \ldots, m\}, \{(1, i) \mid 2 \leq i \leq m + 1\});$

$L_0 \quad = V;$

$L_1 \quad = c(s_1);$

$L_{i+1} = \bigcup_{j=1}^{p} a_j^{j+(i-1)p}, \ 1 \leq i \leq m;$

$$\begin{aligned}
R_1 \;=\; &\{(c(s_k); \mathrm{out}/c(s_k)bh_+(j + (i-1)p)a_j; \mathrm{in});\\
&(c(s_k)b; \mathrm{out}/c(s+k)b^2; \mathrm{in}), (c(s_k)b^2; \mathrm{out}/c(s_k)b^3; \mathrm{in}),\\
&(c(s_k)b^3 h_+(j + (i-1)p); \mathrm{out}/c(s_q); \mathrm{in}) \mid\\
&\qquad (s_k, \gamma_j^+, s_q) \in I,\ 1 \le j \le p,\ 1 \le i \le m\}\cup\\
&\{(c(s_k); \mathrm{out}/c(s_k)bh_-(j + (i-1)p; \mathrm{in}),\\
&(c(s_k)b; \mathrm{out}/c(s_k)b^2; \mathrm{in}), (c(s_k)b^2; \mathrm{out}/c(s_k)b^3; \mathrm{in}),\\
&(c(s_k)b^3 h_-(j + (i-1)p)a_j; \mathrm{out}/c(s_q); \mathrm{in}),\\
&(h_-(j + (i-1)p); \mathrm{out}/a_1^{3g}; \mathrm{in}),\\
&(c(s_k); \mathrm{out}/c(s_k)b^4 h_{=0}(j + (i-1)p); \mathrm{in}),\\
&(c(s_k)b^4; \mathrm{out}/c(s_k)b^5; \mathrm{in}),\\
&(c(s_k)b^5 h_{=0}(j + (i-1)p; \mathrm{out}/c(s_t); \mathrm{in}) \mid\\
&\qquad (s_k, \gamma_j^-, s_q, s_t) \in I,\ 1 \le j \le p,\ 1 \le i \le m\}\cup\\
&\{(c(s_f); \mathrm{out}), (b^r; \mathrm{out}/a_1^{3g}; \mathrm{in}), (a_1^r; \mathrm{out}/a_1^{3g}; \mathrm{in})\};\\
R_{e+1} \;=\; &\{(h_+(j + (i-1)p)a_e; \mathrm{in}), (h_+(j + (i-1)p); \mathrm{out}) \mid\\
&\qquad (s_k, \gamma_j^+, s_q) \in I,\ 1 \le j \le p,\ 1 \le i \le m\}\cup\\
&\{(a_j; \mathrm{out}/h_-(j + (i-1)p); \mathrm{in}), (h_-(j + (i-1)p); \mathrm{out}),\\
&(a_j; \mathrm{out}/(h_0(j + (i-1)p); \mathrm{in}) \mid\\
&\qquad (s_k, \gamma_j^-, s_q, s_t) \in I,\ 1 \le j \le p,\ 1 \le i \le m\} \text{ for } 1 \le e \le m.
\end{aligned}$$

The functions c, h_+, h_-, $h_{=0}$ and the constants r and g are defined in the following.

The functions h_+, h_- and $h_{=0}$, all three from $\{\gamma_1, \ldots, \gamma_n\}$ to the multiset $\{b\} \to \mathbb{N}_+$, encode the registers γ_i, $1 \le i \le n$, present in the instructions of M:

$$\begin{aligned}
h_+(\gamma_i) &= a_1^{(6n+2i)} b^{r-(6n+2i)} && \text{if } (s_k, \gamma_i^+, s_q) \in I,\\
h_-(\gamma_i) &= a_1^{(8n+2i)} b^{r-(8n+2i)} && \text{if } (s_k, \gamma_i^-, s_q, s_t) \in I,\\
h_{=0}(\gamma_i) &= a_1^{(10n+2i)} b^{r-(10n+2i)} && \text{if } (s_k, \gamma_i^-, s_q, s_t) \in I,
\end{aligned}$$

where $r = 24n + 2$, $1 \le k, q, t \le |S|$.

These encodings are such that the number of b symbols always exceeds that of a_1 symbols and that the number of a_1 symbols is unique for each encoding through the tree functions. The number of a_1 symbols present in these encodings ranges from $6n + 2$ to $12n$ while the number of b ranges from $12n + 2$ to $18n$. The function of the a_1 symbol is similar to that of the $\star$ symbol in the proof of Theorem 5.15.

The function $c : S \to (\{b\} \to \mathbb{N}_+)$, with

$$c(s_i) = b^{19n(2|S|-i)}$$

encodes the states in S into unique multisets over $\{b\}$. The constant g is equal to $|c(s_{|S|})| + 1$.

The content of register $j + (i-1)p$ is represented by the number of occurrences of symbol a_1 present in compartment $i + 1$; the symbol b is used to encode instruction of M. Too many occurrences of a_1 in the skin compartment allows the P system Π to never halt. $\square$

As a consequence of this theorem and of the fact that register machines with three registers can generate $\mathbb{N}$ RE (see Section 3.4), we have:

Corollary 5.5 $\mathbb{N}\,O_3P_3(sym_*, anti_*) = \mathbb{N}\,O_2P_4(sym_*, anti_*) = \mathbb{N}$ RE.

5.9   Accepting systems

In this section we denote by $\mathbb{N}\,aO_sP_m(sym_p, anti_q)$ the family of sets accepted by P systems with symport/antiport using s symbols, having degree at most m, using symport of weight p and antiports of weight q.

Theorem 5.17 $\{ki \mid i \in \mathbb{N}_+\} \in \mathbb{N}\,aO_1P_1(sym_*, anti_*)$ *for any* $k \in \mathbb{N}_+$.

Proof   We define the accepting P system with symport/antiport

$$\Pi = (\{a\}, (\{0,1\}, \{(0,1)\}), \{a\}, L_1, \{(a^k; \text{out}), (a; \text{out}/a; \text{in})\}, 1).$$

The multiset L_1 defines the input. The application of rule $(a^k; \text{out})$ allows groups of k symbols to pass from the skin compartment to the environment, while rule $(a; \text{out}/a; \text{in})$ 'keeps busy' the remaining symbols in the skin compartment. The system halts only if the input had a multiple of k symbols. In this case rule $(a; \text{out}/a; \text{in})$ is no longer used. $\square$

Theorem 5.18 $\mathbb{N}\,aO_2P_1(sym_*, anti_*) \supsetneq \mathbb{N}$ REG.

Proof   *Part I:* $(\mathbb{N}\,aO_2P_1(sym_*, anti_*) \supseteq \mathbb{N}$ REG$)$ Let us consider K, K_0, K_1 and k from Proposition 3.1. We define the P system with symport/antiport $\Pi = (\{a,b\}, (\{0,1\}, \{(0,1)\}), \{a,b\}, L_1, R_1, 1)$ with
$$R_1 = \{1: (a^i; \text{out}/b^2; \text{in}) \mid i \in K_0\} \cup \{2: (a^i; \text{out}/b^3; \text{in}) \mid i \in K_1\} \cup$$
$$\{3: (a^k b^3; \text{out}/b^3; \text{in}), 4: (b^4; \text{out}/b^4; \text{in}), 5: (a; \text{out}/a; \text{in})\}.$$

The multiset L_1 defines the input. The P system can check if the number of a's initially present in L_1 belongs to either K_0, rule 1 is used, or to $\{i + jk \mid i \in K_1,\ j \in \mathbb{N}_+\}$, rules 2 and 3 are used. The application of rule 1 allows us to check if the number of a's initially present in L_1 belongs to K_0. Rules 2 and 3 are used to check if the number of a's initially present in L_1 belongs to $\{i + jk \mid i \in K_1,\ j \in \mathbb{N}_+\}$.

The a's not used by any of these rules are 'kept busy' by rule 5. In case more than three occurrences of b pass into the skin compartment, then rule 4 is continuously applied so that Π never halts.

Part II: ($\mathbb{N}$ $aO_2P_1(sym_*, anti_*)$ can accept non-semilinear sets) Let us consider $K = \{2^n \mid n \geq 0\}$. We define the accepting P system with symport/antiport $\Pi = (\{a, b\}, (\{0, 1\}, \{(0, 1)\}), \{a, b\}, L_1, R_1, 1)$ with:
$R_1 = \{(a^2; out/a; in), (a; out/b; in), (b^2; out/b^2; in)\}$.

The multiset L_1 defines the input. During each transition the number of occurrences of a in the skin compartment can be halved. If at any transition there is an odd number of occurrences of a in the skin compartment, then one occurrence of b passes from the environment to the skin compartment. Eventually only one occurrence of a remains in the skin compartment. In this configuration one b passes into this compartment. If at least two occurrences of b are present in the skin compartment, then the system Π never halts.

The system has a finite set of configurations only if the number of occurrences of a in the skin compartment is always even, that is, the input was an element of K. □

Theorem 5.19 $\mathbb{N}$ $aO_1P_3(sym_*, anti_*)$ *contains non-semilinear sets.*

Proof We define the P systems with symport/antiport

$$\Pi = (\{a\}, \mu, \{a\}, L_1, \phi, \phi, R_1, R_2, R_3, 1)$$

with
$\mu = (\{1, 2, 3\}, \{(0, 1), (1, 2), (2, 3)\});$
$R_1 = \{(a^2; out/a; in)\};$
$R_2 = \{(a; in)\};$
$R_3 = \{(a^2; out), (a^2; in)\}.$

The multiset L_1 defines the input. The P system Π divides the number of occurrences of a in the skin compartment by half in any configuration until a single a is left. If the number of occurrences of a in the initial configuration was a power of 2, then Π reaches a configuration in which one a is left in this compartment and it passes into compartment 2 halting in this way the computation. If the number of occurrences of a in the initial configuration was not a power of 2, then there will be a configuration with an odd number of occurrences of a in compartment 1. This allows one occurrence of a to pass into compartment 2. A similar configuration will be present later on in the computation. The presence of two occurrences of a in compartment 2 allows the continuous application of rules $(aa; out), (aa; in) \in R_3$.

We can then say that Π can accept $\{2^n \mid n \in \mathbb{N}\}$, a non-semilinear set. □

Considering the proof of Theorem 5.13 we have:

Corollary 5.6 *P systems with symport/antiport, 1 symbol and degree $n+3$ accept what partially blind register machines with n registers accept.*

Actually, this model of P systems is more powerful from a computational point of view than any partially blind register machine. This is a consequence of Theorem 5.19 and from the fact that a partially blind register machine cannot accept $\{2^n \mid n \in \mathbb{N}\}$.

If one allows *comp*, the compartment containing the input in the initial configuration, to be composite, then

Theorem 5.20 $\mathbb{N} \, aOP_2(sym_1, anti_1) = \mathbb{N} \, RE$.

5.10   Bounded systems

One question that was raised in Membrane Computing is whether or not the deterministic version of a model of membrane systems is less powerful from a computational point of view than the non-deterministic one. This problem was tackled by Oscar H. Ibarra using bounded P systems with symport/antiport.

A *bounded P system with symport/antiport* has only rules of the form $(x; out/y; in)$ with $|y| = |x| = 1$ or 2. These systems are equivalent to restricted register machines.

Lemma 5.1 *Bounded P systems with symport/antiport of degree 1 can accept the sets of numbers accepted by restricted register machines.*

Proof Let $M = (S, I, s_1, s_f)$ be a restricted register machine with n registers $\{\gamma_1, \ldots, \gamma_n\}$ and let $S = \{s_1, \ldots, s_{|S|}\}$ be the set of states. We define a bounded P system with symport/antiport $\Pi = (V, (\{0, 1\}, \{(0, 1)\}), V \setminus \{s_1, x\}, L_1, R_1, 1)$ by

$$V = S \cup \bar{S} \cup \bar{\bar{S}} \cup \{o_1, \ldots, o_n, \bar{o}_1, \ldots, \bar{o}_n, x, \star\};$$
$$L_1 = \{s_1\} \cup \bigcup_{i=1}^{n} o_i^{val(\gamma_i)};$$
$$R_1 = \{1 : (s_i o_p; out/s_j o_q; in), 2 : (s_i x; out/\bar{o}_p \bar{s}_i; in);$$
$$3 : (o_p \bar{o}_p; out/\star^2; in), 4 : (\bar{s}_i; out/\bar{\bar{s}}_i; in);$$
$$5 : (\bar{\bar{s}}_i \bar{o}_p; out/s_k x; in) \mid (s_i, \gamma_p^-, \beta_q^+, s_j, s_k) \in I\} \cup$$
$$\{6 : (\star; out/\star; in)\}$$

where $\bar{S} = \{\bar{s} \mid s \in S\}$, $\bar{\bar{S}} = \{\bar{\bar{s}} \mid s \in S\}$ and $val(\gamma_p)$ is the content of register γ_p, $1 \leq p \leq n$, in the initial configuration of M.

The presence of the symbol s_i in the skin compartment of Π indicates that the P system simulates the register machine M being in state s_i. The number of occurrences of symbols o_p, $1 \leq p \leq n$, in the skin compartment indicates the content of the register γ of the register machine. If in a certain configuration the symbols s_i and o_p are present in the skin compartment, then rule 1 can be applied only if $(s_i, \gamma_p^-, \beta_q^+, s_j, s_k) \in I$. This rule simulated the subtraction of 1 from register γ, the addition of 1 to register β and the passage from state i to state j. In this configuration also rule 2 can be applied. This allows an occurrence of $\bar{o}_p$, to pass into the skin compartment. If an occurrence of o_p is also present in the skin compartment, then rule 3 and 6 are applied, followed by a continuous application of rule 6. This means that the P system never halts. If instead no occurrence of o_p is present in the skin compartment, then rules 4 and 5 are applied.

The computation of Π halts when s_f, the symbol associated with the final state of the register machines, is in the skin compartment. $\qquad \square$

Lemma 5.2 *A restricted register machine can accept any set of numbers accepted by bounded P systems with symport/antiport.*

Proof This proof follows from that of Lemma 4.3, Part II, and from the fact that rules $(x; \text{out}/y; \text{in})$ present in the restricted membrane system are such that $|y| = |x|$. $\qquad \square$

Given a linear-bounded register machine it is possible to define an equivalent restricted register machine. This means that deterministic and non-deterministic bounded P systems with symport/antiport are equivalent if and only if deterministic and non-deterministic linear-bounded register machines are equivalent. As indicated in Section 3.4 this is a long-standing problem in computational complexity theory.

5.11 Restricted models and semilinear sets

In this section we study restricted models of P systems with symport/antiport with respect to their capability to generate or accept semilinear sets.

Definition 5.2 *A* simple P systems with symport/antiport *of degree m*

$$\Pi = (V, \mu, L_0, L_1, \ldots, L_m, R_1, \ldots, R_m)$$

is such that all the components are equivalent to those indicated in Definition 5.1. Moreover:

$V = W \cup \{a\}$, $a \notin W$;

$\mu = (Q, E)$ *with* $Q = \{0, 1, \ldots, m\}$ *is a cell-tree of depth 2, that is, the skin compartment is the parent of all the remaining compartments except the environment. The children compartments of the skin compartment are called input compartments;*

in the initial configuration

the input compartments $\{2, \ldots, m\}$ *contain the input* a^{p_i}, $p_i \in \mathbb{N}$, *for* $2 \leq i \leq m$;

the compartments $\{1, \ldots, m\}$ *contain* w_i, $1 \leq i \leq m$, *multisets over* W;

the environment contains a finite multiset over W, *each element of the multiset having a finite multiplicity;*

the rules associated with R_i, $1 \le i \le m$ are symport and antiport rules over V with the restriction that for rules of the kind $(x; in)$ and $(y; out/x; in)$, x does not contain a's. This means that the number of a's in the compartments $\{1, \ldots, m\}$ cannot increase.

Configuration, computation and *weight* are defined as in Section 5.2. We say that the vector $(p_2, \ldots, p_m)$ is *accepted* by Π if this system halts when a^{p_i}, $p_i \in \mathbb{N}$, $2 \le i \le m$, is given as input. This set of vectors is denoted by $\mathbb{N}^{m-1}(\Pi)$.

We call *decreasing register machines* those having n registers, all of them input registers, such that these registers can only be decremented. If M is a decreasing register machine with n registers, then the set of vectors accepted by Π is denoted by $\mathbb{N}^n(M)$.

Theorem 5.21 *Let* $A \subseteq \mathbb{N}_+^k$, $k \in \mathbb{N}_+$. *The following statements are equivalent:*

(i) $A \in \mathsf{SLS}_k$;

(ii) A *is accepted by a reversal-bounded register machine with a free register;*

(iii) A *is accepted by a reversal-bounded register machine;*

(iv) A *is accepted by a decreasing register machine.*

Proof *((iv) implies (iii))* This follows from the definitions.

((iii) implies (ii)) This follows from the definitions.

((i) implies (iv)) Let A be a semilinear set. From its definition it is easy to define a decreasing register machine M with k registers. As M is non-deterministic it is sufficient to define M when A is a linear set. So, let $A = \{v \mid v = v_0 + \sum_{i=1}^{t} v_i p_i,\ p_i \in \mathbb{N}_+,\ v_0, v_i \in \mathbb{N}_+^k\} \subseteq \mathbb{N}_+^k$, with $v_i = (v_{i_1}, \ldots, v_{i_k})$, $0 \le i \le t$. If $val(\gamma_1), \ldots, val(\gamma_k)$ is the input of M, then the register machine first decreases the content of the input registers of $v_{0_1}, \ldots, v_{0_k}$, respectively. Then M decreases the content of the registers of $v_{j_1}, \ldots, v_{j_k}$ of p_q, $1 \le j, q \le k$, where p_q is a value non-deterministically chosen. If all the registers record 0 in the same configuration, then M accepts the input.

((ii) implies (i)) This trivially follows from the fact that a bounded language $L \subseteq a_1^* \ldots a_k^*$ over an alphabet V, $a_1, \ldots, a_k \in V$, is accepted by a non-deterministic finite automaton with reversal-bounded registers and one unrestricted register; then $\{(n_1, \ldots, n_k) \in \mathbb{N}_+^k \mid a_i^{n_1} \ldots a_k^{n_k} \in L\}$ is semilinear. □

Theorem 5.22 *The set of numbers accepted by simple P systems with symport/antiport is equivalent to that accepted by reversal-bounded register machines with a free register.*

Proof *Part I: (Such P systems can accept the set of numbers accepted by these register machines)* Let $M = (S, I, s_1, s_f)$ be a reversal-bounded register machine with a free register with n registers. Considering the proof of Theorem 5.21 we can assume that M is a decreasing register machine with n registers accepting $N^k(M) \subseteq \mathbb{N}^k$. The initial content of the n registers γ of M is $val(\gamma_1), \ldots, val(\gamma_n)$. We consider a simple P system $\Pi = (V, \mu, L_0, L_1, \ldots, L_{n+1}, R_1, \ldots, R_{n+1})$ with

$V = S \cup \{a, b_1, b_2, b_3\}$;

$\mu = (\{0, 1, \ldots, n+1\}, \{(0, 1)\} \cup \{(1, j+1) \mid 1 \le j \le n\})$;

$L_0 \quad = S \cup \{b_2, b_3\}$;

$L_1 \quad = \{s_1, b_1\}$;

$L_{j+1} = \{a^{val(\gamma_j)}\}, \ 1 \le j \le n$;

$R_1 \quad = \{1 : (sab_3; out/vb_1; in) \mid (s, \gamma^-, v) \in I\} \cup$

$\qquad \{2 : (s_f b_1; out), 3 : (b_1; out/b_2; in), 4 : (b_2; out/b_3; in),$

$\qquad \ 5 : (b_3; out/b_3; in)\}$;

$R_{j+1} = \{6 : (s_j; in), 7 : (sab_3; out/vb_1; in) \mid (s, \gamma_j^-, v) \in I\}, 1 \le j \le n$.

The presence of $s \in S$ in the skin compartment indicates that Π can simulate M being in state s. The simulation of an instruction of the kind $(s, \gamma_i^-, v) \in I$ is performed by the application of rules (6, 3), (7, 4), 1, where rules 6 and 3 and 7 and 4 are applied in parallel. When b_3 is in the skin compartment rule 5 can be applied and its subsequent application allows the system to never halt. This happens if a subtraction from an empty register is attempted. When the symbol s_f is present in the skin compartment, then rule 2 is applied and Π halts.

Part II: (Such register machines can accept the set of numbers accepted by these P systems) We give an informal description of such a register machine M with $n + 1$ registers simulating a simple P systems Π with symport/antiport. Without loss of generality we assume that the first n registers are reversal-bounded and the $(n+1)$-th register is the free one. If we consider Definition 5.2, then in the initial configuration the input registers of M contain $p_1, \ldots, p_k$, the input of Π. The free register keeps track of the number of a's present in the skin compartment. The strings $w_1, \ldots, w_m$ and the set of rules $R_1, \ldots, R_m$ of Π are stored in the finite control S of M together with the distribution of symbols in W in Π. The machine M simulates each non-deterministic maximally parallel transition of Π using several configurations. Clearly the number of occurrences of a in Π cannot be increased.

Rules of the kind $(x; out)$ or $(x; out/y; in)$ with x containing multiple occurrences of a are simulated as follows. Let d be the maximum number of a's present in these rules. For each compartment in Π the finite control of M keeps track of the first d a's in the compartment while using the register associated with that particular compartment for the remaining a. In order to check if one of the previously indicated rules can be applied, then it is simply enough to examine the finite control.

In a maximally parallel transition some, possibly all, input compartments of Π can allow some a's to pass into the skin compartment and the skin compartment can let some other a's pass to the environment. The number of occurrences of these a's is unrelated so the free register can be incremented and decremented an unbounded number of times during the computation. This is the reason why a free register is needed. It follows that M can simulate Π. $\square$

Considering the last two theorems we have:

Theorem 5.23 *Let $A \subseteq \mathbb{N}^k$, $k \in \mathbb{N}_+$. The statements in Theorem 5.21 and:*

      v) A is accepted by a simple P system with symport/antiport

are equivalent.

From the known results on semilinear sets we have:

Corollary 5.7 *Let $k \in \mathbb{N}_+$. Then*

(i) *the class of subsets of $\mathbb{N}_+^k$ accepted by simple P systems with symport/antiport is closed under union, intersection and complementation;*

(ii) *the membership, disjointness, containment and equivalence problems for simple P systems with symport/antiport accepting subsets of $\mathbb{N}_+^k$ are decidable.*

For the next results we need to introduce a model of register machines.

A *cascade register machine* $M = (S, I, s_1, s_f)$ with n registers $\gamma_1, \ldots, \gamma_n$ is a register machine having instructions of the form $(s, \gamma_i^-, \gamma_{i+1}^+, v)$, $1 \leq i \leq n-1$, $(s, \gamma^{=0}, v, w)$ and (s, γ^-, v) with $s, v, w \in S$ and $s \neq s_f$.

A *configuration* of a cascade register machine M with n registers is given by an element in the $(n+1)$-tuple $S \times \mathbb{N}^n$. Given two configurations $(s, val(\gamma_1), \ldots, val(\gamma_n))$, $(s', val(\gamma_1'), \ldots, val(\gamma_n'))$ (where $val : \{\gamma_1, \ldots, \gamma_n\} \to \mathbb{N}$ is the function returning the content of a register) we define a *computational step* as $(s, val(\gamma_1), \ldots, val(\gamma_n)) \vdash (s', val(\gamma_1'), \ldots, val(\gamma_n'))$ and:

if $(s, \gamma_i^-, \gamma_{i+1}^+, v) \in I$, $1 \leq i \leq n-1$, and $val(\gamma_i) \neq 0$, then $s' = v$, $\gamma_i' = val(\gamma_i) - 1$ and $\gamma_{i+1}' = val(\gamma_{i+1}') + 1$, $\gamma_j' = val(\gamma_j)$, $1 \leq j \leq n$, $i \neq j$, $i+1 \neq j$.

Informally: in state s if the content of register γ_i is greater than 0, then subtract 1 from register γ_i, add 1 to register γ_{i+1} and change the state into v.

if $(s, \gamma^{=0}, v, w) \in I$ and $val(\gamma) = 0$, then $s' = v$, otherwise $s' = w$. In either case $\gamma_i' = val(\gamma_i)$, $1 \leq i \leq n$.

Informally: in state s if the content of register γ is 0, then change the state into v, otherwise into w.

if $(s, \gamma^-, v) \in I$ and $val(\gamma) > 0$, then $s' = v$, $\gamma_i' = val(\gamma_i) - 1$ and $\gamma_j' = val(\gamma_j)$, $1 \le j \le n$, $i \ne j$.

Informally: in state s if the content of register γ is greater than 0, then subtract 1 from register γ and change the state into v.

The reflexive and transitive closure of $\vdash$ is denoted by $\vdash^*$. The definitions of *computation, halting* and *ending* are similar to those given in Section 3.4. The set $N(M) = \{val(\gamma_1) \mid (s_1, val(\gamma_1), 0, \dots, 0) \vdash^* (s_f, 0, \dots, 0)\}$ consists of the numbers accepted by M.

Theorem 5.24 *A deterministic cascade register machine with three registers can simulate a deterministic register machine with two registers.*

Proof Let $M = (S, I, s_1, s_f)$ be a deterministic register machine with two registers γ_1 and γ_2. We define $M' = (S', I', s_1, s_f)$ as a deterministic cascade register machine with three registers γ_1', γ_2' and γ_3'. The operations on γ_1 and γ_2 are simulated by operations on γ_2' and γ_3', respectively, as indicated in the following:

instructions in M	instructions in M'
(s, γ_1^+, v)	$(s, \gamma_1'^-, \gamma_2'^+, v)$
(s, γ_2^+, v)	$(s, \gamma_1'^-, \gamma_2'^+, s')$
	$(s', \gamma_2'^-, \gamma_3'^+, v)$
(s, γ_1^-, v, w)	$(s, \gamma_2'^{=0}, w, s')$
	$(s', \gamma_2'^-, v)$
(s, γ_2^-, v, w)	$(s, \gamma_3'^{=0}, w, s')$
	$(s', \gamma_3'^-, v)$

with $s' \notin S$.

If during the simulation the content of γ_1' cannot be decreased, then M' never reaches a configuration with s_f as state. It should be clear that M' simulates M. $\square$

Corollary 5.8 *The emptiness problem for deterministic cascade register machines with three registers is undecidable.*

Proof This follows from the undecidability of the halting problem for register machines with two registers and Theorem 5.24. $\square$

Theorem 5.25 *Deterministic cascade register machines with three registers can accepts non-semilinear sets.*

Proof We define $M = (S, I, s_1, s_f)$ as a deterministic cascade register machine with three registers γ_1, γ_2 and γ_3. The machine M is such that $(s_1, n, 0, 0)$, $n \in \mathbb{N}$, is its initial configuration. In the following we sketch how M works.

(i) if γ_1 is 0, then M does not reach a configuration with s_f as state;

(ii) with the application of two instructions of the kind $(s, \gamma_i^-, \gamma_{i+1}^+, v)$, M configures the registers so as to contain $\gamma_1 = n - 1$, $\gamma_2 = 0$ and $\gamma_3 = 1$;

(iii) if $\gamma_1 = 0$, then M stops (in a state different from s_f);

(iv) set $k = 1$;

(v) starting with $\gamma_1 = n - (1 + 3 + \ldots + (2k - 1))$, $\gamma_2 = 0$ and $\gamma_3 = 2k - 1$

  (a) M decrements γ_3 and γ_1 by 1 and increments γ_2 by 1;

  (b) if $\gamma_3 \neq 0$ go to (1);

  (c) M decrements the content of γ_1 by 2 and increments γ_2 by 2;

  (d) M decrements γ_2 by 1 while incrementing γ_3 by 1;

  (e) if $\gamma_2 \neq 0$ go to (4);

(vi) if $\gamma_1 = 0$ before (5) is completed then M stops (in a state different from s_f);

(vii) if $\gamma_1 = 0$ after (5), then M halts, otherwise $k = k + 1$ and go to (1).

The values of the registers when k is incremented are: $\gamma_1 = n - (1 + 3 + \ldots + (2k - 1) + (2k + 1)) = n - (k + 1)^2$, $\gamma_2 = 0$ and $\gamma_3 = 2k + 1$. So M can accept the set $\{p^2 \mid p \in \mathbb{N}_+\}$ which is a non-semilinear set. $\qquad \square$

It is interesting to note that if in a cascade register machine M register γ_1 cannot be tested for 0, then either $\mathsf{N}(M) = \emptyset$ or there is a $p \in \mathbb{N}$ such that $\mathsf{N}(M) = \{q \mid q \geq p, \, q \in \mathbb{N}\}$. So, cascade register machines lacking a zero-test on γ_1 can accept only semilinear sets.

We are now ready to introduce *cascade P systems with symport/antiport*, very similar to simple P systems with symport/antiport, the only difference being the underlying cell-tree μ. For these systems $\mu = (\{0, 1, \ldots, m\}, \{(0, 1)\} \cup \{(i + 1, i) \mid 1 \leq i \leq m - 1\})$, that is, a unary-tree.

Theorem 5.26 *Let $A \subseteq \mathbb{N}$ and $k \in \mathbb{N}_+$. The set A is accepted by a cascade P system with symport/antiport of degree k if and only if it is accepted by a cascade register machine with k registers.*

This proof is quite complex and we refer interested readers to the original paper cited in Section 5.17.

If in the definition of cascade P systems with symport/antiport we consider $V = W \cup \{a_1, \ldots, a_r\}$, then we obtain *extended cascade P systems with symport/antiport*. The initial configuration of these systems is such that $w_1 a_1^{p_1} \ldots a_r^{p_r}$ is present in compartment 1 (the input compartment) and w_i is present in compartment i, $1 \leq i \leq m$. For these systems the following results are known.

Theorem 5.27 *A set $A \subseteq \mathbb{N}_+^k$ is accepted by an extended cascade P system with symport/antiport of degree 1 if and only if $A \in \mathsf{SLS}_k$.*

Theorem 5.28 *Extended cascade P systems with symport/antiport of degree 2 can accept non-semilinear sets.*

Theorem 5.29 *The emptiness problem for extended cascade P systems with symport/antiport of degree 2 is undecidable.*

Results equivalent to those presented in this section can be obtained if simple P systems with symport/antiport and their modifications are used as generators of tuples.

5.12 Symport/antiport of multiset rewriting rules

The models of P systems with symport/antiport considered in the previous sections allow symbols to pass from one compartment to another. What if instead there are multiset rewriting rules passing from one compartment to another?

Definition 5.3 *A P system with symport/antiport of multiset rewriting rules of degree m is a construct*

$$\Pi = (V, T, \mu, L_1, \ldots, L_m, P_1, \ldots, P_m, A, B, R_1, \ldots, R_m, comp)$$

where

$V, \mu, L_1, \ldots, L_m$ *and comp are as in Definition 5.1;*

$T \subseteq V$ *is the set of terminal symbols;*

$P_i \subseteq V^+ \times V^*$, $1 \leq i \leq m$, *are finite sets of multiset rewriting rules;*

$A = \{a_1, \ldots, a_{|A|}\}$ *is a finite set of labels;*

$B : \bigcup_{i=1}^m P_i \to A$ *is a bijective function associating multiset rewriting rules to labels;*

R_i, $1 \leq i \leq m$, *are sets of symport and antiport rules of the kind $(x; in),(x; out)$ and $(y; out/x; in)$ with $x, y : A \to \{0, 1\}$.*

The multiset rewriting rules P_i associated with compartment i, $1 \leq i \leq m$, can change the multiset M_i over V associated with vertex i in the following way: if $\alpha \to \beta \in P_i$ and $\alpha \subseteq M_i$, then M_i changes into $M_i' = M_i \setminus \alpha \cup \beta$.

The symport and antiport rules R_i associated with compartment i, $1 \leq i \leq m$, can change the sets of multiset rewriting rules P_i and P_j associated with vertices i and j, respectively, in μ, where i is the parent of j, in the following way:

if $(x; \text{in}) \in R_j$ and $supp(x) \subseteq B(P_i)$, then P_i changes into $P_i' = P_i \setminus supp(x)$ while P_j changes into $P_j' = P_j \cup supp(x)$;

if $(x; \text{out}) \in R_j$ and $supp(x) \subseteq B(P_j)$, then P_j changes into $P_j' = P_j \setminus supp(x)$ while P_i changes into $P_i' = P_i \cup supp(x)$;

if $(y; \text{out}/x; \text{in}) \in R_j$, $supp(x) \subseteq B(P_i)$ and $supp(y) \subseteq B(P_j)$, then P_i changes into $P_i' = P_i \setminus supp(x) \cup supp(y)$ and P_j changes into $r_j' = P_j \cup supp(x) \setminus supp(y)$.

A *configuration* of a P system Π with symport/antiport of multiset rewriting rules of degree m is given by the $2m$-tuple $(M_1, \ldots, M_m, S_1, \ldots, S_m)$ with M_i, $1 \leq i \leq m$, multisets over V and $S_j : A \rightarrow \{0, 1\}$, $1 \leq j \leq m$. The $2m$-tuple $(L_1, \ldots, L_m, P_1, \ldots, P_m)$ is called the *initial configuration*. For two configurations $(M_1, \ldots, M_m, S_1, \ldots, S_m)$ and $(M_1', \ldots, M_m', S_1', \ldots, S_m')$ of Π we write $(M_1, \ldots, M_m, S_1, \ldots, S_m) \Rightarrow (M_1', \ldots, M_m', S_1', \ldots, S_m')$ to denote a *transition* from $(M_1, \ldots, M_m, S_1, \ldots, S_m)$ to $(M_1', \ldots, M_m', S_1', \ldots, S_m')$ that is, the application of one multiset of the multiset rewriting rules and symport and antiport rules associated with each compartment under the requirement of maximal parallelism. If in one configuration a symbol is subject to a multiset rewriting rule, then that multiset rewriting rule cannot be subject to a symport or antiport in the same transition. The reflexive and transitive closure of $\Rightarrow$ is denoted by $\Rightarrow^*$.

A *computation* is a sequence of transitions between configurations of a system Π starting from the initial configuration $(L_1, \ldots, L_m, P_1, \ldots, P_m)$. If a computation is finite, then the last configuration is called *final*. A finite computation *halts* when the multiset of applicable symport and antiport rules and multiset rewriting rules have empty support.

Let $\Pi = (V, T, \mu, L_1, \ldots, L_m, P_1, \ldots, P_m, A, B, R_1, \ldots, R_m, comp)$ be a P system with symport/antiport of multiset rewriting rules. We define $N^{|T|}(\Pi) = \{(M_{comp}(t_1), \ldots M_{comp}(t_{|T|})) \mid (L_1, \ldots, L_m, P_1, \ldots, P_m) \Rightarrow^* (M_1, \ldots, M_m, S_1, \ldots, S_m)$ is a halting computation of $\Pi\}$.

The *weight* of a rule is defined as in Section 5.2.

Example 5.4 A P system with symport/antiport of multiset rewriting rules of degree 2, using only antiports of weight 1 generating $\{a^{2^n} \mid n \in \mathbb{N}\}$.

We define the P system with symport/antiport of multiset rewriting rules

$$\Pi_4 = (V, T, \mu, L_1, L_2, P_1, P_2, A, B, R_1, R_2, 2)$$

with:

$V = \{a, b\}$;

$T = \{a\}$;

$\mu = (\{0, 1, 2\}, \{(0, 1), (1, 2)\})$;
$L_1 = \phi$;
$L_2 = \{b\}$;
$P_1 = \{b \rightarrow a\}$;
$P_2 = \{b \rightarrow bb\}$;
$A = \{a_1, a_2\}$;
$B(b \rightarrow a) = a_1$;
$B(b \rightarrow bb) = a_2$;
$R_1 = \emptyset$;
$R_2 = \{(a_2; \text{out}/a_1; \text{in})\}$.

From the initial configuration only the multiset rewriting rule $b \rightarrow bb \in R_2$ can be applied an unbounded number of times doubling every time the number of b's present in compartment 2. At any transition the antiport rule $(a_2; \text{out}/a_1; \text{in}) \in R_2$ can be applied so that $b \rightarrow bb$ passes to compartment 1 and $b \rightarrow a$ passes to compartment 2. Only this last multiset rewriting rule can then be applied so that all the b's are replaced by a's. When this happens the computation halts. We can then say that Π_4 generates $\{a^{2^n} \mid n \in \mathbb{N}\}$. $\diamond$

Example 5.5 A P system with symport/antiport of multiset rewriting rules of degree 2, using only symports of weight 2 generating $\{a^{2^n} \mid n \in \mathbb{N}\}$.

We define the P system with symport/antiport of multiset rewriting rules

$$\Pi_5 = (V, T, \mu, L_1, L_2, P_1, P_2, A, B, R_1, R_2, 2)$$

with:
$V = \{a, b, c\}$;
$T = \{a\}$;
$\mu = (\{0, 1, 2\}, \{(0, 1), (1, 2)\})$;
$L_1 = \phi$;
$L_2 = \{b\}$;
$P_1 = \{b \rightarrow a\}$;
$P_2 = \{b \rightarrow bb, c \rightarrow c\}$;
$A = \{a_1, a_2, a_3\}$;
$B(b \rightarrow a) = a_1$;
$B(b \rightarrow bb) = a_2$;
$B(c \rightarrow c) = a_3$;
$R_1 = \emptyset$;
$R_2 = \{(a_2 a_3; \text{out}), (a_1 a_3; \text{in})\}$.

From the initial configuration only the multiset rewriting rule $b \rightarrow bb \in R_2$ can be applied an unbounded number of times doubling every time the number of b's present in compartment 2. At any transition the antiport rule $(a_2 a_3; \text{out}) \in$

R_2 can be applied and then maximal parallelism forces $(a_1a_3; \text{in}) \in R_2$ to be applied. In this way the multiset rewriting rule $b \to a$ is in compartment 2 and its application allows all the b's to be replaced by a's. When this happens the computation halts. We can then say that Π_5 generates $\{a^{2^n} \mid n \in \mathbb{N}\}$. $\diamond$

Theorem 5.30 *A P system with symport/antiport of multiset rewriting rules having degree 2, using antiports of unbounded weight and no symports, can generate the Parikh images of* ET0L *languages.*

Proof Let $R = (V, H, \alpha, T)$ be an ET0L system of the kind considered in Lemma 3.1 generating $L(R)$. The system R is such that $V = \{v_1, \ldots, v_{|V|}\}$ and $H = \{H_1, H_2\}$. We define a P system $\Pi = (W, T, \mu, L_1, L_2, P_1, P_2, A, B, R_1, R_2, 2)$ with symport/antiport of multiset rewriting rules such that:

$$W = V \cup T \cup \{\star\};$$
$$\mu = (\{0, 1, 2\}, \{(0, 1), (1, 2)\});$$
$$L_1 = \phi;$$
$$L_2 = \bigcup_{v \in V}\{v^{|\alpha|_v}\};$$
$$P_1 = H_1 \cup P';$$
$$P' = \{v \to \star \mid v \in V \setminus T\};$$
$$P_2 = H_1 \cup \{\star \to \star\};$$
$$A = \{a_i \mid 1 \le i \le |H_1| + |H_2| + |V| + 1\};$$

$$B(x) = \begin{cases} a_i & \text{for } 1 \le i \le |H_1|, \ x \in H_1; \\ a_j & \text{for } |H_1| + 1 \le j \le |H_1| + |H_2|, \ x \in H_2; \\ a_p & \text{for } |H_1| + |H_2| + 1 \le p \le |A| - 1, \ x \in V; \\ a_{|A|} & \text{for } x = \star \to \star; \end{cases}$$

$$R_1 = \emptyset;$$
$$R_2 = R_2' \cup R_2'';$$
$$R_2' = \{(x; \text{out}/y; \text{in}) \mid x = \cup_{w \in H_1} B(w), \ y = \cup_{z \in H_2} B(z)\} \cup$$
$$\qquad \{(x; \text{out}/y; \text{in}) \mid x = \cup_{w \in H_2} B(w), \ y = \cup_{z \in H_1} B(z)\};$$
$$R_2'' = \{(x; \text{out}/y; \text{in}) \mid x = \cup_{w \in H_1} B(w), \ y = \cup_{z \in P'} B(z)\} \cup$$
$$\qquad \{(x; \text{out}/y; \text{in}) \mid x = \cup_{w \in H_2} B(w), \ y = \cup_{z \in P'} B(z)\}.$$

The P system works as follows. Because of the rules in R_2' the multiset rewriting rules in H_1 and H_2 pass into and out of compartment 2. In this way the simulation of the multiset rewriting rules in these two tables can be performed. This simulation ends when the rules in R_2'' are applied. When this happens the multiset rewriting rules in P' pass into compartment 2 (and no multiset rewriting rule in H_1 or H_2 is then present in this compartment). If some symbol in $V \setminus T$ is present in compartment 2 when also the multiset rewriting rules in T' are present there, then these symbols are replaced by $\star$. The continuous application of $\star \to \star$ allows Π to never halt.

The P system Π halts only if no symbol in $V \setminus T$ is present in compartment 2 when the multiset rewriting rules in P' pass into this compartment. In this case compartment 2 contains $\Psi_T(\mathsf{L}(R))$. The weight of the antiport rules present in Π is the maximum of the cardinalities of the tables in R and $|P'|$. This maximum is unbounded. We can say that Π generates $\Psi_T(\mathsf{L}(R))$. $\qquad\square$

It is important to notice that the previous theorem indicates precise bounds on the weight of the rules in case of specific ET0L systems. If, for instance, we consider a T0L system such that each table contains only one production, then the weight of the antiport rules is 1 (this is because $|P'| = 0$).

It is possible to bound the weight of the rules of the P system considered in the previous theorem by keeping their computational power unchanged at the cost of introducing priorities between rules and increasing the number of compartments. Let Π be a P system with symport/antiport of multiset rewriting rules with priorities between rules and let R' and R'', $R' \cap R'' = \emptyset$ be two sets of multiset rewriting rules in Π. The sets R' and R'' can contain rules associated with different compartments in Π. If all the rules in R' have higher priority than all the rules in R'', then for a transition, first the rules in R' and then the rules in R'' are chosen in a non-deterministic, maximally parallel way. This means that no rule in R'' is applied if this would not allow a rule in R' to be applied.

Differently from what is indicated by Definition 5.3 in the P systems considered in the following theorem and in Theorem 5.32 the final compartment is composite. This issue is discussed in Section 5.16.

Theorem 5.31 *P systems with symport/antiport of multiset rewriting rules with priorities between multiset rewriting rule, having degree 5, using antiports of weight 2 and no symports, can generate the Parikh images of* ET0L *languages.*

Proof Let $R = (V, H, \alpha, T)$, $H = \{H_1, H_2\}$, be an ET0L system of the kind considered in Lemma 3.1 generating $\mathsf{L}(R)$. Let $P' = \{v \to \star \mid v \in V \setminus T\}$ and $s = max(\{|H_1|, |H_2|, |P'|\})$. We increase the cardinality of H_1, H_2 and P' to s with multiset rewriting rules of the kind $d_i \to d_i$, $d_i \notin V$ and $1 \leq i \leq k$. So we assume we have sets of multiset rewriting rules H_1', H_2' and P'' all with the same cardinality s. Moreover, the multiset rewriting rules of the kind $x \to y$, $x \in V \setminus T$, present in H_1' and H_2' are replaced by $x \to d'y$ and $x \to d''y$, respectively with d', $d'' \notin V$.

We define a P system with symport/antiport of multiset rewriting rules with priorities between multiset rewriting rules $\Pi = (W, T, \mu, L_1, \ldots, L_5, P_1, \ldots, P_5, A, B, R_1, \ldots, R_5, 3)$ such that:

$$W = V \cup T \cup \{d_i \mid 1 \leq i \leq k\} \cup \{d', d'', \star\};$$
$$\mu = (\{0, 1, \ldots, 5\}, \{(i, i+1) \mid 0 \leq i \leq 4\});$$
$$L_1 = L_5 = \phi;$$
$$L_2 = L_4 = \{\star\};$$
$$L_3 = \{\alpha\};$$
$$P_1 = P_5 = \{\star \to \star\};$$
$$P_2 = H_2';$$
$$P_3 = H_1' \cup \{d' \to \epsilon, d'' \to \epsilon, \star \to \star\};$$
$$P_4 = P'';$$
$$A = \{a_i \mid 1 \leq i \leq |H_1'| + |H_2'| + |P''| + 3\};$$

$$B(x) = \begin{cases} a_i & \text{for } 1 \leq i \leq |H_1'|, \ x \in H_1'; \\ a_j & \text{for } |H_1'| + 1 \leq j \leq |H_1'| + |H_2'|, \ x \in H_2'; \\ a_p & \text{for } |H_1'| + |H_2'| + 1 \leq p \leq |A| - 3, \ x \in P''; \\ a_{|A|-2} & \text{for } x = d' \to \epsilon; \\ a_{|A|-1} & \text{for } x = d'' \to \epsilon; \\ a_{|A|} & \text{for } x = \star \to \star; \end{cases}$$

$$R_1 = \emptyset;$$
$$R_2 = \{(xy; \text{out}/a_{|A|}; \text{in}) \mid x \in \cup_{w \in H_1'} B(w), \ y \in \cup_{z \in H_2'} B(z)\};$$
$$R_3 = \{(a_i; \text{out}/a_j; \text{in}), (a_j; \text{out}/a_i; \text{in}) \mid B(x) = a_i, \ x \in H_2', \ a_j = a_{i-s}\};$$
$$R_4 = \{(a_i; \text{out}/a_j; \text{in}) \mid B(x) = a_i, \ x \in P'', \ a_j = a_{i-2s}\};$$
$$R_5 = \{(a_{|A|}; \text{out}/x, y; \text{in}) \mid x \in \cup_{w \in H_1'} B(w), \ y \in \cup_{z \in P''} B(z)\}.$$

Rules in R_5 have higher priority than rules in R_4 and rules in R_2 have higher priority than rules in R_3.

The simulation of R takes place in compartment 3. In the initial configuration the multiset rewriting rules in H_1' can be simulated on α in compartment 3. The d' symbols introduced by the application of such multiset rewriting rules are removed by $d' \to \epsilon \in P_3$ (ϵ being the empty string). The multiset rewriting rules in H_2' and P'' are initially in compartments 2 and 4, respectively. The multiset rewriting rules in H_2' can pass from compartment 2 into compartment 3 by the application of the antiports in R_3. At the same time the multiset rewriting rules in H_1' can pass from compartment 3 into compartment 2. This exchange has to be done in such a way that the multiset rewriting rules in the two sets are not mixed.

If some multiset rewriting rules of H_1' and H_2' are present in compartment 2 in the same configuration, then one of the rules in R_2 (having higher priority than the rules in R_3) is applied. This leads to $\star \to \star$ passing into compartment 2. As $\star$ is present in this compartment, then the continuous application of $\star \to \star$ allows Π to never halt. The exchange of multiset rewriting rules in H_2' with those in H_1' is performed in a similar way. The d'' symbols introduced by the application of such multiset rewriting rules are removed by $d'' \to \epsilon \in P_3$.

When the multiset rewriting rules in H_1' are in compartment 3 the rules in R_4 can be applied, too. These rules exchange the multiset rewriting rules in P'', initially present in compartment 4, with the multiset rewriting rules in H_1'. In case this exchange leads to multiset rewriting rules in the two sets to be mixed, then the application of rules in R_5 allows Π to never halt (similarly to what was indicated before, notice that the rules in R_5 have higher priority than the rules in R_4).

If some symbols in $V \setminus T$ are present in compartment 2 when the rules in P'' pass into this compartment, then $\star$ is introduced and the continuous application of $\star \rightarrow \star$ allows Π to never halt.

The P system Π halts only if no symbol in $V \setminus T$ is present in compartment 3 when the multiset rewriting rules in P'' pass into this compartment. In this case compartment 3 contains $\Psi_T(\mathsf{L}(R))$. The weight of the antiport rules present in Π is the maximum cardinality of the tables in R. The system Π uses only antiports of weight 2. We can say that Π generates $\Psi_T(\mathsf{L}(R))$. $\qquad\square$

It is possible to further change Definition 5.3 to allow the presence of catalysts in the multiset rewriting rules. These kinds of multiset rewriting rules are described in Section 6; here we simply say that a catalyst is a symbol needed for the application of a multiset rewriting rule but which is not affected by the application of any multiset rewriting rule. The next theorem, also having a composite final compartment, considers the matrix grammars of Definition 3.3.

Theorem 5.32 *P systems with symport/antiport of multiset rewriting rules with catalysts having degree 3, using only antiports of weight 2 can generate the Parikh images of the languages generated by a matrix grammar with appearance checking in Z-binary normal form.*

Proof Let $G_M = (N, T, S, M, F)$ be a grammar such as that of Definition 3.3. We define a P system with symport/antiport of multiset rewriting rules with catalysts $\Pi = (W, T, \mu, L_1, L_2, L_3, P_1, P_2, P_3, A, B, R_1, R_2, R_3, 2)$ such that:

$W = N \cup T \cup \{c, d, e_1, e_2, e_3, e_4, e_5\} \cup \{f_r \mid 1 \leq r \leq k - 1\}$;

$\mu = (\{0, 1, 2, 3\}, \{(0, 1), (1, 2), (2, 3)\})$;

$L_1 = \phi$;

$L_2 = \{S_X, S_A\}$;

$L_3 = N_1 \cup N_2 \cup \{Z\}$;

$P_1 = S_1 \cup S_2 \cup S_3 \cup S_6$;

$P_2 = S_4$;

$P_3 = S_5$;

$S_1 = \{X \rightarrow f_r Y d, \ cA \rightarrow cf_r ad \mid M_r = (X \rightarrow Y, \ A \rightarrow a) \in M, 1 \leq r \leq s\}$;

$S_2 = \{X \rightarrow f_r Y d, \ A \rightarrow f_r \star \mid M_r = (X \rightarrow Y, \ A \rightarrow \star) \in M,$
$$s + 1 \leq r \leq k - 1\};$$

$S_3 = \{e_3 \to e_3, \, e_4 \to e_4, \, e_5 \to e_5\} \cup \{f_r \to \epsilon \mid 1 \le r \le k - 1\};$
$S_4 = \{e_1 \to e_1, \, d \to \epsilon\};$
$S_5 = \{e_2 \to e_2, \, d \to \star, \, \star \to \star\};$
$S_6 = \{Z \to d \mid M_k = (Z \to \epsilon) \in M\};$
$A = \{a_i \mid 1 \le i \le \sum_{j=1}^{6} |S_j|\};$

$$B(x) = \begin{cases} a_i & \text{for } 1 \le i \le |S_1|, \, x = X \to f_r Y d \in S_1 \cup S_2; \\ a_j & \text{for } |S_1| + 1 \le j \le |S_1| + |S_2|, \, x = cA \to c f_r a d \in S_1 \text{ or} \\ & \qquad x = A \to f_r \star \in S_2; \\ a_p & \text{for } |S_1| + |S_2| + 1 \le p \le \sum_{q=1}^{5} |S_q|, \, x \in S_3 \cup S_4 \cup S_5; \\ a_{|A|} & \text{for } x = Z \to d \in S_6; \end{cases}$$

$R_1 = \emptyset;$
$R_2 = \{(B(e_1 \to e_1); \text{out}/a_i a_j; \text{in}), \, (a_i a_j; \text{out}/B(e_1 \to e_1); \text{in}) \mid$
$\qquad\qquad 1 \le i \le |S_1|, \, |S_1| + 1 \le j \le |S_1| + |S_2|\} \cup$
$\qquad \{(B(e_1 \to e_1); \text{out}/B(Z \to d)B(e_3 \to e_3); \text{in}),$
$\qquad (B(e_3 \to e_3); \text{out}/B(e_4 \to e_4); \text{in});$
$\qquad (B(e_4 \to e_4)B(Z \to d); \text{out}/B(e_5 \to e_5); \text{in})\},$
$R_3 = \{(B(e_2 \to e_2); \text{out}/B(Z \to d); \text{in})\} \cup$
$\qquad \{(B(e_2 \to e_2); \text{out}/a_i; \text{in}) \mid 1 \le i \le |S_1| + |S_2|\} \cup$
$\qquad \{(B(e_2 \to e_2); \text{out}/a_i; \text{in}) \mid 1 \le i \le |S_1|\}.$

The multiset rewriting rules present in S_1, S_2 and S_6 derive from those of G_M. They have been modified so as to generate, when applied, the symbol d, that is, used to check if a multiset rewriting rule has been applied. The symbols f_r, $1 \le r \le k - 1$, present in the multiset rewriting rules in S_1 and S_2 are used to identify rules in a unique way. Such f_r symbols are removed by $f_r \to \epsilon \in S_3$, σ being the empty string. The sets of multiset rewriting rules S_1, S_2, S_3 and S_6 are initially present in compartment 1.

The simulation takes place in compartment 2 where, in the initial configuration S_X, S_A and the catalyst c are present.

Using one rule of the kind $(B(e_1 \to e_1); \text{out}/a_i a_j; \text{in}) \in R_2$ two multiset rewriting rules associated with a matrix in G_M can pass into compartment 2 while $e_1 \to e_1$ passes into compartment 1. When the multiset rewriting rules having labels a_i and a_j are in compartment 2 they can be either applied (possibly in subsequent transitions) or passed back to compartment 1 because of the application of $(a_i a_j; \text{out}/B(e_1 \to e_1); \text{in}) \in R_2$. The multiset rewriting rules $e_1 \to e_1$ is only used to guarantee that only one pair of multiset rewriting rules at one time passes into compartment 2.

Let us assume that two multiset rewriting rules present in matrices of type (ii) (see Definition 3.3) having label a_{i_1} and a_{j_1} pass into compartment 2 but only the multiset rewriting rules having label a_{i_1} can be applied. Because of maximal parallelism the multiset rewriting rule having label a_{j_1} passes into compartment 3 while $e_2 \rightarrow e_2$ passes into compartment 2. This occurs because of the application of one of rules of the kind $(B(e_2 \rightarrow e_2); \text{out}/a_i; \text{in}) \in R_3$. When a multiset rewriting rule having label a_i is in compartment 3 it is applied. In this way the symbol d is generated. The subsequent application of $d \rightarrow \star$ and $\star \rightarrow \star$ (repeatedly) both in P_3 allows Π to never halt.

Let us assume now that the two multiset rewriting rules having labels a_{i_1} and a_{j_1} passing into compartment 2 are associated with productions present in matrices of type (iii). If $B(A \rightarrow f_r\star) = a_{j_1}$ is applied but $B(X \rightarrow f_rY) = a_{i_1}$ is not applied, then Π behaves in a way similar to what has just been indicated. If instead $A \rightarrow f_r\star$ is not applied while $X \rightarrow f_rY$ is applied, then, after the application (possibly in subsequent transitions) of $X \rightarrow f_rY$, both multiset rewriting rules pass into compartment 1. In this way the appearance checking is simulated.

The P system Π can halt after $(B(e_1 \rightarrow e_1); \text{out}/B(Z \rightarrow d)B(e_3 \rightarrow e_3); \text{in}) \in R_2$ is applied. If when this happens Z is not present in compartment 2, then $(B(e_2 \rightarrow e_2); \text{out}/B(Z \rightarrow d); \text{in}) \in R_3$ is applied followed by the application of $d \rightarrow \star$ and $\star \rightarrow \star$ (repeatedly) both in P_3. In this way Π never halts. If instead Z is present in compartment 2, then $Z \rightarrow d$ and $(B(e_3 \rightarrow e_3); \text{out}/B(e_4 \rightarrow e_4); \text{in}) \in R_2$ are applied. This is followed by the application of $(B(e_4 \rightarrow e_4), B(Z \rightarrow d); \text{out}/B(e_5 \rightarrow e_5); \text{in}) \in R_2$. When this happens the P system Π halts with the Parikh image of $\mathsf{L}(G_M)$ present in compartment 2. We can say that Π generates $\Psi_T(\mathsf{L}(G_M))$. $\square$

5.13 Tissue P systems

In the remainder of the present chapter we consider membrane systems with symport/antiport having a cell-graph as underlying structure. To fully understand the motivation behind the research presented in the following we have to recall a few aspects in the history of Membrane Computing. As we said in Chapter 1 Membrane Computing originated by observations of the functioning of single cells. For this reason the first models had a cell-tree as underlying structure. The study of cell-graphs as underlying structures arrived later in the field. The models of membrane systems considering cell-graphs were either inspired by biological processes or they simply originated by mathematical abstractions.

Tissue P systems are among the models having a cell-graph as underlying structure inspired by biological reality. Under this name a broad range of models has been studied, some of these inspired by neural interaction (see Chapter 7).

The following section indicates one biological reality defining cell–cell interaction different from neural interaction.

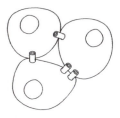

Fig. 5.5 Gap junctions between cells

5.13.1 Biological motivations

Cells can be connected to each other so as to form tissues. Our skin, guts and muscles are examples of tissues. Cell junctions can be classified into three categories:

occluding: prevent small molecules from passing from one side of the tissue to the other;

anchoring: mechanically attach cells to other cells;

communicating: allow cells to pass chemicals or electrical signals between adjacent cells.

This last forms of junction have been of inspiration to a model of membrane systems: *tissue P system.*

Gap junctions allow small chemicals to pass directly from the cytoplasm of one cell to that of another. This passage is performed by transmembrane proteins, called *channel-forming* proteins, present on the plasma membrane. When two of these proteins present on the plasma membrane of two cells are aligned, then a channel is established. This is depicted in Fig. 5.5.

These channel-forming proteins allow the passage of molecules with a small diameter such as ions, sugars, amino acids, etc., but they do not allow the passage of bigger molecules, such as proteins, for instance. Gap junctions are not continuously open, rather these junctions are dynamical structures that can close or open themselves in response to changes in the cells.

5.13.2 Basic definitions

Definition 5.4 *A* (generating) tissue P system with symport/antiport *of degree* m *is a construct*

$$\Pi = (V, \mu_R, L_0, L_1, \ldots, L_m, comp)$$

where:

V is an alphabet;

$\mu_R = (Q, E, lr)$ *is an asymmetric edge-labelled directed multigraph (a cell-graph)*
underling Π where:

$Q \subseteq \mathbb{N}$ *contains* vertices. *For simplicity we define* $Q = \{0, 1, \ldots, m\}$. *Each*
vertex in Q defines a cell *of Π.*

$E \subseteq Q \times Q$ *defines directed labelled* edges *between vertices.*

$lr : Q \times Q \rightarrow R$ *is the labelling relation where the elements of R are of*
the kind $(x; in), (x; out)$ *called* symport *rules, or* $(y; out/x; in)$*, called*
antiport *rules, with x and y multisets over V with a finite and not-*
empty support. This means that $(\phi; in), (\phi; out)$ *and* $(\phi; out/\phi; in)$ *are*
not allowed.

L_i, $0 \le i \le m$, *are multisets over V. All the elements in L_0 have infinite*
multiplicity while the elements in $L_1, \ldots, L_m$ have finite multiplicity.

comp $\in Q$ defines the output cell.

If $(i, j) \in E$ and $lr(i, j) = r$, then we write (i, j, r), a *labelled edge*. The
use of 'asymmetric' edge-labelled directed multigraphs as underlying structure
can be misleading since, as we will see in the following, the application of an
antiport on such edges allows symbols to move in two directions. We chose to
indicate 'asymmetric' for consistency with what is present in the literature of
tissue P systems with symport/antiport. We will discuss this further at the end of
Section 5.15.

If $(i, j, r) \in E$, then the application of r can change the multisets associated with the multisets M_i and M_j over V, associated with vertices i and j,
respectively, in the following way:

if $r = (x; in)$ and $x \in M_j$, then M_i changes into $M_i' = M_i \cup x$ while M_j changes
into $M_j' = M_j \setminus x$;

if $r = (x; out)$ and $x \in M_i$, then M_i changes into $M_i' = M_i \setminus x$ while M_j changes
into $M_j' = M_j \cup x$;

if $r = (y; out/x; in)$, $y \in M_i$ and $x \in M_j$, then M_i changes into $M_i' = M_i \setminus y \cup x$
while M_j changes into $M_j' = M_j \setminus x \cup y$.

In general, if a multiset x is subtracted from M_i and united to M_j we say
that x *passes* from cell (vertex) i to cell (vertex) j.

A *configuration* of a tissue P system with symport/antiport of degree m is
given by the $(m+1)$-tuple $(M_0 \setminus L_0, M_1, \ldots, M_m)$ of multisets over V associated
with the cells of Π. Note that the configuration does not record the symbols in
the environment occurring with infinite multiplicity as they are invariant to any
configuration. The $(m+1)$-tuple $(\phi, L_1, \ldots, L_m)$ is called the *initial configuration*.
For two configurations $(M_0 \setminus L_0, M_1, \ldots, M_m)$ and $(M_0' \setminus L_0, M_1', \ldots, M_m')$ of Π
we write $(M_0 \setminus L_0, M_1, \ldots, M_m) \Rightarrow (M_0' \setminus L_0, M_1', \ldots, M_m')$ to denote a *transition*
from $(M_0 \setminus L_0, M_1, \ldots, M_m)$ to $(M_0' \setminus L_0, M_1', \ldots, M_m')$ that is, the application

of a multiset of rules associated with each compartment under the requirement
of maximal parallelism.

A *computation* is a sequence of transitions between configurations of a
system Π starting from the initial configuration $(\phi, L_1, \ldots, L_m)$. If a computation is finite, then the last configuration is called *final*. In this case we
say that the system *halts*. The result of a finite computation is given by the
multiset of symbols present in the output cell *comp* when the computations
halts.

The set $N(\Pi)$ denotes the set of numbers generated by a tissue P system Π
with symport/antiport. The family of sets $N(\Pi)$ generated by tissue P systems
with symport/antiport of degree at most m, using symports of weight at most p
and antiports of weight at most q, is denoted by $N \, OtP_m(sym_p, anti_q)$. The family
of sets $N(\Pi)$ generated by tissue P systems with symport/antiport of degree at
most m, using symports of weight at most p, antiports of weight at most q and
at most s symbols, is denoted by $N \, O_stP_m(sym_p, anti_q)$. When p, q or m are not
bounded then they are replaced by $*$.

5.14   From previous proofs

It should be clear that there is a straightforward link between P systems
with symport/antiport and tissue P systems with symport/antiport. Moreover,
because of the underlying structure, all the results about P systems with
symport/antiport are valid for tissue P systems with symport/antiport.

This is indeed the case for the systems considered in the proofs of Theorems
5.1, 5.2, 5.10, 5.11 and Corollary 5.3. We have:

Theorem 5.33 $N \, OtP_1(sym_1, anti_1) \subseteq N \, FIN.$

Theorem 5.34 $N \, OtP_1(sym_2, anti_0) \subseteq N \, FIN.$

Theorem 5.35 $N \, O_1tP_1(sym_*, anti_*) = N \, FIN.$

Theorem 5.36 $N \, O_1tP_2(sym_*, anti_*) \supseteq N \, REG$

Theorem 5.37 $N \, O_1tP_4(sym_*, anti_*) \supseteq N \, REG.$

Also the link between tissue P systems with symport/antiport and P/T systems follows directly from that between P systems with symport/antiport and
P/T systems.

5.15 New proofs

In this section we prove results concerning tissue P systems with symport/antiport not resembling any of the results on P systems with symport/antiport.

Theorem 5.38 $\mathbb{N} \, O_1 tP_7(sym_*, anti_*) = \mathbb{N} \, RE$.

Proof Let $M = (S, I, s_1, s_f)$ be a register machine with three registers such that one of the registers is only incremented. We define a tissue P system Π with symport/antiport, degree 7 and using only one symbol able to simulate M. The P system is $\Pi = (\{a\}, \mu_R, L_0, L_1, \ldots, L_7, 1)$ with

$$\mu_R = (Q, E, lr);$$
$$Q = \{0, 1, \ldots, 7\};$$
$$E = \{1 : (4, 2, (a^2; \mathrm{out}/a^6; \mathrm{in})), 2 : (4, 7(a^3; \mathrm{out})), 3 : (5, 3, (a^6; \mathrm{out}/a^{18}; \mathrm{in}));$$
$$\quad 4 : (5, 7, (a^7; \mathrm{out})), 5 : (6, 1, (a^1; \mathrm{out})), 6 : (6, 2, (a^4; \mathrm{out})),$$
$$\quad 7 : (6, 2, (a^2; \mathrm{out}/a^6; \mathrm{in})), 8 : (6, 3, (a^{12}; \mathrm{out})),$$
$$\quad 9 : (6, 3, (a^6; \mathrm{out}/a^{18}; \mathrm{in})), 10 : (6, 7, (a^1; \mathrm{out})),$$
$$\quad 11 : (7, 0, (a^2; \mathrm{out}/a^2; \mathrm{in}))\} \cup$$
$$\{12 : (6, 0, (a^{c(p)}; \mathrm{out}/a^{2c^{(1)}(p)}; \mathrm{in})),$$
$$\quad 13 : (6, 0, (a^{c^{(1)}(p)}; \mathrm{out}/a^{2c^{(2)}(p)-c(p)+1}; \mathrm{in})),$$
$$\quad 14 : (6, 0, (a^{c^{(3)}(p)}; \mathrm{out}/a^{2c(k)-c^{(2)}(p)-1}; \mathrm{in})),$$
$$\quad 15 : (6, 4, (a^{c(p)}; \mathrm{out})), 16 : (6, 4, (a^{c^{(1)}(p)}; \mathrm{out}/a^{c(p)-1}; \mathrm{in})),$$
$$\quad 17 : (6, 4, (a^{c^{(1)}(p)+1}; \mathrm{out})), 18 : (6, 5, (a^{c(p)}; \mathrm{out})),$$
$$\quad 19 : (6, 5, (a^{c^{(1)}(p)}; \mathrm{out}/a^{c(p)-1}; \mathrm{in})), 20 : (6, 5, (a^{c^{(1)}(p)+1}; \mathrm{out})) \mid$$
$$\qquad\qquad\qquad\qquad (s_p, \gamma_i^-, s_q, s_k) \in I, \ i = 2, 3\} \cup$$
$$\{21 : (6, 0, (a^{2c(p)-1}; \mathrm{out}/a^{2c(q)}; \mathrm{in})) \mid (s_p, \gamma_1^+, s_q) \in I\} \cup$$
$$\{22 : (6, 0, (a^{2c(p)-4}; \mathrm{out}/a^{2c(q)}; \mathrm{in})) \mid (s_p, \gamma_2^+, s_q) \in I\} \cup$$
$$\{23 : (6, 0, (a^{2c(p)-12}; \mathrm{out}/a^{2c(q)}; \mathrm{in})) \mid (s_p, \gamma_3^+, s_q) \in I\} \cup$$
$$\{24 : (6, 0, (a^{2c(p)-2}; \mathrm{out}/a^{2c^{(1)}(p)}; \mathrm{in})),$$
$$\quad 25 : (6, 0, (a^{2c^{(1)}(p)}; \mathrm{out}/a^{2c(q)}; \mathrm{in})) \mid (s_p, \gamma_2^-, s_q, s_k) \in I\} \cup$$
$$\{26 : (6, 0, (a^{2c(p)-6}; \mathrm{out}/a^{2c^{(1)}(p)}; \mathrm{in})),$$
$$\quad 27 : (6, 0, (a^{2c^{(1)}(p)+18}; \mathrm{out}/a^{2c(q)}; \mathrm{in})) \mid (s_p, \gamma_3^-, s_q, s_k) \in I\} \cup$$
$$\{28 : (6, 0, (a^{2c(f)}; \mathrm{out})\},$$
$$L_0 = \{a\};$$
$$L_1 = \phi;$$
$$L_2 = \{a^2\};$$
$$L_3 = \{a^6\};$$
$$L_4 = L_7 = \{a\};$$
$$L_5 = \{a^5\};$$
$$L_6 = \{a^{2c(0)}\}.$$

This proof requires the use of a unique-sum set $U_{3|S|}$ with $3|S|$ elements. Different multiplicities of a, where the multiplicities are elements in $U_{3|S|}$, are associated with the elements in S by the functions c, $c^{(1)}$ and $c^{(2)}$ all from S to $U_{3|S|}$. The exact definition of the three functions is irrelevant for the proof.

The cells in Π have specific functions. Cells 1, 2 and 3 are associated with the registers γ_1, γ_2 and γ_3 of M, respectively, γ_1 being the register that is only incremented. Cells 4 and 5 are used to test if cells 2 and 3, respectively, indicate that the registers γ_2 and γ_3 contain 0. Cell 6 has overall control of what takes place in Π, while in cell 7 rule 11 can be applied forever, letting Π never halt.

The content of the three registers of M is encoded in different ways in cells 1, 2 and 3. It is $val(\gamma_1)$ in cell 1, $4val(\gamma_2) + 2$ in cell 2 and $12val(\gamma_3) + 6$ in cell 3 (where val is the function returning the content of a register). Notice that the presence of no a in cell 1, a^2 in cell 2 and a^6 in cell 6 in the initial configuration indicates that the registers are empty. If Π simulates M being in state p, then $a^{2c(p)}$ is present in cell 6. Informally we can say that $c(p)$ occurrences of a are used to simulate the state change of M, while the remaining $c(p)$ occurrences of a are used to simulate an operation on the registers of M. Initially $a^{2c(0)}$, where $c(0)$ is the element in $U_{3|S|}$ associated with the initial state s_0, is present in cell 6.

During the computation of Π the passage of one occurrence of a in cell 7 allows rule 11 to be applied forever so that Π does not halt. The simulation performed by the P system Π is strongly based on the use of a unique-sum set and on the property that for such sets none of the elements can be obtained as a linear combination of the remaining elements in the set. During the computation of Π, different occurrences of the symbol a are present in the skin compartment. Only unique (maximally parallel) sequences of multisets of applied rules allow Π to simulate instructions of M. Any other maximally parallel sequences of multiset of applied rules allow one occurrence of a to pass into cell 7. This can occur with the application of any of rules 2, 4 or 10.

In the following rules, numbers between parentheses indicate that the associated rules are applied in parallel.

The simulation of instructions of the kind (s_p, γ_1^+, s_q) is performed by the application of rules $(5, 21)$. The simulation of instructions of the kind (s_p, γ_2^+, s_q) and (s_p, γ_3^+, s_q) is performed by the application of $(6, 22)$ and $(8, 23)$, respectively. Notice that, in accordance with what was said before, the application of these rules allows the number of occurrences present in cells 1, 2, and 3 to be increased by 1, 4 and 12 occurrences of a, respectively.

The simulation of instructions of the kind $(s_p, \gamma_2^-, s_q, s_k)$, starting when $a^{2c(p)}$ is present in cell 6, is performed in the following way. If $val(\gamma_2) > 0$, then in cell 2 there will be at least six occurrences of a so that the rules $(7, 24)$, 25 can be applied. Notice that the application of rule 7 lets four occurrences of a pass from cell 2 to cell 6, simulating in this way the subtraction of 1 from γ_2. Moreover, after the application of rule 25, cell 6 contains $a^{2c(q)}$ simulating M being in state s_q. The rules (12, 15), (1, 13, 16), (2, 14, 17) can be also applied when $a^{2c(p)}$ is present in cell 6, but in this case the subsequent repeated application of rule 11 allows Π to never halt.

If instead $val(\gamma_2) > 0$, then in cell 2 there are two occurrences of a. The rules (12, 15), (13, 16), (14, 17) can be applied letting $a^{2c(k)}$ be present in compartment 6 after the application of (14, 17). Notice that the simulation of the passage from state s_p to state s_k allows occurrences of a to pass into and out of cell 4. If $a^{2c(p)}$ is present in cell 6 and two occurrences of a are present in cell 2, then also the rules (12, 15), (1, 13, 16), (2, 14, 17) can be applied, but in this case the subsequent repeated application of rule 11 allows Π to never halt.

The simulation of instructions of the kind $(s_p, \gamma_3^-, s_q, s_k)$ is similar to that just described.

The system Π can halt when $a^{2c(f)}$ is present in compartment 1. If rule 28 is applied, then Π halts. $\qquad\square$

Theorem 5.39 $\mathbb{N} \, O_2 tP_3(sym_*, anti_*) = \mathbb{N} \, \mathrm{RE}$.

Proof Let $M = (S, I, s_1, s_f)$ be a register machine with three registers such that one of the registers is only incremented. We define a tissue P system Π with symport/antiport, degree 3 and using only two symbols able to simulate M. The P system is $\Pi = (\{a, b\}, \mu_R, L_0, L_1, L_2, L_3, 1)$ with

$\mu_R = (Q, E, lr)$;

$Q = \{0, 1, 2, 3\}$;

$E = \{1 : (1, 0, (c(s_f); out)), 2 : (1, 2, (b^2; out)), 3 : (2, 0, (ab; out/b^9; in)),$
$\quad 4 : (2, 0, (b^2; out/b^9; in)), 5 : (2, 3, (b^2; in)), 6 : (3, 0, (ab; out/b^2; in))\} \cup$
$\{7 : (1, 0, (c(s_p); out/c(s_q)a; in)) \mid (s_p, \gamma_1^+, s_q) \in I\} \cup$
$\{8 : (1, 0, (c(s_p); out/c(s_p^{(1)}), a; in)), 9 : (1, 0(c(s_p^{(2)}); out/c(s_q); in)),$
$\quad 10 : (1, 2, (c(s_p^{(1)}), a; out)), 11 : (3, 0, (c(s_p^{(1)}); out/c(s_p^{(2)}); in)),$
$\quad 12 : (3, 1(c(s_p^{(2)}); out)) \mid (s_p, \gamma_i^+, s_q) \in I, \; i = 2, 3\} \cup$
$\{13 : (2, 3, (c(s_p^{(1)}); out)) \mid (s_p, \gamma_2^+, s_q) \in I\} \cup$
$\{14 : (2, 3, (c(s_p^{(1)}); out)) \mid (s_p, \gamma_3^+, s_q) \in I\} \cup$
$\{15 : (1, 0, (c(s_p); out/c(s_p^{(1)})b; in)),$
$\quad 16 : (1, 0, (c(s_p^{(2)}); out/c(s_p^{(3)}); in)),$
$\quad 17 : (1, 0, (c(s_p^{(4)}); out/c(s_k); in)), 18 : (1, 0, (c(s_p); out/c(s_p^{(5)}); in)),$
$\quad 19 : (1, 0, (c(s_p^{(6)}); out/c(s_q); in)), 20 : (1, 2(c(s_p^{(1)}), b; out)),$
$\quad 21 : (1, 2, (c(s_p^{(3)}); out)), 22 : (1, 2, (c(s_p^{(5)}); out)),$
$\quad 23 : (3, 0, (c(s_p^{(1)}); out/c(s_p^{(2)}); in)),$
$\quad 24 : (3, 0, (c(s_p^{(3)})b); out/c(s_p^{(4)}); in),$
$\quad 25 : (3, 0, (c(s_p^{(5)})a; out/c(s_p^{(6)}); in)), 26 : (3, 1, (c(s_p^{(2)}); out)),$

$$27 : (3, 1, (c(s_p^{(4)}); \text{out})), 28 : (3, 1, (c(s_p^{(6)}); \text{out})) \mid$$
$$(s_p, \gamma_i^-, s_q, s_k) \in I, \ i = 2, 3\} \cup$$
$$\{29 : (2, 3, (c(s_p^{(1)})a; \text{out})), 30 : (2, 3, (c(s_p^{(3)})b; \text{out})),$$
$$31 : (2, 3, (c(s_p^{(5)})a; \text{out})) \mid s_p \gamma_2 s_q s_k \in I\} \cup$$
$$\{32 : (3, 2, (c(s_p^{(1)}); \text{out})), 33 : (2, 3, (c(s_p^{(3)}); \text{out})),$$
$$34 : (2, 3, (c(s_p^{(5)}); \text{out})) \mid s_p \gamma_3 s_q s_k \in I\};$$

$L_0 = \{a, b\};$

$L_1 = \{c(s_0)\};$

$L_2 = L_3 = \phi.$

This proof follows the lines of the proof of Theorem 5.15. Similar to that proof, also here there is a function c and a constant g defined in the following.

We define the sets $S^{(i)} = \{s^{(i)} \mid s \in S\}$, $1 \le i \le 6$ and $\bar{S} = S \cup \bigcup_{i=1}^{6} S^{(i)} = \{\bar{s}_1, \ldots, \bar{s}_{|\bar{S}|}\}$. The function $c : \bar{S} \to (\{b\} \to \mathbb{N}_+)$ is such that:

$|c(\bar{s}_i)| + |c(\bar{s}_j)| > |c(\bar{s}_{|\bar{S}|})|$, $1 \le i, j \le |\bar{S}|$, $i \ne j$, that is, the sum of the occurrences of b's returned by c for two elements in $\bar{S}$ has to be bigger that the number of b's returned by c for $\bar{s}_{|\bar{S}|}$;

$|c(\bar{s}_{i+1})| - |c(\bar{s}_i)| \ge 3$, $1 \le i \le |\bar{S}| - 1$, the number 3 depends on the defined rules: an application of rules that is not a simulation of an instruction of M has to leave in cell 1 at least three occurrences of b.

The c function is defined as: $c(\bar{s}_i) = 3i + 3 \cdot 7|I| = 3i + 21|I|$, $1 \le i \le |\bar{S}|$. The number 7 before is due to the fact that in $\bar{S}$ there are seven symbols $(s, s^{(1)}, \ldots, s^{(6)})$ associated with each $s \in S$.

The constant g has to be so large that if g occurrences of b are present in cell 1, then the system never halts. We chose $g = 2c(s_{|\bar{S}|})$.

During a simulation the number of occurrences of a present in cell i represents the content of register γ_i, $i = 1, 2, 3$. The presence of $c(s)$, $s \in S$, in cell 1 indicates the simulation of M being in state s. In the initial configuration $c(s_0)$ is present in cell 1. In case a sequence of applied rules does not follow the simulation of an instruction of M, then the P system Π never halts due to the continuous application of rule 4.

As this proof is similar to that of Theorem 5.15, we do not indicate the details of it, but simply the sequence of applied rules simulating instructions of M. The simulation of instructions of the kind (s_p, γ_1^+, s_q) is performed by the application of rule 7. The simulation of instructions of the kind (s_p, γ_2^+, s_q) is performed by the application of rules 8, 10, 13, 11, 12, 9. Similarly for the simulation of rules of the kind (s_p, γ_3^+, s_q). The simulation of instructions of the kind $(s_p, \gamma_2^-, s_q, s_k)$ if $val(\gamma_2) > 0$ is performed by the application of rules 18, 22, 29, 25, 28, 19; while if $val(\gamma) = 0$ by the application of rules 15, 20, 30, 11, 26, 16, 21, 33, 24, 27, 17. Similarly for the simulation of rules of the kind $(s_p, \gamma_3^-, s_q, s_k)$.

When $c(s_f)$ is present in cell 1, then the application of rule 1 halts the computation of Π. $\square$

With a proof similar to the just given one it is possible to prove:

Theorem 5.40 $\mathbb{N} O_4 t P_2(sym_*, anti_*) = \mathbb{N} RE$.

If only two symbols and one compartment are considered, then:

Theorem 5.41 $\mathbb{N} O_2 t P_1(sym_*, anti_*) = \mathbb{N} REG$.

Proof *Part I: ($\mathbb{N} REG \subseteq \mathbb{N} O_2 t P_1(sym_*, anti_*)$)* Similar to the proof of Theorem 5.11.

Part II: ($\mathbb{N} O_2 t P_1(sym_*, anti_*) \subseteq \mathbb{N} REG$) We show that a tissue P system with symport/antiport of degree 1 and more than one symbol cannot generate more than $\mathbb{N} REG$. As it is known that $\mathbb{N} REG = \mathbb{N} MAT$, then we only have to show that $\mathbb{N} O_2 t P_1(sym_*, anti_*) \subseteq \mathbb{N} MAT$. Let $\Pi = (V, \mu_R, L_0, L_1, 1)$ be a tissue P system with symport/antiport of degree 1 with

$V = \{v_1, \ldots, v_{|V|}\}$;
$\mu_R = (Q, E, lr)$ where:
$\quad Q = \{0, 1\}$;
$\quad E = \{(1, 0, r) | r \in R\}$.

We consider the set $\bar{V}$ and we define the function $bar : V \to \bar{V}$ such that $bar(v) = \bar{v}$, $v \in V$. We also define $bar : (V \to \mathbb{N}) \to \bar{V}^+$ such that if X is a multiset over V and $X(v) = n$; then $bar(v) = \bar{v} \ldots \bar{v}$ where $\bar{v}$ is repeated n times.

We define a matrix grammar $G_M = (N, V, S, M)$ with:

$N = \{S, X, Y\} \cup \bar{V}$;
$M = \bigcup_{i=1}^{7} M_i$;
$M_1 = (S \to X bar(L_1))$;
$M_2 = (X \to Y)$;
$M_3 = (Y \to \epsilon)$;
$M_4 = \bigcup_{(x;out) \in R}(X \to X, [\bar{v}_1 \to \epsilon]^{x(v_1)}, \ldots, [\bar{v}_i \to \epsilon]^{x(v_{i|supp(x)|})})$;
$M_5 = \bigcup_{(y;in) \in R}(X \to X \bar{w}_{j_1}^{y(v_{j_1})} \ldots \bar{w}_{j_{|supp(y)|}}^{y(v_{j_{|supp(y)|}})})$;
$M_6 = \bigcup_{(x;out/y;in) \in R}([\bar{v}_1 \to \epsilon]^{x(v_1)}, \ldots, [\bar{v}_i \to \epsilon]^{x(v_{i|supp(x)|})},$
$\qquad\qquad\qquad X \to X \bar{w}_{j_1}^{y(v_{j_1})} \ldots \bar{w}_{j_{|supp(y)|}}^{y(v_{j_{|supp(y)|}})})$;
$M_7 = \bigcup_{i=1}^{|V|}(Y \to Y, \bar{v}_i \to v_i)$.

In the above $[\alpha \to \beta]^{\gamma}$ in a matrix denotes that the production $\alpha \to \beta$ is present γ times in that matrix.

The matrix grammar G_M starts simulating Π applying (the production in the) matrix M_1. After this the matrices of type M_4, M_5 and M_6 can be applied. This corresponds to the simulation of rules of the kind $(x; out), (y; in)$

and $(x; \text{out}/y; \text{in})$, respectively. The fact that only one of these matrices can be applied in one sentential form reflects the fact that only one rule per configuration can be applied in Π. The derivation of G_M can result in a string over T only if the application of matrix M_2 is followed by several applications of matrix M_7 and finally by the application of matrix M_3.

It should be clear that $\mathbb{N} \, O_2 t P_1(\text{sym}_*, \text{anti}_*) \subseteq \mathbb{N} \, \text{MAT}$. $\square$

The only results concerning tissue P systems with a bounded weight we mention in this section are:

Theorem 5.42 $\mathbb{N} \, OtP_2(sym_1, anti_1) = \mathbb{N} \, \text{RE}$.

and

Theorem 5.43 $\mathbb{N} \, OtP_2(sym_2, anti_0) = \mathbb{N} \, \text{RE}$.

We consider the proofs of these theorems to be rather convoluted, and for this reason we omit them.

As we said in Section 5.13.2 the use of 'asymmetric' edge-labelled directed multigraphs as underlying structure for tissue P systems with symport/antiport can be misleading. Anyhow the use of edge-labelled directed multigraphs (that is, not 'asymmetric') as underlying structure leads to a few new results.

The families $\mathbb{N} \, Ot'P_m(\text{sym}_p, \text{anti}_q)$ and $\mathbb{N} \, O_s t'P_m(\text{sym}_p, \text{anti}_q)$ are equivalent to $\mathbb{N} \, OtP_m(\text{sym}_p, \text{anti}_q)$ and $\mathbb{N} \, O_s tP_m(\text{sym}_p, \text{anti}_q)$, respectively, but with an edge-labelled directed multigraph as underlying structure of the relative tissue P systems with symport/antiport. The known results are:

Theorem 5.44 $\mathbb{N} \, O_1 t'P_6(sym_*, anti_*) \ = \ \mathbb{N} \, O_3 t'P_2(sym_*, anti_*) \ = \ \mathbb{N} \, O_5 t'P_1 (sym_*, anti_*) = \mathbb{N} \, \text{RE}$.

5.16   Final remarks and research topics

We summarise the results presented in Section 5.5 and Section 5.8 and then we briefly discuss them.

	results Section 5.5		results Section 5.8
5.1	$\mathbb{N} \, OP_1(sym_1, anti_1) \subseteq \mathbb{N} \, \text{FIN}$	5.10	$\mathbb{N} \, O_1P_1(sym_*, anti_*) = \mathbb{N} \, \text{FIN}$
5.2	$\mathbb{N} \, OP_1(sym_2, anti_0) \subseteq \mathbb{N} \, \text{FIN}$	5.11	$\mathbb{N} \, O_1P_2(sym_*, anti_*) \supseteq \mathbb{N} \, \text{REG}$
5.3	$\mathbb{N} \, OP_1(sym_1, anti_2) = \mathbb{N} \, \text{RE}$	5.12	$\mathbb{N} \, O_2P_1(sym_*, anti_*) \supseteq \mathbb{N} \, \text{REG}$
5.1	$\mathbb{N}_1 \, OP_1(sym_0, anti_2) = \mathbb{N}_1 \, \text{RE}$	5.3	$\mathbb{N} \, O_1P_4(sym_*, anti_*) \supseteq \mathbb{N} \, \text{REG}$
5.4	$\mathbb{N}_7 \, OP_1(sym_3, anti_0) = \mathbb{N}_7 \, \text{RE}$	5.4	$\mathbb{N} \, O_5P_1(sym_*, anti_*) = \mathbb{N} \, \text{RE}$
5.5	$\mathbb{N}_1 \, OP_2(sym_1, anti_1) = \mathbb{N}_1 \, \text{RE}$	5.4	$\mathbb{N} \, O_4P_2(sym_*, anti_*) = \mathbb{N} \, \text{RE}$
5.6	$\mathbb{N}_1 \, OP_2(sym_2, anti_0) = \mathbb{N}_1 \, \text{RE}$	5.5	$\mathbb{N} \, O_3P_3(sym_*, anti_*) = \mathbb{N} \, \text{RE}$
		5.5	$\mathbb{N} \, O_2P_4(sym_*, anti_*) = \mathbb{N} \, \text{RE}$

If we look at the results concerning P systems with symport/antiport with no restrictions on the number of symbols (left column in the previous table), then we notice that the addition of one compartment allows these systems to make a rather big computational jump: from $\mathbb{N}$ FIN to $\mathbb{N}$ RE (see Theorems 5.1 and 5.5 and Theorems 5.2 and 5.6). A similar feature does not seem obvious in P systems with symport/antiport having restrictions on the number of symbols. In this case it is more the number of different symbols that plays a role (see Theorems 5.10, 5.12 and Corollary 5.4).

For this reason we formulate the following:

Suggestion for research 5.1 *Understand the reason of such computational jumps for P systems with symport/antiport with or without restrictions on the number of symbols.*

The previous suggestion for research aims at investigations similar to those described in Chapter 4. What are the processes that one extra compartment allows us to have if no restriction on the number of symbols is present? What are the processes that extra symbols allow us to have if there are restrictions on the number of symbols?

There are many 'gaps' in the previous table, that is, combinations of numbers of symbols, degrees and weights for which no results are known or the known results are not equivalences but inclusions. So, the suggestion for research concerning 'filling the gaps' is natural (for this as for any other model of computation). We want to explicitly mention one of these ideas:

Suggestion for research 5.2 *Study the computational power of $\mathbb{N} O_1 P_3(sym_*, anti_*)$ and $\mathbb{N} O_1 P_4(sym_*, anti_*)$.*

Theorem 5.13 is definitely unusual. The P systems considered there require the presence of maximal parallelism to simulate partially blind register machines. This may not be in line with what was stated by Theorem 4.3. We say 'may not be' because other elements, not considered in Theorem 4.3, could be relevant in the simulation of partially blind register machines. These elements are implicitly present in Theorem 4.3 but not in the P systems considered in Theorem 5.13. Another unusual component in this last theorem is the fact that the sum of unique-sum sets grows in an exponential way (and Theorem 3.4 states that there are no unique-sum sets having a smaller sum). This is the only result of this kind we are aware of requiring a variable growth in this way.

We can foresee two possibilities here:

Suggestion for research 5.3 *Either improve Theorem 5.13 so that neither maximal parallelism nor unique-sum sets are needed or research the links between the number of symbols, the possible processes (as studied in Chapter 4) and the growth of some variable to the computational power of (P) systems.*

Theorem 5.14 is very close to Theorem 5.13. We think that P systems with symport/antiport, one symbol and without priorities between rules can simulate register machines. This *conjecture* is due to Theorem 5.38 indicating that tissue P systems with symport/antiport and one symbol can simulate register machines.

Suggestion for research 5.4 *Proof or disprove the previous conjecture.*

The model of P systems with symport/antiport considered in Section 5.12 is certainly of interest. Moreover, some of the results reported there do not have an elementary final compartment.

Suggestion for research 5.5 *Study further P systems with symport/antiport of productions having elementary final compartments.*

Such study can also consider the introduction of further measures of descriptional complexity such as, for instance, the kind of productions.

The next table summarises the results concerning tissue P systems with symport/antiport.

5.33	$N\,OtP_1(sym_1, anti_1) \subseteq N\,FIN$		5.35	$N\,O_1tP_1(sym_*, anti_*) = N\,FIN$
5.34	$N\,OtP_1(sym_2, anti_0) \subseteq N\,FIN$		5.36	$N\,O_1tP_2(sym_*, anti_*) \supseteq N\,REG$
5.42	$N\,OtP_2(sym_1, anti_1) = N\,RE$		5.37	$N\,O_1tP_4(sym_*, anti_*) \supseteq N\,REG$
5.43	$N\,OtP_2(sym_2, anti_0) = N\,RE$		5.38	$N\,O_1tP_7(sym_*, anti_*) = N\,RE$
			5.39	$N\,O_2tP_3(sym_*, anti_*) = N\,RE$
			5.40	$N\,O_4tP_2(sym_*, anti_*) = N\,RE$
			5.41	$N\,O_2tP_1(sym_*, anti_*) = N\,REG$
			5.44	$N\,O_1t'P_6(sym_*, anti_*) = N\,RE$
			5.44	$N\,O_3t'P_2(sym_*, anti_*) = N\,RE$
			5.44	$N\,O_5t'P_1(sym_*, anti_*) = N\,RE$

We were not able to find common patterns between the results concerning P systems with symport/antiport and tissue P systems with symport/antiport. If we compare the results listed in the two tables given in the present section, then we realise that the majority of them are very similar (this was already indicated in Section 5.14). How do we compare the few exceptions? For instance, we know that $N\,OtP_2(sym_1, anti_2) = N\,RE$. Shall we compare this results with $N\,OP_1(sym_1, anti_2) = N\,RE$ (suggesting that if the weight of the symport and antiport is fixed, then tissue P systems need one more compartment) or with $N_1\,OP_2(sym_1, anti_1) = N_1\,RE$ (suggesting that if the number of compartments is fixed, then tissue P systems can have symports of smaller weight)?

We believe that the only message that can be deduced from these results is that when symports and antiports are used the topology of the compartments does not influence much the computational power of the resulting devices. As opposed to this, the results in Section 4.4 clearly show that the underlying topology can influence the computing power. Our understanding of this lets us think that the use of symports and antiports presented in this chapter is 'too powerful', so as to mask the differences in topology. Different topologies could influence the computational power if the use of symports and antiports were 'less powerful'.

Suggestion for research 5.6 *Study models of P systems and tissue P systems with 'weaker' symports and antiports.*

For instance, only symports can be allowed in tissue P systems. Moreover, deterministic systems can be studied. In Section 6.5 we will see how determinism can influence the computational power of a system. There the number of catalysts bounds the maximum number of rules that can be applied in a configuration. In a similar way in a tissue P system the number of edges bounds the number of rules that can be applied in a configuration. One study in this direction saw the introduction of *conditional uniports*: only one symbol at one time can pass from one compartment to another only if another symbol, called the *activator*, is present in one of the two compartments. These studies only considered cellgraphs as underlying structure.

5.17 Bibliographical notes

Membrane systems with symport/antiport were introduced in [203] in 2002 and one year later the Institute for Scientific Information, [129], reported this paper as 'citation leader in the domain'.

Theorem 5.1 was originally proved in [16], while Theorem 5.2 was originally proved in [88]. In [203] it was proved that $N \, OP_2(sym_2, anti_2) = N \, RE$; the improved result, Theorem 5.3, was independently obtained in [88] and [69].

The study of P systems with symport/antiport with antiports of weight 0 has a longer history. In [171] the equalities

$$N \, OP_2(sym_5, anti_0) = N \, OP_3(sym_4, anti_0) = N \, OP_5(sym_3, anti_0) = N \, RE$$

were proved. In [88] it was proved that $N \, OP_{13}(sym_1, anti_3) = N_{13} \, RE$ and that $N \, OP_2(sym_3, anti_0) = N \, RE$. Theorem 5.4, stating that $N_7 \, OP_1(sym_3, anti_0) = N_7 \, RE$, was originally proved in [9].

Minimal cooperation for P systems with symport/antiport was considered already in [203] where it was proved that $N \, OP_5(sym_2, anti_1) = N \, RE$. The first proof of the computational completeness of P systems with symport/antiport with weight at most 1 was given in [27]. The number of compartments of these systems was then reduced to six [141], five [29], and four [79, 167]. In [241] it was proved that $N_5 \, OP_3(sym_1, anti_1) = N_5 \, RE$. To the best of our knowledge this proof introduced a new technique to perform simulations. This technique starts with a phase in which the simulating system 'loads' itself with symbols used during the subsequent simulation. The number and kind of 'loaded' symbols is independent of the initial configuration of the simulating system. In the case of membrane systems this 'loading' consists of symbols passing from the environment to the skin compartment.

P systems with symport/antiport generating vectors of numbers were introduced in [10]. In this case a system with three compartments, symports and antiports of weight 1 can generate the vector of recursively enumerable numbers without any additional symbol in the final compartment. Two compartments are sufficient if one considers a terminal alphabet [9].

Theorem 5.5, stating that $N_1 \ OP_2(sym_1, anti_1) = N_1 \ RE$, was originally proved in [14] and preceded by $N_3 \ OP_2(sym_1, anti_1) = N_3 \ RE$ proved in [8]. Theorem 5.6 was originally proved in [14]. This last result was preceded by $N \ OP_4(sym_2, anti_0) = N \ RE$ in [88] and $N_6 \ OP_2(sym_2, anti_0) = N_6 \ RE$ in [9]. If, instead, one considers vectors of numbers or a terminal alphabet, then two compartments are sufficient for computational completeness [10]. Theorem 5.7 is original to this monograph and it subsumes the only result concerning P systems with symport/antiport operating in asynchronous mode presented in [79].

Trace languages for P systems with symport/antiport were initially considered in [130]. The results presented in Section 5.6 are mainly from [88] but adapted to the results obtained after that publication. Incoming languages, as defined at the end of Section 5.6, for membrane systems, were considered in [56,70].

Timed P systems with symport/antiport were studied in [184].

With the exception of Theorem 3.4 and Theorem 5.13 all the results of Section 5.8 are from [5]. These results were preceded by [213] showing that $N \ O_3P_4(sym_*, anti_*) = N \ RE$ and [3] proving that $N \ O_5P_1(sym_*, anti_*) = N \ RE$. Theorem 5.13 is original to this monograph and it was preceded by [124] defining a P system with symport/antiport with only one symbol and degree $2m + 3$ simulating a partially blind register machine. Similarly to Theorem 5.14 in [124] it is also proved that if priorities are used then the P systems with symport/antiport with only one symbol and degree $2m + 3$ can simulate register machines.

Theorem 5.17 is from [5]. The proof of Theorem 5.20 can be found in [15]. The remaining results in Section 5.9 are from [124, 125]. In [124] it is also proved that a partially blind register machine cannot accept $\{2^n \mid n \in N\}$.

The question of whether or not the deterministic version of a model of membrane systems is less powerful from a computational point of view than the non-deterministic one was raised in [211] and it was addressed in [74, 116]. The answer to this question given by Oscar H. Ibarra was reported in [122]. What was indicated in Section 5.10 is different from the original answer, but inspired by it. Also on this issue, see Section 6.5.

In [122] and [120] results on models of P systems with symport/antiport not presented in this chapter can be found. With regret, we decided not to introduce these results as they require a considerable number of concepts not otherwise used in the present monograph. In particular, [120] introduces accepting P systems with symport/antiport having an input tape and gives a characterisation of CS in terms of this introduced model. Moreover, in the same paper, a model of P systems with symport/antiport inducing an infinite hierarchy of accepted sets of numbers is described. In Section 8.6 we describe models of membrane systems inducing infinite hierarchies.

Section 5.11 is entirely based on [125]. In Theorem 5.21 the proof that item 2 implies item 1 is based on a result in [115].

Section 5.12 is entirely based on [45]. Our comment following the proof of Theorem 5.30 subsumes a result presented in [45] but not recalled in the present chapter.

Other interesting publications related to P systems with symport/antiport are [9, 118].

Tissue P systems were introduced in [174] where the passage of symbols from cell to cell is not based on symports and antiports.

The proofs of Theorem 5.42 and Theorem 5.43 can be found in [17]. Theorem 5.42 improves the result $N\ OtP_3(sym_1, anti_1) = N\ RE$ presented in [243]. Theorem 5.38 was originally proved in [71] while Theorems 5.39, 5.40, and 5.41 were originally proved in [6]. The proofs related to Theorem 5.44 can be found in [6, 71]. Tissue P systems with conditional uniports were considered in [245], while in [230] the number of rules is considered as a measure of complexity.

In addition to [174], other publications about tissue P systems inspired by neural interactions are [75, 200, 201, 252]; see also what is cited in Section 7.9.

Tissue P systems based on strings were studied in [55, 152, 157, 183] and other models of tissue P systems can be found in [4, 26, 28]. A graphical simulator for tissue P systems was described in [33].

5.17.1 About the notation

The notation used in the present chapter is widespread in Membrane Computing:

antiport rules of the kind $(x; out/y; in)$ are sometimes denoted by x/y. In the same fashion, symport rules of the kind $(x; in)$ and $(x; out)$ are denoted by ϕ/x and x/ϕ, respectively (see, for instance, [5]);

the family $N\ OP_m(sym_p, anti_q)$ is sometimes denoted by $N\ OP_m(sym_p)$ or by $N\ OP_m(anti_q)$ if $q = 0$ or $p = 0$, respectively; similarly for the other families of sets of numbers defined in the present chapter (see, for instance, [8]);

in [88] the family $N\ OP_m(sym_p, anti_q)$ was denoted by $N\ PP_m(sym_p, anti_q)$.

6 Catalysts

A specific kind of chemical reaction, biochemical reactions involving catalysts, inspired the model of membrane systems considered in this chapter.

Some of these devices do not have several of the features shared by the majority of the models of membrane system. For instance the environment, a compartment with symbols occurring in an infinite multiplicity, is not present. Often also the underlying topological structure is absent, that is, it 'collapses' to just one compartment (so, no passage of symbols between compartments). The resulting membrane systems are then multiset rewriting systems.

6.1 Biological motivations

It is very likely that readers know the meaning of

$$(6.1) \qquad 2H_2 + O_2 \rightarrow 2H_2O$$

It indicates a *chemical reaction* between two molecules of hydrogen and one molecule of oxygen resulting in the creation of two molecules of water.

The chemical reactions taking places in living organisms are called *biochemical reactions*. One example of biochemical reactions occurring in living cells is the binding of a restriction enzyme to a DNA molecule creating two DNA molecules (Section 9.1.2). The number of different types of biochemical reactions taking place in a cell is on the order of some thousands. Reactions can take place at different *reactions rates*, that is, at different speeds. This depends on the reaction itself and on some external factors such as temperature, pH, etc.

The vast majority of the reactions taking place in a cell involve catalysts: molecules taking part in a reaction, not consumed by it and accelerating the reaction rate.

In general, a reaction involving a catalyst can be indicated by

$$(6.2) \qquad C + A \rightarrow C + B$$

where C represents the catalyst, A represents the molecule(s) to which the catalyst binds and B represent the molecule(s) produced by the reaction. Notice that the catalyst is present on the left and the right side of the arrow. The proper term to refer to A is a *substrate* while that to refer to B is the *product*. If, for instance, we consider the binding of a restriction enzyme to a DNA molecule (Section 9.1.2), then the enzyme is the catalyst C, the DNA molecule is the substrate A and the two resulting DNA molecules are the product B.

It is important to say that enzymes are catalysts that can increase specific reaction rates by orders of magnitude.

6.2 Basic definitions

Definition 6.1 *A* (generating) P system with catalysts *of degree* m *is a construct*

$$\Pi = (V, C, \mu, L_1, \ldots, L_m, R_1, \ldots, R_m, comp)$$

where:

V *is an alphabet;*

$C \subset V$ *is a set of* catalysts;

$\mu = (Q, E)$ *is a cell-tree underlying* Π *where;*

    $Q \subset \mathbb{N}_+$ *contains* vertices. *For simplicity we define* $Q = \{1, \ldots, m\}$. *Each vertex in* Q *defines a* compartment *of* Π.

    $E \subseteq Q \times Q$ *defines arcs between vertices, denoted by* (i, j), $i, j \in Q, i \neq j$.

L_i, $1 \leq i \leq m$, *are multisets over* V. *The elements in* $L_1, \ldots, L_m$ *have finite multiplicity.*

R_i, $1 \leq i \leq m$, *are finite sets of* rules *of the kind:* $a \to \alpha$ *or* $ca \to c\alpha$ *with* $a \in V \setminus C$, $c \in C$ *and* $\alpha \in ((V \setminus C) \times \{here, in, out\})^*$, *where the set* $\{here, in, out\}$ *contains* target indicators.

$comp \in Q$ *is an elementary compartment called the* output compartment.

The symbols in $V \setminus C$ are called *non-catalysts*. Rules of the kind $a \to \alpha$ abstract chemical reactions as (6.1), while rules of the kind $ca \to c\alpha$ abstract chemical reactions as (6.2) in the previous section. Vertex 1 defines the skin compartment of μ. A *configuration* of Π is an m-tuple $(M_1, \ldots, M_m)$ of multisets over V associated with the compartments of Π. The m-tuple $(L_1, \ldots, L_m)$ is called the *initial configuration*.

Let $\Pi = (V, C, \mu, L_1, \ldots, L_m, R_1, \ldots, R_m, comp)$ be a P system with catalysts and let j, i and k be three vertices in μ such that i is the parent of j and j is the parent of k. Moreover, let R_j be the set of rules associated with compartment j and let M_j, M_i and M_k be multisets over V, associated with vertices j, i and k, respectively, such that $a \to \alpha \in R_j$ ($ca \to c\alpha \in R_j$), $a \in M_j$ ($c, a \in M_j$), $\alpha = (\alpha_1, tar_1), \ldots, (\alpha_{|\alpha|}, tar_{|\alpha|})$ with $\alpha_p \in V \setminus C$ and $tar_p \in \{here, in, out\}$, $1 \leq p \leq |\alpha|$. Without loss of generality, we assume that $tar_1 = \ldots = tar_{p_1} = here$, $tar_{p_1+1} = \ldots = tar_{p_2} = in$ and $tar_{p_2+1} = \ldots = tar_{|\alpha|} = out$. The application of $\to \alpha$ ($ca \to c\alpha \in R_j$) changes:

M_j into $M'_j = M_j \setminus \{a\} \cup \{\alpha_1, \ldots, \alpha_{p_1}\}$;

M_i into $M'_i = M_i \cup \{\alpha_{p_2+1}, \ldots, \alpha_{|\alpha|}\}$;

M_k into $M'_k = M_k \cup \{\alpha_{p_1+1}, \ldots, \alpha_{p_2}\}$.

The use of target indicators is an abstraction of the passage of molecules across membranes (see Section 5.1). In the following the target indicator *here* is omitted. For two configurations $(M_1, \ldots, M_m)$ and $(M'_1, \ldots, M'_m)$ of Π we write $(M_1, \ldots, M_m) \Rightarrow (M'_1, \ldots, M'_m)$ to denote a *transition* from $(M_1, \ldots, M_m)$ to $(M'_1, \ldots, M'_m)$, that is, the application of a multiset of rules associated with each compartment under the requirement of maximal parallelism. The reflexive and transitive closure of $\Rightarrow$ is denoted by $\Rightarrow^*$.

A *computation* is a sequence of transitions between configurations of a system Π starting from the initial configuration $(L_1, \ldots, L_m)$. If a computation is finite, then the last configuration is called *final* and we say that the system *halts*. Let $\Pi = (V, C, \mu, L_1, \ldots, L_m, R_1, \ldots, R_m, comp)$ be a P systems with catalysts. We define:

$N(\Pi) = \{|M_{\text{comp}}| \mid (L_1, \ldots, L_m) \Rightarrow^* (M_1, \ldots, M_m)$ is a halting computation
$\qquad\qquad$ of $\Pi\}$;

$N_{-c}(\Pi) = \{\sum_{i=1}^{|V|} M_{\text{comp}}(v_i) \mid v_i \in V \setminus C, 1 \le i \le |V|$ and $(L_1, \ldots, L_m) \Rightarrow^*$
$\qquad\qquad (M_1, \ldots, M_m)$ is a halting computation of $\Pi\}$;

$N^{|V|}(\Pi) = \{(M_{\text{comp}}(v_1), \ldots, M_{\text{comp}}(v_{|V|})) \mid (L_1, \ldots, L_m) \Rightarrow^* (M_1, \ldots, M_m)$ is
$\qquad\qquad$ a halting computation of $\Pi\}$;

$N_{-c}^{|V \setminus C|}(\Pi) = \{(M_{\text{comp}}(v_1), \ldots, M_{\text{comp}}(v_{|V \setminus C|})) \mid v_i \in V \setminus C, 1 \le i \le |V|$ and
$\qquad\qquad (L_1, \ldots, L_m) \Rightarrow^* (M_1, \ldots, M_m)$ is a halting computation of $\Pi\}$.

Moreover,

$\mathbb{N} \, OP_m(\text{cat}_p) = \{N(\Pi) \mid \Pi$ is a P systems with catalysts of degree at most m
$\qquad\qquad$ and using at most p catalysts$\}$;

$\mathbb{N} \, OP_{m,-c}(\text{cat}_p) = \{N_{-c}(\Pi) \mid \Pi$ is a P systems with catalysts of degree at most
$\qquad\qquad m$ and using at most p catalysts$\}$;

$\mathbb{N}^q \, OP_m(\text{cat}_p) = \{N^{|V|}(\Pi) \mid \Pi$ is a P systems with catalysts of degree at most
$\qquad\qquad m$ and using at most p catalysts and q non-catalysts$\}$;

$\mathbb{N}^{q-p} \, OP_{m,-c}(\text{cat}_p) = \{N_{-c}^{|V \setminus C|}(\Pi) \mid \Pi$ is a P systems with catalysts of degree
$\qquad\qquad$ at most m and using at most p catalysts and q non-catalysts$\}$.

So, for instance, if in a halting computation of such a P system compartment *comp* has two catalysts c_1 and c_2, three occurrences of a non-catalyst a and four occurrences of a non-catalyst b, then:

9 is the output if symbols counted with their multiplicity are considered;

7 is the output if non-catalysts counted with their multiplicity are considered;

$(2, 3, 4)$ is the output if vectors of multiplicities of symbols are considered; $(3, 4)$ is the output if vectors of multiplicities of non-catalysts are considered.

In a similar way *accepting P systems with catalysts* can be defined. These systems are such that the compartment *comp*, called in this case the *input compartment*, contains in the initial configuration the input, that is, a multiset over V. If such a system halts, then it is said to accept the input. Following the notation given in the list above, the families of sets or vectors of numbers accepted by P systems with catalysts of degree at most m, using at most p catalysts and at most q non-catalysts, are denoted by $\mathbb{N} \, \text{aOP}_m(\text{cat}_p), \mathbb{N} \, \text{OP}_{m,-c}(\text{cat}_p), \mathbb{N}^q \, \text{aOP}_m(\text{cat}_p)$ and $\mathbb{N}^{q-p} \, \text{aOP}_{m,-c}(\text{cat}_p)$, respectively.

A P system with catalysts is called a *purely catalytic P system* if every rule involves a catalyst, that is, if no rule of the kind $a \to \alpha$, $a \in V \backslash C$, $\alpha \in (V \backslash C)^*$ is present. Following the denotations given above, the families of generating and accepting purely catalytic P systems of degree at most m, using at most p catalysts and at most q non-catalysts are denoted by $\mathbb{N} \, \text{OP}_m(\text{pcat}_p)$, $\mathbb{N} \, \text{OP}_{m,-c}(\text{pcat}_p)$, $\mathbb{N}^q \, \text{OP}_m(\text{pcat}_p)$, $\mathbb{N}^{q-p} \, \text{OP}_{m,-c}(\text{pcat}_p)$, $\mathbb{N} \, \text{aOP}_m(\text{pcat}_p)$, $\mathbb{N} \, \text{aOP}_{m,-c}(\text{pcat}_p)$, $\mathbb{N}^q \, \text{aOP}_m(\text{pcat}_p)$ and $\mathbb{N}^{q-p} \, \text{aOP}_{m,-c}(\text{pcat}_p)$, respectively.

Some variants of purely catalytic P systems are introduced in the following sections.

It is important to notice that in these definition the environment, that is, the compartment parent of the skin compartment, is not present. This is because it is not needed in the systems considered in the present chapter. The original definition of P systems with catalysts considered the presence of the environment (and of priority between rules) but subsequent studies let us realise that this component was not always needed. More details are presented in Section 6.8.

6.3 Examples

The rules present in P systems with catalysts are very similar to the kind of productions present in context-free grammars. It should be clear then that this kind of grammar can be simulated by P systems with catalysts.

Example 6.1 A P system with catalysts of degree 2 generating $\mathbb{N}$ CF.

Let $G = (N, T, S, P)$ be a context-free grammar. We define the P system with catalysts

$$\Pi_1 = (V, C, \mu, L_1, L_2, R_1, R_2, 2)$$

with

$$V = N \cup T \cup C;$$
$$C = \{c\};$$
$$\mu = (\{1,2\}, \{(1,2)\});$$
$$L_1 = \{c, S\};$$
$$L_2 = \phi;$$
$$R_1 = \{cA \to c\alpha_1 \dots \alpha_p(\alpha_{p+1}, in) \dots (\alpha_q, in) \mid A \to \alpha_1 \dots \alpha_q \in P,$$
$$\alpha_1, \dots, \alpha_p \in N, \alpha_{p+1}, \dots, \alpha_q \in T\} \cup$$
$$\{A \to A \mid A \in N\};$$
$$R_2 = \emptyset.$$

The P system has two compartments, compartment 2, the final one, is elementary, as requested by Definition 6.1. There is only one catalyst: c. As there are rules not involving c, then Π_1 is not a catalytic P system. In the initial configuration the skin compartment contains the catalyst c and the start symbol S, while compartment 2 is empty.

Only the skin compartment contains rules. Some of these are a rewriting of the productions $A \to \alpha_1 \dots \alpha_q$ present in the grammar G. Without loss of generality in the definition of these rules we consider $\alpha_1 \dots \alpha_p$ non-terminal symbols and $\alpha_{p+1} \dots \alpha_q$ terminal symbols. The remaining rules let symbols in N be rewritten into themselves. Notice that the target indicator *here* is omitted.

The system operates by simulating one production of G per unit time. Every non-terminal symbol remains in the skin compartment while the terminal symbols pass into compartment 2. If the system Π_1 reaches a configuration in which no non-terminal symbols are present in the skin compartment, then it halts, otherwise not. When Π_1 halts, the catalyst remains in the skin compartment while compartment 2 contains only terminal symbols.

It should be clear that Π_1 simulates G. We can then say that $\mathsf{N}(\Pi_1) \supseteq \mathsf{N}\,\mathsf{CF}$. ◇

Example 6.2 A purely catalytic P system of degree 2 generating non-negative even numbers.

We define the purely catalytic P system

$$\Pi_2 = (\{a, b, c\}, \{c\}, (\{1, 2\}, \{(1, 2)\}), \{c, b\}, \phi, R_1, \emptyset, 2)$$

with $R_1 = \{cb \to cb(a, in)(a, in), cb \to c\}$.

The P system has two compartments, compartment 2, the final one, is elementary, as requested by Definition 6.1. There is only one catalyst, c, and it occurs in all the rules present in the system. This makes Π_2 a purely catalytic P system. In the initial configuration the skin compartment contains the catalyst c and the symbol b, while compartment 2 is empty.

Only the skin compartment contains rules. If the rule $cb \to cb(a, in)(a, in)$ is applied, then two occurrences of a pass into compartment 2 while b remains in the skin compartment (the target indicator *here* related to the symbol b is omitted). This rule can be applied an unbounded number of times. The system halts when the rule $cb \to c$ is applied. In this configuration the skin compartment contains only the catalyst c while compartment 2 contains an even number of a's or no a. No a is present in compartment 2 only if the only applied rule is $cb \to c$. We can then say that Π_2 generates all even numbers. $\diamond$

If in the previous two examples the catalysts are not considered in the final multiset of symbols indicating the output of the system, then there is no need for two compartments. Accordingly, the rules are changed such that no *in* target indicator is used.

Example 6.3 A P system with catalysts of degree 1 accepting non-negative even numbers.

We define the P system with catalysts of degree 1

$$\Pi_3 = (V, C, \mu, L_1, R_1, 1)$$

with

$$V = \{a, a'_1, a''_1, a'_2, a''_2, \star\} \cup C;$$
$$C = \{c_1, c_2\};$$
$$\mu = (\{1\}, \emptyset);$$
$$L_1 = \{c_1 c_2\} \cup \{a^n \mid n \in \mathbb{N}\};$$
$$
\begin{aligned}
R_1 = \{ \ &1: c_1 a \to c_1 a'_1 a''_1, \quad 2: c_2 a \to c_2 a'_2 a''_2, \quad 3: c_1 a'_2 \to c_1, \\
&4: c_2 a'_1 \to c_2 \qquad\quad 5: c_1 a''_2 \to c_1, \qquad 6: c_2 a''_1 \to c_2, \\
&7: c_1 a''_1 \to c_1 \star, \quad\ \ 8: c_2 a''_2 \to c_2 \star, \quad 9: c_1 a'_1 \to c_a \star, \\
&10: c_2 a'_2 \to c_2 \star, \quad\ 11: a'_1 \to \star, \qquad\quad 12: a'_2 \to \star, \\
&13: \star \to \star \}.
\end{aligned}
$$

In order to facilitate the explanation, rules have been numbered. The P system has only one compartment; this means that it is a (multiset) rewriting system with no need to specify any compartment. The reason we indicate μ is to be consistent with Definition 6.1.

In the initial configuration the catalysts c_1 and c_2 are in the skin compartment (metaphorically) together with n occurrences of the symbol a, the input of Π. The P system works by removing pairs of a's. If n is an even number, then Π can halt, otherwise it never halts because of the continuous application of rule 13. This rule is applied only if the symbol $\star$ is introduced in the system. This can also happen if n is an even number.

In the following, numbers in between parenthesis indicate that the relative rules are applied in parallel. If at least two a's are present in the system then the following sequence of rules can be applied: (1, 2), (3, 4), (5, 6). This can be repeated until no two a's are present in the system. When this happens the system halts.

Other sequences of rules that can be applied if at least two a's are present in the system are: (1, 2), (5, 4, 12); (1, 2), (3, 6, 11) and (1, 2), (1, 4, 11). If only one a is present, then the sequences of rules that can be applied are: 1, (4, 7); 2, (3, 8); 1, (6, 9) and 2, (5, 10). In all these cases the symbol $\star$ is introduced. If $n = 0$, then no rule is applied. We can then say that Π_3 accepts non-negative even numbers. ◇

6.4 P systems with catalysts and P/T systems

In this section we show how the study of the computational power of P systems with catalysts can be facilitated by the results in Section 4.4.

Lemma 6.1 *The building blocks* join *and* fork *can be simulated by P systems with catalysts of degree 1 and with two catalysts.*

Proof A *fork* building block, see Fig. 4.15, can be simulated by a rule of the kind $a \to b_1 b_2$ where $\{a\}$ is the input set and $\{b_1, b_2\}$ is the output set of the transition in the *fork*.

A *join* building block, see Fig. 4.15, can be regarded as a rule of the kind $a_1 a_2 \to b_1$ where $\{a_1, a_2\}$ is the input set and $\{b_1\}$ is the output set of the transition in the *join*. In general, this kind of rule does not belong to P systems with catalysts, but they can be simulated by sets of rules (very similar to those of Example 6.3) in these systems.

We define such a P system with $V = \{a_1, a_1', a_1'', a_2, a_2', a_2'', b_1, c_1, c_2, \star\}$, $C = \{c_1, c_2\}$ and the rules

$$1 : c_1 a_1 \to c_1 a_1' a_1'', \quad 2 : c_2 a_2 \to c_2 a_2' a_2'', \quad 3 : c_1 a_2' \to c_1, \quad 4 : c_2 a_1' \to c_2,$$
$$5 : c_1 a_2'' \to c_1 b_1, \quad 6 : c_2 a_1'' \to c_2, \quad 7 : c_1 a_1'' \to c_1 \star, \quad 8 : c_2 a_2'' \to c_2 \star,$$
$$9 : c_1 a_1' \to c_1 \star, \quad 10 : c_2 a_2' \to c_2 \star, \quad 11 : a_1' \to \star, \quad 12 : a_2' \to \star,$$
$$13 : \star \to \star.$$

In the following, numbers in between parenthesis indicate that the relative rules are applied in parallel. Only one sequence of application of the previous rules can lead to the simulation of the *join*. Any other application leads to the introduction of the symbol $\star$ and the subsequent continuous application of rule 13 allows the P system to never halt.

If at least one occurrence of a_1 and a_2 is present in the compartment of the P system, then the application of (1, 2), (3, 4), (5, 6) allows it to have one occurrence of b_1. From the same initial configuration the application of (1, 2), (11, 12) allows the system to never halt. If instead only one occurrence of a_1 and no occurrence of a_2 is present, then the application of either 1, (4, 7), or 1, (6, 9) or 1, (6, 11) allows the system to never halt. Similarly if only one occurrence of a_2 and no occurrence of a_1 is present.

If more than one occurrence of a_1 and only one occurrence of a_2 are present, then the application of (1, 2), (1, 4, 12) lets the system never halt. Similarly if more than one occurrence of a_2 and only one occurrence of a_1 are present.

This proves the lemma.   □

We notice a few things with respect to the proof of the previous lemma:

(i) The simulation of *fork* can be performed in one transition while the simulation of a *join* requires at least three transitions.

(ii) In some cases (such as the application of (1, 4, 12)) three rules are applied in parallel. If we want to define a purely catalytic P system, then a third catalyst has to be introduced.

(iii) Because of the rules $c_1 a_1 \rightarrow c_1 a_1' a_1''$ and $c_2 a_2 \rightarrow c_2 a_2' a_2''$ we say that c_1 is *associated with* a_1 and that c_2 is *associated with* a_2.

(iv) The simulation of a *join* requires both catalysts. This means that if we consider following this simulation and the given P system has only two catalysts, then at most one *join* per unit time can be simulated.

(v) The rules $c_1 a_1'' \rightarrow c_1 \star$ and $c_2 a_2'' \rightarrow c_2 \star$ are present to introduce the symbol $\star$ in case one of the rules $c_1 a_1 \rightarrow c_1 a_1' a_1''$ and $c_2 a_2 \rightarrow c_2 a_2' a_2''$, respectively, is not applied. This can only happen only if a_1 and a_2, respectively, are not present. But this means that if in a P system the symbol a_1, for instance, is certainly present, then there is no need for the rule $c_1 a_1 \rightarrow c_1 \star$. Similarly, if the symbol a_1 can be present at most once, then there is no need to have the rule $a_1' \rightarrow \star$.

(vi) The simulation of a *join* is non-deterministic as, for instance, when a_1 and a_2 are present, then several sequences of rules can be applied.

If we consider the proof of Theorem 4.2, Part I, then we realise that Lemma 6.1 provides the necessary conditions to prove that P systems with catalysts can simulate register machines. The sufficient condition is provided showing that the composition of *join* and *fork* as depicted in Fig. 4.1 can be simulated.

In the following we call a *0-test net* the net depicted in Fig. 4.1 and a *0-test P/T system* any P/T system having the 0-test net as underlying net and either $\{p_s\}$ or $\{p_s, p_\gamma = k\}$, $k \in \mathbb{N}_+$, as initial configuration. The maximal strategy configuration graph of a 0-test P/T system is then denoted by $MSCG$ (*0-test*).

Lemma 6.2 *P systems of degree 1 and with two catalysts can weakly simulate* $MSCG$(0-test).

Proof If we consider the $MSCG(\textit{0-test})$ depicted in Fig. 4.7(c) (associated with a P/T system having the net depicted in Fig. 4.1 as underlying net), then we realise that each firing contains at most one transition of a *join*. This means that the simulation of *join* used in the proof of Lemma 6.1 can be used. Moreover, we notice that in the 0-test P/T system some places (p_γ) can have more than one token while others (all the remaining places in that net) can have at most one token. Let us now assume that the number of tokens in a place p in the 0-test P/T system is represented in the P system by the number of occurrences of a symbol p.

We associate (as indicated in item (iii) in the list on page 140) catalyst c_1 with the place p_γ. This means that the rules $c_1 p_\gamma \rightarrow c_1 p'_\gamma p''_\gamma$ and $c_1 p''_\gamma \rightarrow c_1 \star$ are in the P system. As the place p_γ can have more than one token, then the catalyst c_1 has to be 'kept busy' during the simulation of the $MSCG(\textit{0-test})$. If not, then the presence of more than one token in p_γ would lead to the application of $c_1 p_\gamma \rightarrow c_1 p'_\gamma p''_\gamma$ when c_1 is not 'busy'. This, in turn, could imply the application of $c_1 p''_\gamma \rightarrow c_1 \star$ and the consequent impossibility to simulate the $MSCG(\textit{0-test})$.

Now we are ready to list the rules needed for this simulation:

$1 : c_1 p_s \rightarrow c_1 p_1,$ $\quad 2 : c_1 p_\gamma \rightarrow c_1 p'_\gamma p''_\gamma,$ $\quad 3 : c_2 p_1 \rightarrow c_2 p'_1 p''_1,$ $\quad 4 : c_1 p'_1 \rightarrow c_1$
$5 : c_2 p'_\gamma \rightarrow c_2,$ $\quad 6 : c_1 p''_\gamma \rightarrow c_1 p_3,$ $\quad 7 : c_2 p''_\gamma \rightarrow c_2,$ $\quad 8 : c_2 p''_1 \rightarrow c_2 p_4,$
$9 : c_1 p'_\gamma \rightarrow c_1 \star,$ $\quad 10 : c_2 p'_1 \rightarrow c_2 \star,$ $\quad 11 : p'_\gamma \rightarrow \star,$ $\quad 12 : c_1 p_3 \rightarrow c_1 p_v,$
$13 : c_1 p_4 \rightarrow c_1 p_w,$ $\quad 14 : \star \rightarrow \star.$

The simulation of the $MSCG(\textit{0-test})$ in case there are no tokens in p_γ is performed by the application of 1, 3, (4, 8), 13. In case there is at least one token in p_γ, then the simulation is performed by 1, (2, 3), (4, 5), (6, 7), 12. Other applications of rules, similar to those in the proof of Lemma 6.1, are possible, but they all lead to the introduction of $\star$. It should be clear that the simulation is a weak one. $\qquad\square$

The previous two lemmas tell us that P systems with catalysts can simulate register machines. So, if somebody were strictly interested in this quest, then the research could stop here. Anyhow, these two lemmas do not tell all the details: how can such P systems simulate register machines?, how many compartments are needed? and so on. Considering what we said until now these are not difficult questions to answer.

The content of the different registers of the register machine can be represented by the number of occurrences of different symbols, while the states can be represented by other different symbols. The only restriction present in this simulation is given by the catalyst c_1. As said in the proof of Lemma 6.2 this catalyst has to be 'kept busy'. This means that if when the symbol s_f, associated with the final state of the register machine, is present in the compartment where the symbols representing the values of the registers are also present, then the P system would introduce the $\star$ and never halt. We can overcome this limitation in two ways.

If we consider a generating register machine, then we can consider one such machine only increasing the value of the output computer. In this case the P system would have a cell-tree with two compartments as underlying structure: $\mu = (\{1, 2\}, \{(1, 2)\})$. Compartment 1, the final compartment, would collect the symbols associated with the output register by rules of the kind $c_1 s \rightarrow c_1 v(\gamma, in)$ (simulating instructions of the kind (s, γ^+, v)).

If instead we consider accepting register machines, then one compartment is enough. In the initial configuration the P system would have in its compartment: the number of occurrences of symbols associated with the input register, the two catalysts and s_1 the symbol associated with the initial state of the register machine. The P system can halt only if a simulation of the register machine has been performed. In all halting configurations the two catalysts remain in the compartment (s_f, the symbol associated with the final state of the register machine, can be removed by this compartment by $c_1 s_f \rightarrow c_1$).

We can then state:

Theorem 6.1 $\mathbb{N}_2 \, \mathrm{aOP}_1(cat_2) = \mathbb{N}_2 \, \mathrm{aOP}_1(pcat_3) = \mathbb{N}_2 \, \mathrm{RE}$;
$\mathbb{N} \, \mathrm{OP}_2(cat_2) = \mathbb{N} \, \mathrm{OP}_2(pcat_3) = \mathbb{N} \, \mathrm{aOP}_{1,-c}(cat_2) = \mathbb{N} \, \mathrm{aOP}_{1,-c}(pcat_3) = \mathbb{N} \, \mathrm{RE}$.

It is because of item (ii) in the list on page 140 that the purely catalytic P systems mentioned in the previous theorem have three registers.

From Theorem 4.3 we then know that

Theorem 6.2 *The sets* $\mathbb{N}_2 \, \mathrm{aOP}_1(cat_2)$, $\mathbb{N}_2 \, \mathrm{aOP}_1(pcat_3)$, $\mathbb{N} \, \mathrm{OP}_2(cat_2)$, $\mathbb{N} \, \mathrm{OP}_2$ $(pcat_3), \mathbb{N} \, \mathrm{aOP}_{1,-c}(cat_2)$ *and* $\mathbb{N} \, \mathrm{aOP}_{1,-c}(pcat_3)$ *generated by systems operating in asynchronous mode are equal to the set generated by partially blind register machines.*

A *join* building block can be regarded as a rule of the kind $ca \rightarrow c$, $c \in C$ and $a \in V \setminus C$. If we consider this together with Theorem 4.6, then we obtain

Corollary 6.1 *Accepting purely catalytic P systems with only rules of the kind* $ca \rightarrow c$, $c \in C$ *and* $a \in V \setminus C$ *accept only finite sets.*

Let us change Definition 6.1 so that rules of the kind $a \rightarrow \alpha$ can be present only if rules of the kind $cb \rightarrow ca$ are also present and let us call the resulting P systems with catalysts *restricted*. Then, as a consequence of Lemma 6.1, Lemma 6.2, Theorem 6.1 and Theorem 4.4 we obtain:

Corollary 6.2 *Restricted P systems with catalysts of degree 2 and two catalysts and restricted purely catalytic P systems of degree 2 and three catalysts can simulate restricted register machines.*

6.5 Deterministic P systems with catalysts of degree 1

In the present and in the following section we consider P systems with catalysts of degree 1, that is, multiset rewriting systems operating under maximal parallelism. Given one such device $\Pi = (V, C, L, R)$, where μ and *comp* have been omitted for obvious reasons, its rules are of the kind $a \to \alpha$ and $ca \to c\alpha$ with $a \in V$, $c \in C$ and $\alpha \in (V \setminus C) \to \mathbb{N}$. Moreover, $\mathbb{C}_\Pi$ denotes the set of configurations of Π. For $M^{(1)}, \ldots, M^{(r+1)} \in \mathbb{C}_\Pi$, $r \in \mathbb{N}_+$ and $\rho^{(1)}, \ldots, \rho^{(r)}$ multisets of rules over R, $M^{(1)} \Rightarrow^{\rho^{(1)}} M^{(2)}$ denote the transition from configuration $M^{(1)}$ to configuration $M^{(2)}$ due to the application of the rules in $\rho^{(1)}$. In a similar way $M^{(1)} \Rightarrow^{\rho^{(1)}} M^{(2)} \Rightarrow^{\rho^{(2)}} \cdots \Rightarrow^{\rho^{(r)}} M^{(r+1)}$ is denoted by $M^{(1)} \Rightarrow^{\rho^{(1)}, \rho^{(2)}, \ldots, \rho^{(r)}} M^{(r+1)}$. We recall that the reflexive and transitive closure of $\Rightarrow$ is denoted by $\Rightarrow^*$ and that, as we consider systems of degree 1, each configuration is a multiset.

Lemma 6.3 *Let $\Pi = (V, C, L, R)$ be a deterministic purely catalytic P system of degree 1, $M^{(1)}, \ldots, M^{(4)} \in \mathbb{C}_\Pi$ and $\rho^{(1)}, \rho^{(2)}$ multisets of rules over R such that $M^{(1)} \Rightarrow^{\rho^{(1)}} M^{(2)}$, $M^{(3)} \Rightarrow^{\rho^{(2)}} M^{(4)}$ and $\rho^{(1)} \subseteq \rho^{(2)}$. If $ca \to c\alpha \in \rho^{(2)} \setminus \rho^{(1)}$, $c \in C$, $a \in V \setminus C$ and $\alpha : (V \setminus C) \to \mathbb{N}$, then $M^{(1)}(a) < M^{(3)}(a)$.*

Proof This follows from Π being deterministic and operating under maximal parallelism. □

Lemma 6.4 *Let $\Pi = (V, C, L, R)$ be a deterministic purely catalytic P system of degree 1, $M^{(1)}, \ldots, M^{(r+1)}, M^{(1)'} \in \mathbb{C}_\Pi$ and $\rho^{(1)}, \ldots, \rho^{(r)}$ be multisets of rules over R such that $M^{(1)} \Rightarrow^{\rho^{(1)}, \ldots, \rho^{(r)}} M^{(r+1)}$ and $M^{(1)'}(v) \geq M^{(1)}(v)$ for each $v \in V$. There are $M^{(2)'}, \ldots, M^{(r+1)'} \in \mathbb{C}_\Pi$ and $\rho^{(1)'}, \ldots, \rho^{(r)'}$ multisets of rules over R such that:*

(i) $M^{(1)'} \Rightarrow^{\rho^{(1)'}, \ldots, \rho^{(r)'}} M^{(r+1)'}$;

(ii) $\rho^{(i)} \subseteq \rho^{(i)'}$ for $1 \leq i \leq r$;

(iii) $M^{(i)}(v) \leq M^{(i)'}(v)$ for each $v \in V$ and for $1 \leq i \leq r + 1$.

Proof By induction on r.

Basis: The case $r = 1$ it is clearly true.

Inductive hypothesis: Let us assume that the hypothesis holds for $r = p$.

Inductive step: We know that $M^{(1)} \Rightarrow^{\rho^{(1)}, \ldots, \rho^{(r)}} M^{(r+1)}$. Because of the inductive hypothesis there are $M^{(1)'}, \ldots, M^{(p+1)'} \in \mathbb{C}_\Pi$ and $\rho^{(2)'}, \ldots, \rho^{(p)'}$ multisets of rules over R such that $M^{(1)'} \Rightarrow^{\rho^{(1)'}, \ldots, \rho^{(p)'}} M^{(p+1)'}$ satisfying (ii) and (iii). Since by the inductive hypothesis $M^{(p)'}(v) \geq M^{(p)}(v)$ for all $v \in V$, then all the rules in ρ_p can be applied to $M^{(p)'}$. Thus $M^{(p)'} \Rightarrow^{\rho^{(p)'}} M^{(p+1)'}$ is such that $\rho^{(p)} \subseteq \rho^{(p)'}$. If we now consider Lemma 6.3, then a rule of the kind $ca \to c\alpha \in \rho^{(p)'} \setminus \rho^{(p)}$ implies $M^{(p)'}(a) > M^{(p)}(a)$. This means that the application of this additional rule uses only one extra occurrence of the symbol a in $M^{(p)'}$ but not in $M^{(p)}$. Thus $\rho^{(p+1)} \leq \rho^{(p+1)'}$ and the hypothesis holds for $r = p + 1$. □

Lemma 6.5 *Let* $\Pi = (V, C, L, R)$ *be a deterministic purely catalytic P system of degree* 1. $M^{(1)}, M^{(2)}, M^{(3)} \in \mathbb{C}_\Pi$ *and* $\rho^{(1)}, \ldots, \rho^{(r)}$ *multisets of rules over* R. *If* $M^{(1)} \Rightarrow^{\rho^{(1)}, \ldots, \rho^{(r)}} M^{(2)}$ *and* $M^{(2)} \Rightarrow^{\rho^{(1)}, \ldots, \rho^{(r)}} M^{(3)}$. *Then from configuration* $M^{(1)}$ *the application of* $\rho^{(1)}, \ldots, \rho^{(r)}$ *repeats forever.*

Proof By contradiction let $M^{(p)}$ be the first configuration and $\rho^{(q)}$ the first multiset of rules over R such that $M^{(1)} \Rightarrow^{\rho^{(1)}, \ldots, \rho^{(r)}} M^{(2)} \Rightarrow^{\rho^{(1)}, \ldots, \rho^{(r)}} M^{(3)} \cdots$ $\Rightarrow^{\rho^{(1)}, \ldots, \rho^{(r)}} M^{(p)} \Rightarrow^{\rho^{(1)}, \ldots, \rho^{(q)'}, \ldots \rho^{(r)'}} M^{(p+1)}$ and $\rho^{(q)} \neq \rho^{(q)'}$, $p \geq 3$ and $1 \leq q < r$.

Let us consider the three sequences of configurations: $M^{(p-2)} \Rightarrow^{\rho^{(1)}, \ldots, \rho^{(q-1)}} M^{(p-2)'} \Rightarrow^{\rho^{(q)}, \ldots, \rho^{(r)}} M^{(p-1)}$, $M^{(p-1)} \Rightarrow^{\rho^{(1)}, \ldots, \rho^{(q-1)}} M^{(p-1)'} \Rightarrow^{\rho^{(q)}, \ldots, \rho^{(r)}} M^{(p)}$ and $M^{(p)} \Rightarrow^{\rho^{(1)}, \ldots, \rho^{(q-1)}} M^{(p)'} \Rightarrow^{\rho^{(q)'}, \ldots, \rho^{(r)'}} M^{(p+1)}$. From Lemma 6.4 we know that $\rho^{(q)} \subseteq \rho^{(q)'}$ holds (note that $M^{(1)'}$ in Lemma 6.4 is $M^{(2)}$ in the present proof). If we assume that $\rho^{(q)} \neq \rho^{(q)'}$, then a rule of the kind $ca \rightarrow c\alpha$ belongs to $\rho^{(q)'} \setminus \rho^{(q)}$. Thus, because of Lemma 6.3, $M^{(p)'}(x) > M^{(p-1)'}(x)$.

As the sequence of rules $\rho^{(q)}, \ldots, \rho^{(r)}, \rho^{(1)}, \ldots, \rho^{(q-1)}$ is applied to both $M^{(p-2)'}$ and $M^{(p-1)'}$, then $M^{(p)'}(a) > M^{(p-1)'}(a) > M^{(p-2)'}(a)$. Moreover, as $\rho^{(q)}$ was applied to $M^{(p-1)'}$, then the set $\rho^{(q)} \cup \{ca \rightarrow c\alpha\}$ can be applied to $\rho^{(p-1)}$. A contradiction. □

Theorem 6.3 *Let* $\Pi = (V, C, L, R)$ *be a deterministic purely catalytic P system of degree* 1. *The following three statements are equivalent:*

(i) Π *does not halt;*

(ii) *there are* $M, M' \in \mathbb{C}_\Pi$ *such that* $L \Rightarrow^* M \Rightarrow^* M'$ *and* $M'(v) \geq M(v)$ *for each* $v \in V$;

(iii) *all the sequences of transitions of* Π *have a periodic subsequence* $\rho^{(p+1)}, \ldots, \rho^{(q)}$ *of applied rules, that is,* $L \Rightarrow^{\rho^{(1)}, \ldots, \rho^{(p)}(\rho^{(p+1)}, \ldots, \rho^{(q)})^\omega} \cdots$. *This means that the application of a finite number of multisets of rules* $\rho^{(1)}, \ldots, \rho^{(p)}$ *is followed by a repetition* (ω) *of the application of a finite number of multisets of rules* $\rho^{(p+1)}, \ldots, \rho^{(q)}$.

Proof *((i) implies (ii))* If Π does not halt, then certainly there are two configurations $M, M' \in \mathbb{C}_\Pi$ such that $M(v) \leq M'(v)$ for each $v \in V$. If this were not the case, then the number of elements in the configurations monotonically decreases so that sooner or later the system halts because it runs out of symbols. A contradiction.

((ii) implies (iii)) Let $\rho^{(1)}, \ldots, \rho^{(r)}$ be the multisets of rules such that $M \Rightarrow^{\rho^{(1)}, \ldots, \rho^{(r)}} M'$ and $M(v) \leq M'(v)$ for each $v \in V$. Because of Lemma

6.4 there are $\rho_j^{(1)}, \ldots, \rho_j^{(r)}$, $1 \le j \le q$, $q \in \mathbb{N}_+$, multisets of rules and $M^{(1)'}, \ldots,$ $M^{(q)'} \in \mathbb{C}_\Pi$ such that $M \Rightarrow^{\rho^{(1)}, \ldots, \rho^{(r)}} M' \Rightarrow^{\rho_1^{(1)}, \ldots, \rho_1^{(r)}} M^{(1)'} \Rightarrow^{\rho_2^{(1)}, \ldots, \rho_2^{(r)}} \ldots$ Moreover, for each $1 \le t \le r$ and $q \in \mathbb{N}_+$, $\rho_q^{(t)} \subseteq \rho_{q+1}^{(t)}$ and $M^{(q)'}(v) \le M^{(q+1)'}(v)$ for each $v \in V$. As the maximum number of rules applied in a configuration is at most $|C|$, then for all $1 \le t \le r$ there must be a q_t such that $\rho_{q_t}^{(t)} = \rho_{q_t+1}^{(t)} = \cdots$. Let $q' = max(\{q_t \mid 1 \le t \le r\})$ such that $\rho_{q'}^{(t)} = \rho_{q'+1}^{(t)}$ for all $1 \le t \le r$. So, $M^{(q')'} \Rightarrow^{\rho_{q'}^{(1)}, \ldots, \rho_{q'}^{(r)}} M^{(q'+1)'} \Rightarrow^{\rho_{q'}^{(1)}, \ldots, \rho_{q'}^{(r)}} \ldots$. By choosing $\rho^{(p+t)'} = \rho_{q'}^{(t)}$, $1 \le t \le r$ Lemma 6.5 guarantees that $\rho^{(p+1)'}, \ldots, \rho^{(p+r)'}$ repeats forever.

((iii) implies (i)) Trivially true. $\qquad\square$

Corollary 6.3 *If $\Pi = (V, C, L, R)$ is a deterministic purely catalytic P system of degree 1 with $V \setminus C = \{a_1, \ldots, a_k\}$, then $\{(M(a_1), \ldots, M(a_k)) \mid L \Rightarrow^* M\} \in$ SLS_k.*

Proof This follows from item (iii) of Theorem 6.3. $\qquad\square$

Corollary 6.4 *Let $\Pi = (V, C, L, R)$ be a deterministic purely catalytic P system of degree 1. It is decidable to determine whether Π halts.*

Proof From Theorem 6.3 we know that Π does not halt if and only if there are $M, M' \in \mathbb{C}_\Pi$ such that $L \Rightarrow^* M \Rightarrow^* M'$ and $M'(v) \ge M(v)$ for each $v \in V$. The system Π can then be simulated and if M and M' are found, then Π does not halt, otherwise it does. $\qquad\square$

Theorem 6.4 *The set of numbers accepted by deterministic purely catalytic P systems of degree 1 is recursive.*

Proof Considering Corollary 6.4 there is a deterministic procedure telling us if a certain deterministic purely catalytic P systems of degree 1 accepts a given number or not. $\qquad\square$

If we consider the previous theorem together with $\mathbb{N}_2 \, aOP_1(pcat_3) = \mathbb{N}_2 \, RE$ and $\mathbb{N} \, aOP_{1,-c}(pcat_3) = \mathbb{N} \, RE$ from Theorem 6.1, then we see that the presence of non-determinism increases the computational power of purely catalytic P systems. A result similar to Theorem 6.4 holds also for deterministic P systems with catalyst of degree m, $m \in \mathbb{N}_+$.

Theorem 6.5 *The set of numbers accepted by deterministic P systems with catalysts of degree 1 is recursive.*

Proof We show this by proving that deterministic purely catalytic P systems of degree 1 and deterministic P systems with catalysts of degree 1 are equivalent.

Part I: (Deterministic purely catalytic P systems of degree 1 can simulate deterministic P systems with catalysts of degree 1) Let $\Pi = (V, C, L, R)$ be a deterministic P system with catalysts of degree 1. The set R is equal to $R_1 \cup R_2$ with R_1 having only rules of the kind $ca \to c\alpha$ and R_2 having only rules of the kind $a \to \alpha$, with $a \in V$, $c \in C$ and $\alpha : (V \setminus C) \to \mathbb{N}$. Moreover, we assume that a unique number from 1 to $|R_2|$ is associated with each rule in R_2. We define $\Pi' = (V \cup C', C \cup C', L \cup C', R')$, a purely catalytic P system of degree 1, with:

$C' = \{c_i' \mid 1 \le i \le |R_2|\}$, $C' \cap V = \emptyset$;

$R' = R_1 \cup \{c_i'a \to c_i'\alpha \mid a \to \alpha \in R_2$ and $1 \le i \le |R_2|$ is the number associated with $a \to \alpha\}$.

The system Π' is clearly deterministic and it can simulate Π.

Part II: (Deterministic P systems with catalysts of degree 1 can simulate deterministic purely catalytic P systems of degree 1) This follows from the fact that a deterministic purely catalytic P system of degree 1 is a special case of a deterministic P system with catalysts of degree 1.   □

Now we generalise the previous theorem:

Theorem 6.6 *The set of numbers accepted by deterministic P systems with catalysts of degree m, $m \in \mathbb{N}_+$, is recursive.*

Proof We show this by proving that deterministic purely catalytic P systems of degree 1 and deterministic P systems with catalysts of degree $m > 1$ are equivalent.

Part I: (Deterministic purely catalytic P systems of degree 1 can simulate deterministic P systems with catalysts of degree $m > 1$) Let $\Pi = (V, C, L, R)$ be a deterministic P systems with catalysts of degree $m > 1$. We define $\Pi' = (V', C', L', R')$, a P system with catalysts of degree 1, with

$V' = \{a^{(i)} \mid a \in V, 1 \le i \le m\}$;

$C' = \{a^{(i)} \mid a \in C, 1 \le i \le m\}$;

$L'(a^{(i)}) = L_i(a)$ for $1 \le i \le m$;

R' is such that if in Π

  compartment i is the parent of compartment j and compartment j is the parent of compartment k;

  $ca \to c\alpha \in R_j$ with $\alpha \in (V \setminus C)^*$, $\alpha = (\alpha_1, tar_1) \dots (\alpha_{|\alpha|}, tar_{|\alpha|})$ and $\alpha_1, \dots, \alpha_{|\alpha|} \in V$;

  without loss of generality, $tar_1 = \dots = tar_{p_1} = here$, $tar_{p_1+1} = \dots = tar_{p_2} = in$ and $tar_{p_2+1} = \dots = tar_{|y|} = out$;

  then $c^{(j)}a^{(j)} \to c^{(j)}\alpha_1^{(j)} \dots \alpha_{p_1}^{(j)} \alpha_{p_1+1}^{(k)} \dots \alpha_{p_2}^{(k)} \alpha_{p_2+1}^{(i)} \dots \alpha_{|\alpha|}^{(i)} \in R'$ (similarly for $a \to \alpha \in R_j$).

The idea at the base of the description of Π' is that the symbols in Π are rewritten with the indication of the compartment they are in (using a superscript with the number of the compartment between parenthesis). The system Π' is deterministic and it simulates Π.

Part II: (Deterministic P systems with catalysts of degree $m > 1$ can simulate deterministic purely catalytic P systems of degree 1) To the deterministic purely catalytic P systems of degree 1 we add a 'parent' compartment with the empty multiset initially associated with it. In this way a deterministic P system with catalysts of degree 2 is obtained where the inner membrane is equivalent to the deterministic purely catalytic P systems of degree 1. □

Part I in the proof of the previous theorem is very important as it can be generalised to many, if not all, models of membrane systems. The resulting system, multisets rewriting systems, lose then all the features characterising membrane systems. This point will be discussed further in Section 6.7.

The introduction of priorities between rules allows deterministic purely catalytic P systems to simulate register machines. We recall that if *priorities* are present, then a rule of the kind $a \to \alpha$ can be applied only if no rule of the kind $a \to \alpha'$ having higher priority can be applied.

Theorem 6.7 *Deterministic purely catalytic P systems of degree 1 with priorities can simulate register machines.*

Proof This derives from the proof of Lemma 6.2 if:

  rule 1 has higher priority than rule 2;

  rule 4 has higher priority than rule 6;

  rule 5 has higher priority than rule 7;

  rule 12 has the highest priority between all rules.

These priorities allow the simulation of the 0-test P/T system to be deterministic.
□

6.6   Catalytic P systems of degree 1

Given a purely catalytic P system of degree 1 $\Pi = (V, C, L, R)$ we say that a non-catalyst $a \in V \setminus C$ is *evolutionary* if there is at least one rule $ca \to c\alpha \in R$, $\alpha : (V \setminus C) \to \mathbb{N}$, otherwise a is said to be *non-evolutionary*.

Theorem 6.8 *Communication-free vector addition systems (VASs) and purely catalytic P systems of degree 1 and one catalyst can simulate each other.*

Proof *Part I: (Catalytic P systems of degree 1 and one catalyst can simulate communication-free VASs)* Let us consider a communication-free VAS $W = (v_0, Y)$

of dimension k with $v_0 = (v_{0_1}, \ldots, v_{0_k})$. We define a purely catalytic P system of degree 1 and one catalyst $\Pi = (V, \{c\}, L, R)$ with:

$V = \{a_1, \ldots, a_k, d, c\}$;
$L(a_i) = v_{0,i}$ for $1 \le i \le m$, and $L(c) = L(d) = 1$;

R is such that for each $(w_1, \ldots, w_k) \in Y$:

if $w_i \ge 0$, $1 \le i \le k$, then $cd \to cda_1^{w_1} \ldots a_k^{w_k} \in R$;
if there is $p \in \{1, \ldots, k\}$ such that $w_p = -1$ and $w_q \ge 0$, $1 \le q \le k$, $q \ne p$,
    then $ca_p \to ca_1^{w_1} \ldots a_{p-1}^{w_{p-1}} a_{p+1}^{w_{p+1}} \ldots a_k^{w_k} \in R$.

Clearly Π simulates W.

Part II: (Communication-free VASs can simulate purely catalytic P systems of degree 1 and one catalyst) Let us assume that $\Pi = (V, \{c\}, L, R)$ is a purely catalytic P systems of degree 1 with $V = \{a_1, \ldots, a_k, c\}$. Moreover, the rules of the kind $ca \to c\alpha \in R$, $a \in \alpha$, are numbered in a unique way with elements in $\{1, \ldots, r\}, r \in \mathbb{N}$. We define $\Pi' = (V', \{c\}, L, R')$, a purely catalytic P systems of degree 1 having rules of the kind $ca \to x\alpha$ with $a \notin \alpha$ that can simulate Π. The system Π' is such that

$$V' = V \cup \{a_1', \ldots, a_r'\};$$

R' is such that

if $ca \to c\alpha \in R$ and $a \notin \alpha$, then $ca \to c\alpha \in R'$;
if $ca \to c\alpha \in R$, $a \in \alpha$ and i is the unique number associated with this rule,
    then $ca \to ca_i'$, $ca_i' \to c\alpha \in R'$.

Clearly Π' simulates Π.
We now define a communication-free VAS $W = (v_0, Y)$ able to simulate systems as Π'. The VAS W is such that:

$$v_0 = (L(a_1), \ldots, L(a_k));$$

Y is such that for each rule $ca_p \to ca_1^{j_1} \ldots a_{p-1}^{j_{p-1}} a_{p+1}^{j_{p+1}} \ldots a_k^{j_k}$, the vector $(j_1, \ldots, j_{p-1}, -1, j_{p+1}, \ldots, j_k) \in Y$.

Clearly V simulates Π'. $\square$

A purely catalytic P system of degree 1 $\Pi = (V, C, L, R)$ is called *simple* if each rule $ca \to c\alpha \in R$ has at most one non-evolutionary symbol in α. The result of the halting computations of the P systems considered in the next theorem is the vector of the multiplicities of non-catalysts present when the computation halts.

Theorem 6.9 *The sets of numbers generated by simple purely catalytic P systems of degree 1, one catalyst and an unbounded number of non-catalysts coincides with semilinear sets.*

Proof *Part I: (Such simple purely catalytic P systems can generate semilinear sets)* Let $A \subseteq \mathbb{N}_+^k$ be a semilinear set with $A = \bigcup_{i=1}^q A_i$ each $A_i = \{v_{i,0} + \sum_{j=1}^k p_{i,j} v_{i,j} \mid p_{i,j} \in \mathbb{N}_+, \ v_{i,j} \subseteq \mathbb{N}_+^k\}$ being a linear set. Let $v_{i,0} = (v_{i,0_1}, \ldots, v_{i,0_k})$ and similarly for $v_{i,j}$. We define a simple purely catalytic P system of degree 1 $\Pi = (V, \{c\}, L, R)$ with

$$
\begin{aligned}
V &= \{a_1, \ldots, a_k, b_1, \ldots, b_p, d_1, \ldots, d_p, e, c\}; \\
L &= \{c, e\}; \\
R &= R_i' \cup R_i'' \cup R_i''' \cup R_i'''', \ 1 \leq i \leq q; \\
R_i' &= \{ce \rightarrow cb_i\}; \\
R_i'' &= \{cb_i \rightarrow cd_i a_1^{v_{i,0_1}} \ldots a_k^{v_{i,0_k}}\}; \\
R_i''' &= \bigcup_{j=1}^t \{cd_i \rightarrow cd_i a_1^{v_{i,j_1}} \ldots a_k^{v_{i,j_k}}\}; \\
R_i'''' &= \{cd_i \rightarrow c\}.
\end{aligned}
$$

The rules in R_i' allow the system to select in a non-deterministic way any of the q linear sets. Fixed an i the rules in R_i'' allow us to generate multiple occurrences of $a_1, \ldots, a_k$ symbols depending on $v_{i,0}$ while the rules in R_i'''' allow us to generate multiple occurrences of $v_{i,j}$ repeatedly for $1 \leq j \leq k$. The repeated application of the rule in R_i''' halts when the rule in R_i'''' is applied. It is clear that only one catalyst is used and that, because of the variables q and k, the number of non-catalysts is unbounded.

Part II: (All the sets generated by such purely catalytic P systems are semilinear) We prove this considering Theorem 5.21, that is, we describe a reversal-bounded register machine M able to simulate such a simple purely catalytic P systems. Let Π be one such system. By definition the multiplicity of the catalysts and the evolutionary non-catalysts present in Π is finite. So M can use its finite control to keep track of these multiplicities. The registers of M are associated with the non-evolutionary non-catalyst symbols in Π. From the definition of simple purely catalytic P systems we know that these registers are non-decreasing. So M can simulate Π faithfully. The value of the registers present in M when it halts corresponds to the occurrences of non-evolutionary non-catalysts symbols in Π when it halts. $\square$

Now we are going to consider the following model of a purely catalytic P system:

Definition 6.2 *A* purely catalytic P system with states *is a construct*

$$\Pi = (V, C, S, L, O, s_0)$$

where:

V and C are as in Definition 6.1;

$L : V \to \mathbb{N}$ *is a multiset;*

S *is a set of states,* $S \cap V = \emptyset$;

$O : S \to S \times R$ *is the* operation function *with R a set of* rules *of the kind*
$ca \to c\alpha$, $a \in V \setminus C$, $c \in C$ *and* $\alpha : (V \setminus C) \to \mathbb{N}$;

$s_0 \in S$ *is the* initial state.

If $O(s) = (s', ca \to c\alpha)$, then the tuple $(s, ca \to c\alpha, s')$ is called an *operation*. A *configuration* of Π is a pair (s, M) with $s \in S$ and M multiset over V. The initial configuration is (s_0, L). The application of operations in O can change the state and the multiset of non-catalysts of Π in the following way. If (s', M') is a configuration of Π, then the operation $(s', ca \to c\alpha, s'')$ can be applied only if $a \in M'$. In this case the state of Π changes into s'' and M' changes into $M'' = M' \setminus \{a\} \cup \{\alpha\}$.

For two configurations (s', M') and (s'', M'') of Π we denote by $(s', M') \Rightarrow (s'', M'')$ a *transition* from (s', M') to (s'', M''), that is, the application of a multiset of transitions under the requirement of maximal parallelism. The reflexive and transitive closure of $\Rightarrow$ is denoted by $\Rightarrow^*$.

A *computation* is a sequence of transitions between configurations of a system Π starting from the initial configuration (s_0, L). If a computation is finite, then the last configuration is called *final* and we say that the system *halts*. Let $\Pi = (V, C, S, L, O, s_0)$ be a purely catalytic P system with states and let $(s_0, L) \Rightarrow^* (s, M)$ be a halting computation of Π. The system Π can generate either the number $|M|$ or the vector $(M_{\text{comp}}(v_1), \dots, M_{\text{comp}}(v_{|V \setminus C|}))$ for $v_i \in V \setminus C, 1 \le i \le |V|$.

We start by studying the computational power of purely catalytic P systems with states using one catalyst.

Lemma 6.6 *Let Π be a purely catalytic P system with states. It is possible to define a vector addition system with states (VASS) W such that $\mathbb{C}_W$, the set of configurations of W, is equal to M for each (s, M) configuration of Π.*

Proof Let $\Pi = (V, \{c\}, S, L, O, s_0)$ be a purely catalytic P system with states where $V = \{a_1, \dots, a_k, c\}$, $S = \{s_0, \dots, s_p\}$ and $O : S \to S \times R$. Moreover, each rule in R has a unique number from 1 to $|R|$ associated with it. We define a VASS $W = (v_0, S', H, s_0)$ with:

$v_0 = (L(a_1), \dots, L(a_k))$;

$S' = S \cup \{(s_i, r) \mid s_i \in S, \ 1 \le r \le |R|\}$;

let $O(s_q) = (s_t, ca_u \to ca_1^{j_1} \dots a_k^{j_k})$ for $s_q, s_t \in S$. Then for $1 \le u \le k$

if $j_u = 0$, then $H(s_q) = (s_t, (j_1, \ldots, j_{u-1}, -1, j_{u+1}, \ldots, j_k))$;

if $j_u > 0$, then

$\quad H(s_q) = ((s_q, r), (0, \ldots, 0, -1, 0, \ldots, 0)), -1$ being the value of the u-th element of $(0, \ldots, 0, -1, 0, \ldots, 0)$,

$H((s_q, r)) = (s_t, (j_1, \ldots, j_{u-1}, j_u, j_{u+1}, \ldots, j_k))$ where r is the unique number associated with the rule $ca_u \to ca_1^{j_1} \ldots a_k^{j_k}$.

If should be clear that $\mathbb{C}_W = M$ for each (s, M) configuration of Π. $\qquad \square$

Lemma 6.7 *Every vector addition system with states (VASS) can be simulated by a purely catalytic P system with states.*

Proof Let $W = (v_0, S, H, s_0)$ be a VASS of dimension k with $v_0 = (v_{0_1}, \ldots, v_{0_k})$. Moreover, let the transitions in W be numbered in a unique way from 1 to p. For each transition $i : (s_i, w_i, s_i')$, $1 \leq i \leq p$, in W let q_i be the negated sum of the negative elements in w_i. So, for instance, if $1 : (s_1, (2, -3, -2), s_1')$ and $2 : (s_2, (8, 2, 0), s_2')$ are two transitions in W, then $q_1 = 5$ and $q_2 = 0$.

We define a purely catalytic P system with states $\Pi = (V, \{c\}, S', L, O, s_0)$ with:

$V = \{a_1, \ldots, a_k, d, c\}$;

$S' = S \cup \{(s, i, j) \mid s \in S, 1 \leq i \leq p, 1 \leq j \leq q_i\}$ and we assume s to be the same as $(s, i, 0)$ for $1 \leq i \leq p$;

$L(a_i) = v_{0_i}$ for $1 \leq i \leq k$;

let $H(s) = ((j_1, \ldots, j_k), s')$, then:

$\quad$ if $j_i \geq 0$, $1 \leq i \leq k$, then $O(s) = (s', cd \to cda_1^{j_1} \ldots a_k^{j_k})$;

$\quad$ if some j_i, $1 \leq i \leq k$, are negative, then non-catalysts associated with the negative $j_i's$ are removed one by one. We explain this with an example in which we assume that in a transition having number l in W j_p and j_q are negative while the remaining j_i are positive. Then:

$\quad O(s) = ((s, l, 1), ca_p \to c)$;

$\quad O((s, l, 1)) = ((s, l, 2), ca_p \to c)$;

$\quad \vdots$

$\quad O((s, l, j_{p-1})) = ((s, l, j_p), ca_q \to c)$;

$\quad O((s, l, j_p)) = ((s, l, j_{p+1}), ca_q \to c)$;

$\quad \vdots$

$\quad O((s, l, j_p + j_q - 1)) = (s', ca_q \to ca_1^{j_1} \ldots a_{p-1}^{j_{p-1}} a_{p+1}^{j_{p+1}} \ldots a_{q-1}^{j_{q-1}} a_{q+1}^{j_{q+1}} \ldots a_k^{j_k})$.

It should be clear that Π can simulate W. $\qquad \square$

If we consider the previous two lemmas and Definition 4.7, then

Theorem 6.10 *Catalytic P system with states, vector addition systems with states (VASS) and communication-free VASS can simulate each other.*

Computational completeness is obtained if priorities are added to the operations present in purely catalytic P system with states. Informally: if *priorities* are present, then an operation of the kind $(s, ca \rightarrow c\alpha, s')$ can be applied only if no operation of the kind $(s, ca \rightarrow c\alpha'', s'')$ having higher priority can be applied.

Theorem 6.11 *The computational power of a purely catalytic P system with states and priorities is equivalent to that of register machines.*

Another model of a purely catalytic P system considered in the literature of Membrane Computing is the following.

Definition 6.3 *A* matrix purely catalytic P system *is a construct*

$$\Pi = (V, C, L, M)$$

where:

V, C and L are as in Definition 6.2;
$M = \{M_i \mid 1 \leq i \leq p, \ p \in \mathbb{N}_+\}$ *where each M_i, called a* matrix, *contains rules of the kind $ca \rightarrow c\alpha$, $a \in V \setminus C$, $c \in C$ and $\alpha \in (V \setminus C)^*$.*

Matrix purely catalytic P systems with states *are defined as*

$$\Pi = (V, C, S, L, O, s_0)$$

where:

V, C, S, L and s_0 are as in Definition 6.2;
$O : S \rightarrow S \times M_i$ *where M_i are defined as for matrix purely catalytic P systems.*

In a matrix purely catalytic P system a matrix can be applied only if all the rules present in it can be applied. If not, then the matrix is not applied. Let Π be a matrix purely catalytic P system with states and let $O(s_1) = (s_2, M)$. If Π is in state s_1 and all the rules in M can be applied, then all the rules in M are applied and Π changes state into s_2. The definitions of *configuration*, *computation* and *halt* follow from the above.

Following proof techniques similar to those used for Lemma 6.6 and Lemma 6.7 it is possible to prove that

Theorem 6.12 *A matrix purely catalytic P system, a matrix purely catalytic P system with states and purely catalytic P systems with states can simulate each other.*

The results indicating equivalences between models of purely catalytic P systems, VAS and VASS are important also because of what is known about VAS and VASS. Of particular value are the equivalences to families of semilinear sets and the (un)decidability properties indicated in Theorem 4.8.

6.7 Final remarks and research topics

The results in Section 6.4 show how far reaching the simulation of *join* and *fork* can be in respect to the direct simulation of a specific computational model. We say 'can be' and not 'is' because it is not always the case that some features considered for building blocks can be naturally translated into features of the simulating system (P systems with catalysts, in this chapter). Corollary 6.2 is an example of the difficulties in this 'translation'. How do we translate in natural terms for P systems with catalysts the 'whose underlying net is a composition of sequences of *join* and *fork*' present in the statement of Theorem 4.4?

The definition of restricted P systems with catalysts considered for Corollary 6.2 is a very simple way to perform such a translation, but still it did not allow us to say that the computational power of such P systems is equivalent to that of restricted register machines. For this reason we formulate:

Suggestion for research 6.1 *Define a model of P systems with catalysts having computational power equivalent to that of restricted register machines.*

The answer to this suggestion could either be the proof that restricted register machines can simulate restricted P systems without catalysts or the definition of another natural model for these P systems.

Proving that Lemma 6.1 is optimal in the number of catalysts, that is, that *join* cannot be simulated with only one catalyst, would imply the optimality in the number of catalysts of the theorems and corollaries presented in Section 6.4. For this reason we formulate:

Suggestion for research 6.2 *Find the minimum number of catalysts needed to simulate a join.*

The results in Section 6.6 and Section 6.5 indicate that the presence of some features (as priorities, states, determinism, etc.) have a direct influence on the computing power of P systems with catalysts. Is it possible to formalise these concepts in broader terms, that is, not limited to P systems with catalysts?

Suggestion for research 6.3 *Study how the features considered in Section 6.6 and Section 6.5 relate to other concepts present in formal systems.*

The previous suggestion for research aims to find laws similar to those indicated by Theorem 4.1. We believe that, also in this case, Petri nets could be of help.

Theorem 6.6 is definitely very important. As already said, we can regard it as 'collapsing' P systems with catalysts into multiset rewriting systems. But then why study (membrane) systems with an underlying topology if they can become systems without a topology? Well, we think that the underlying topology plays an important part and that it could be studied with respect to the computational power of a membrane system.

Suggestion for research 6.4 *Study how the topology of the information flow (and by this we mean the passage of symbols between compartments, replacement of a symbol with others, etc.) is related to the computational power of membrane systems.*

6.8   Bibliographical notes

Membrane systems with catalysts were introduced in [207]. In the original definition, target indicators of the kind $in_{k_1}, \ldots, in_{k_p}$ were present. If a P system with catalysts has a compartment j having compartments $k_1, \ldots, k_p$ as children, then the target indicators $in_{k_1}, \ldots, in_{k_p}$ let symbols pass to these children in a specific way. For instance, if compartment 1 has children 2 and 3, then the application of the rule $a \rightarrow (\alpha_1, in_2)(\alpha_2, in_3)$ in compartment 1 allows α_1 to pass into compartment 2 and α_2 to pass into compartment 3, respectively.

Moreover, the first studies on P systems with catalysts used the target indicator *out* and priorities between rules. The *out* indicator is present in Definition 6.1 but none of the systems considered in the present chapter uses it. Priorities between rules are not included in Definition 6.1: they are only considered for Theorem 6.11 and Theorem 6.7 and they are informally defined just before these theorems.

What was presented in Section 6.4 is original to this monograph. The direct (that is, not using building blocks) proof of Theorem 6.1 can be found in [65]. In the same paper also similar proofs for other models of P systems with catalysts can be found.

Section 6.5 is based on the results present in [126] but several of the proofs are original to this monograph. The cited paper gave the first example of a model of a (membrane) system for which the non-deterministic version is universal while the deterministic version is not. Such a study of P systems with catalysts was posed as a problem in [74].

Section 6.6 is entirely based on [119] where it is possible to find also the proof of Theorem 6.11 and Theorem 6.12.

Other publications dealing with P systems with catalysts are: [37,73,235,248].

6.8.1 About the notation

To the best of our knowledge there is no established notation for the sets of numbers or vectors generated or accepted by membrane systems with catalysts. Here we adopted one notation following the style of similar notations present in this monograph.

7 Spiking

The model of membrane systems considered in the present chapter is inspired by a very specific kind of cells: *neurons*. Despite the fact that this model is relatively recent compared to the other ones considered in this monograph, the number of results concerning it is considerable.

7.1 Biological motivations

The *neuron* is the essential building block of the nervous system. In Fig. 7.1 a typical vertebrate neuron is depicted. Neurons are cells specialised in responding to stimuli with *nerve impulses*, that is, electrical discharges, and in conducting nerve impulses over long distances.

In a typical neuron two kinds of extensions depart from the part of the cell, called the *soma*, where the nucleus and the endoplasmatic reticulum are located:

dendrites: several of these are present in a typical neuron and they branch to form *dendrite trees*;

axon: only one such extension is present. An axon can have ramifications called *axon collaterals*.

The morphology of neurons can vary with respect to the size of the soma, the number, length and branching of the dendrites and the length and branching of the axon. Axons, for instance, can be from 0.1 millimetre to 3 metres long. The initial segment of an axon is called an *axon hillock*. Axons terminate with *nerve terminals* (or *boutons*) normally found close to the plasma membrane of cells different from neurons or close to the dendrites or the soma of other neurons.

The location of this point of contact is called a *synapse*; it is here that the transmission of information from one neuron to another takes place.

Axons are specialised for the conduction of a particular type of electrical impulse called an *action potential* or *spike*. An action potential, caused by the voltage dependent opening and closing of certain ion channels (see Section 5.1), is a series of sudden changes in the voltage across the axon membrane. The arrival of an action potential to the nerve terminals causes secretion of chemicals called *neurotransmitters*. Neurotransmitters bind to proteins present on the membrane of the receptor neuron and cause changes in their membrane potential. This potential is called the *synaptic potential* or *synaptic input*. These small electrical impulses

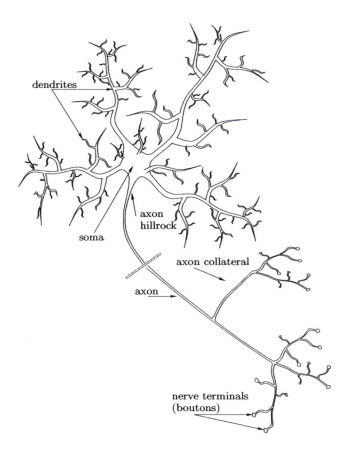

Fig. 7.1 A typical vertebrate neuron

are then conducted towards the soma. So, electrical signals carry information within a neuron, while chemical signals transmit information between neurons.

An electrical signal present on the plasma membrane is more likely to give rise to an action potential at the axon hillock. It is here that the activity of all synaptic potentials is summed and, if a certain threshold is reached, the neuron *fires* an action potential. An action potential is all-or-none. If synaptic potentials are below a certain threshold, then no action potential is produced; if synaptic potentials are above a certain threshold, then the same action potential is produced. This means that it is the frequency of action potentials that encodes different information.

A neuron cannot fire a second action potential until the ion channels have returned to the close conformation. This interval of time in which the neuron is refractory to stimulation is called the *refractory period*.

The way a neuron reacts to a synaptic input depends on the different kinds of ion channels present in some parts of it. Cells with different combinations of ion channels react differently to constant synaptic potentials. Some cells react with a single action potential, others with a constant-frequency sequence of action potentials, others with either an accelerating or decelerating sequences of action potentials. Such sequences are called *action potential trains* or *spiking trains*. Some neurons can even fire without any synaptic potential. There is no general agreement on the time interval between the spikes defining a spiking train.

It is important to say that the firing rate of some neurons can be increased by changes in the strength of the synaptic potentials.

In addition to the electrical and chemical signals the function of the nervous system depends on the topology of the neurons, that is, on how they are interconnected. One axon can have synapses with hundreds of neurons and induce responses to each of them simultaneously. These connections determine the pathway of signals. So, although each neuron is an agent, it is only through cooperation that neurons can fulfil specific tasks.

7.2 Basic definitions

In the following $L(E) \subseteq$ REG denotes the language defined by a regular expression E.

Definition 7.1 *An* extended spiking (neural) P system *of degree m is a construct*

$$(7.1) \qquad \Pi = (V, \mu, L_1, \ldots, L_m, R_1, \ldots, R_m, comp)$$

where:

$V = \{a\}$ *is a singleton alphabet and* a *is called a* spike;

$\mu = (Q, E)$ *is a cell-graph underlying* Π *where:*

> $Q \subset \mathbb{N}$ *contains* vertexes. *For simplicity we define* $Q = \{1, \ldots, m\}$. *Each vertex in Q defines a* neuron *of* Π;

> $E \subseteq Q \times Q$ *defines directed labelled* edges *between vertexes, denoted by* (i, j), $i, j \in Q, i \neq j$. *Each edge in E defines a* synapse;

$L_i : V \to \mathbb{N}$, $1 \leq i \leq m$ *are finite multisets of spikes initially associated with each neuron;*

R_i, $1 \leq i \leq m$, *are sets of rules of the following two forms:*

> $E/a^r \to a^s; d$, *called* spiking rules, *where E is a regular expression over V, $r, s \in \mathbb{N}_+$, $r \geq s$ and s are the spikes generated by the application of the rule, $d \in \mathbb{N}$ and d is called the* delay;

$a^t \rightarrow \epsilon$, *called* forgetting rules, *for* $t \in \mathbb{N}_+$ *and with the restriction that* $t \notin L(E)$ *for any spiking rule of the kind* $E/a^r \rightarrow a^s; d$ *in* R_i:

comp $\in Q$ *denotes the* output neuron.

In the literature of spiking P systems, rules of the kind $E/a^r \rightarrow a^s; d$ with $L(E) = \{a^r\}$, $r \in \mathbb{N}_+$, are denoted by $a^r \rightarrow a^s; d$. In the following we do not use this simplified notation. Neurons having only rules of the kind $E/a^r \rightarrow a^s; d$ with $L(E) = \{a^r\}$, $r \in \mathbb{N}_+$, are called *bounded* and spiking P systems having only bounded neurons are called *bounded* or *finite*. Notice that forgetting rules are a special case of bounded rules having $s = 0$. Neurons having only rules of the kind $a^i(a^k)^*/a^r \rightarrow a^s; d$ where $i, s, d \geq 0$ and $k, j \geq 1$ are called *unbounded* and spiking P systems having only unbounded neurons are called *unbounded*. Neurons having both bounded and unbounded rules are called *general* and spiking P systems having only general neurons are called *general*.

Spiking P systems having only rules of the kind $E/a^r \rightarrow a; d$ or of the kind $a^t \rightarrow \epsilon$ are said to be of *standard* type or *non-extended*.

A spiking P system is assumed to operate in a way synchronised by the ticks (*time units*) of a global clock.

Each neuron can be in one of two states: *active* and *idle* (*closed*). Rules can be applied only in active neurons. The application of a rule of the kind $E/a^r \rightarrow a^s; d$ on a neuron i is composed of two parts:

firing: This occurs in a neuron having at least k spikes associated with it. If $a^k \in L(E)$ and $k \geq r$, then r spikes are removed by the neuron. As a consequence of this the neuron becomes idle for d time units.

spiking: After firing and waiting for d time units s spikes are *sent* (we also say *passed*) to each neuron $j \in Q$ such that $(i, j) \in E$ and j is active.

If $d = 0$, then the neuron does not become idle and it spikes in the same time unit in which it fires.

The application of a rule of the kind $a^s \rightarrow \epsilon$ allows s spikes to be *removed* by neuron i.

It is important to notice that the applicability of a rule depends on the total number of spikes present in a neuron. If, for instance, neuron i contains five spikes and $R_i = \{(aa)^*/a \rightarrow a; 0, a^3/a^3 \rightarrow a; 0, a^2 \rightarrow \epsilon\}$, then none of these rules is applied as $a^5 \notin L((aa)^*)$, $a^5 \notin L(a^3)$ and $a^5 \notin L(a^2)$. However, if the rule $a^5/a^2 \rightarrow a; 0 \in R_i$, then this rule can be applied so that two spikes are removed from i and one spike is generated.

In each time unit at most one (any) rule can be applied in an active neuron. If more than one rule can be applied, then one is non-deterministically chosen.

A *configuration* of a spiking P system Π of degree m is given by the $3m$-tuple $(M_1, f_1, p_1, \ldots, M_m, f_m, p_m)$ where $M_i : V \rightarrow \mathbb{N}$ are multisets of spikes, while $f_i, p_i \in \mathbb{N}$ with f_i being the number of time units neuron i is idle and p_i the number of spikes that neuron i sends when it becomes active again. The $3m$-tuple $(L_1, 0, 0, \ldots, L_m, 0, 0)$ is called the *initial configuration*.

For two configurations $(M_1, f_1, p_1, \ldots, M_m, f_m, p_m)$ and $(M'_1, f'_1, p'_1, \ldots, M'_m, f'_m, p'_m)$ of Π we write $(M_1, f_1, p_1, \ldots, M_m, f_m, p_m) \Rightarrow (M'_1, f'_1, p'_1, \ldots, M'_m, f'_m, p'_m)$ to denote a *transition* from $(M_1, f_1, p_1, \ldots, M_m, f_m, p_m)$ to $(M'_1, f'_1, p'_1, \ldots, M'_m, f'_m, p'_m)$, that is, the application of rules associated with each neuron under the requirement of maximal strategy. The reflexive and transitive closure of $\Rightarrow$ is denoted by $\Rightarrow^*$. The time unit of the initial configuration is 1 and every transition increases the time unit of 1.

For $1 \leq i \leq m$ in a configuration:

if $f_i = 0$ and $p_i = 0$, then neuron i is active;

if $f_i > 1$ and $p_i > 0$, then neuron i is idle and does not fire in the next time unit;

if $f_i = 1$ and $p_i > 0$, then neuron i is idle and it fires in the next time unit.

Let $\Pi = (V, \mu, L_1, \ldots, L_m, R_1, \ldots, R_m, comp)$ be a spiking P system with $\mu = (Q, E)$ and let $c = (M_1, f_1, p_1, \ldots, M_m, f_m, p_m)$ be a configuration of Π. The application of the rules of Π can change c into $(M'_1, f'_1, p'_1, \ldots, M'_m, f'_m, p'_m)$ in the following way. For $1 \leq i \leq m$:

if neuron i is active and

$E/a^r \rightarrow a^s; d \in R_i$, $d > 0$ and this rule is applied, then $M'_i = M_i \setminus \{a^r\}$, $f'_i = d$, $p'_i = s$, $M'_z = M_z$, $f'_z = f_z$ and $p'_z = p_z$ for $1 \leq z \leq m$, $z \neq i$.

Informally: if a neuron is active and a spiking rule with $d > 0$ is applied, then the neuron becomes idle, r spikes are removed from the neuron, and p_i is set to s to denote that when the neuron becomes active it spikes s spikes.

$E/a^r \rightarrow a^s; d \in R_i$, $d = 0$ and this rule is applied, then $M'_i = M_i \setminus \{a^r\}$, $f'_i = f_i$, $p'_i = p_i$, $M'_j = M_j \cup \{a^s\}$, $f'_j = 0$, $p'_j = 0$ for each $j \in Q$, $(i, j) \in E$ and $f_j = 0$. $M'_z = M_z$, $f'_z = f_z$ and $p'_z = p_z$ for $1 \leq z \leq m$, $z \neq i, j$.

Informally: if a neuron is active and a spiking rule with $d = 0$ is applied, then the generated spikes are sent to all the active neurons connected to it.

$a^r \rightarrow \epsilon \in R_i$ and this rule is applied, then $M'_i = M_i \setminus \{a^r\}$, $f'_i = f_i$, $p'_i = p_i$, $M'_z = M_z$, $f'_z = f_z$ and $p'_z = p_z$ for $1 \leq z \leq m$, $z \neq i$.

Informally: if a neuron is active and a forgetting rule is applied, then r spikes are removed from the neuron.

if neuron i is idle and

$f_i > 1$, then $f'_i = f_i - 1$, $M'_i = M_i$, $p'_i = p_i$, $M'_z = M_z$, $f'_z = f_z$ and $p'_z = p_z$ for $1 \leq z \leq m$, $z \neq i$.

Informally: if a neuron is idle and its f parameter is bigger than one, then this parameter is decreased by one.

$f_i = 1$, then $M'_i = M_i$, $f'_i = 0$, $p'_i = 0$, $M'_j = M_j \cup \{a^{p_i}\}$, $f'_j = 0$, $p'_j = 0$ for each $j \in Q$, $(i,j) \in E$ and $f_j = 0$. $M'_z = M_z$, $f'_z = f_z$ and $p'_z = p_z$ for $1 \leq z \leq m$, $z \neq i$.

Informally: if a neuron is idle and its f parameter is equal to one, then the neuron fires p_i spikes to the active neurons connected to it and it becomes active.

A *computation* is a sequence of transitions between configurations of a system Π starting from the initial configuration. A computation *halts* if it reaches a configuration in which all neurons are active and no rule can be applied. Each computation has a *spiking train* associated with it: the sequence of time units in which the output neuron spikes. If in a time unit the output neuron spikes, then the spiking train records that time unit, otherwise not.

Let $\Pi = (V, \mu, L_1, \ldots, L_m, R_1, \ldots, R_m, comp)$ be a spiking P system, let $\mathbb{C}_\Pi$ be the set of configurations of Π and let $c_i \subseteq \mathbb{C}_\Pi$, $i \geq 0$, with $C = c_0 \Rightarrow c_1 \Rightarrow \cdots$ be a computation of Π. The spiking train of C is a vector denoted by $st(C) = (t_1, t_2, \ldots)$ with $t_i \in \mathbb{N}_+$, $i \in \mathbb{N}_+$.

It is possible to associate sets of number with Π in several ways:

(2) considering only the first two spikes:
$$N_2(\Pi) = \{t_2 - t_1 \mid st(C) = (t_1, t_2, \ldots), C \in \mathbb{C}_\Pi\};$$

(k) generalising the first $k \geq 2$ spikes of a computation with at least k configurations:
$$N_k(\Pi) = \{n \mid n = t_i - t_{i-1} \text{ for } 1 \leq i \leq k, \ st(C) = (t_1, t_2, \ldots, t_k, \ldots), \ C \in \mathbb{C}_\Pi\};$$

($\underline{k}$) generalising the first $k \geq 2$ spikes of a computation with k configurations:
$$N_{\underline{k}}(\Pi) = \{n \mid n = t_i - t_{i-1} \text{ for } 1 \leq i \leq k, \ st(C) = (t_1, t_2, \ldots, t_k), C \in \mathbb{C}_\Pi\};$$

(ω) considering all spikes of computations with infinite spike trains:
$$N_\omega(\Pi) = \{n \mid n = t_i - t_{i-1} \text{ for } 1 \geq 1, \ C \in \mathbb{C}_\Pi, \ st(C) = (t_1, t_2, \ldots) \ infinite\};$$

(all) considering all intervals of all computations:
$$N_{\text{all}}(\Pi) = \bigcup_{k \geq 2} N_k(\Pi) \cup N_\omega(\Pi).$$

Two subsets of some of these sets are also of interest:

(h) considering only halting computations. This is valid only for $N_k(\Pi), k \geq 2$ and for $N_{\text{all}}(\Pi)$. The respective subsets are denoted by $N_k^h(\Pi)$ and $N_{\text{all}}^h(\Pi)$;

($\underline{h}$) considering only halting computations such that in the last configuration no spike is associated with any neuron. The respective subsets are denoted by $N_k^{\underline{h}}(\Pi)$ and $N_{\text{all}}^{\underline{h}}(\Pi)$;

(a) considering *alternately* intervals:

$$N^a(\Pi) = \{n \mid n = t_{2k} - t_{2k-1} \text{ for } k \geq 1, \ C \in \mathbb{C}_\Pi, \ st(C) = (t_1, t_2, \ldots)\}.$$

Every second interval of k configurations is ignored. The first interval is considered, the second is ignored, the third is considered, and so on. The respective subsets of $N_k(\Pi)$, $N_\omega(\Pi)$ and $N_{\text{all}}(\Pi)$ are denoted by $N_k^a(\Pi)$, $N_\omega^a(\Pi)$ and $N_{\text{all}}^a(\Pi)$.

It is possible to combine the h to the a restrictions obtaining the sets $N_\alpha^{ha}(\Pi)$ and $N_\alpha^{\underline{h}a}(\Pi)$ for $\alpha \in \{\omega, all\} \cup \{k \mid k \geq 2\}$, as well as $N_{\underline{k}}^{ha}(\Pi)$ and $N_{\underline{k}}^{\underline{h}a}(\Pi)$ for $k \geq 2$.

With $\mathbb{N}\, SN_\alpha^\beta P_m(rule_{p_1}, cons_{p_2}, gen_{p_3}, dlay_{p_4}, forg_{p_5}, outd_{p_6}, p_7)$ we denote the family of numbers generated by spiking P systems with:

$\alpha \in \{all, \omega\} \cup \{k, \underline{k} \mid k \geq 2\}$;

$\beta \in \{h, a, ha, \underline{h}, \underline{h}a\}$ or β is omitted;

degree at most m;

at most p_1 rules in each neuron;

at most p_2 spikes are removed by rules of the kind $E/a^{p_2} \to a^s; d$;

at most p_3 spikes are generated by rules of the kind $E/a^r \to a^{p_3}; d$;

at most p_4 as the longest delay caused by rules of the kind $E/a^r \to a^s; p_4$;

at most p_5 spikes are removed by rules of the kind $a^{p_5} \to \epsilon$;

at most p_6 synapses outgoing from a neuron. Formally, $p_6 = max(|E_i|)$, $i \in Q$
   where $E_i = \{(i, j) \mid j \in Q\}$;

at most p_7 spikes present at any time unit in any neuron.

$\mathbb{N}\, SN_\alpha^\beta P_m(rule_{p_1^+}, cons_{p_2}, gen_{p_3}, dlay_{p_4}, forg_{p_5}, outd_{p_6}, p_7)$ is the subset of $\mathbb{N}\, SN_\alpha^\beta P_m(p_1, p_2, p_3, p_4, p_5, p_6, p_7)$ generated by spiking P systems having only rules of the kind $a^+/a^r \to a^s; d$ or of the kind $a^r/a^r \to a^s; d$.

The parameters $m, p_1, p_2, p_3, p_4, p_5, p_6$ and p_7 are replace by $*$ if they are unbounded. If parameter p_7 is specified, then if a system in such a family reaches a configuration where a neuron has more than p_7 spikes, then the system ends without providing any result.

The presence of so many parameters allows us to define very many classes of spiking P systems. In the following we report the results related to only some of these classes. In Section 7.9 we indicate where other results on this model of P systems can be found.

7.3　Examples

Figures representing spiking P system have neurons represented by rectangles having their label written in **bold** on their right-upper corner. Rules and the initial spikes present in the neurons are indicated inside the respective rectangles. Arrows between rectangles represent the synapses from one neuron to another. Output neurons have one outgoing arrow incoming no neuron.

In the following we use *modules*: groups of neurons with spikes and rules in a spiking P system able to perform a specific task. As a module is part of a bigger system it can have synapses coming from or going to neurons not present in the module itself.

Example 7.1 A non-extended spiking P system Π_1 of degree 3 generating $3k+2$ for $k \in \mathbb{N}_+$.

The formal definition of this system is $\Pi_1 = (\{a\}, \mu, L_1, L_2, L_3, R_1, R_2, R_3, 3)$ with:

$L_1(a) = k$;
$L_2(a) = 0$;
$L_3(a) = 1$;
$R_1 = \{a^+/a \rightarrow a; 2\}$;
$R_2 = \{a^+/a^k \rightarrow a; 1\}$;
$R_3 = \{a^+/a \rightarrow a; 0\}$.

As Π_1 has only rules of the kind $E/a^r \rightarrow a; d$ it is non-extended. This system is depicted in Fig. 7.2.

The initial configuration of Π_1 sees k spikes in neuron 1, one spike in neuron 3 and no spike in neuron 2. All the rules present in Π_1 are spiking rules.

We describe the work of Π_1 considering $k = 2$ and with the help of Table 7.1. The generalisation to any value of $k \in \mathbb{N}_+$ is straightforward.

Table 7.1 indicates the time units and, for each neuron, the number of spikes, the state and the action performed by it. The initial configuration (time unit 1) is then indicated by a^2 in neuron 1 and a in neuron 3. As $a^2 \in L(a^+)$, then neuron 1 fires rule $a^+/a \rightarrow a; 2$. This means that the number of spikes in this neuron is reduced by 1 and the neuron becomes idle for 2 time units. As $a \in L(a^+)$, then

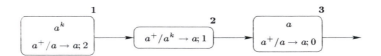

Fig. 7.2 The spiking P system associated with Example 7.1

Table 7.1 Table associated with Example 7.1

time unit	neuron 1	neuron 2	neuron 3
1	a^2 fires	ϕ active	a fires and spikes
2	a idle	ϕ active	ϕ active
3	a idle	ϕ active	ϕ active
4	a spikes and fires	a active	ϕ active
5	ϕ idle	a active	ϕ active
6	ϕ idle	a active	ϕ active
7	ϕ spikes	a^2 fires	ϕ active
8	ϕ active	ϕ idle	ϕ active
9	ϕ active	ϕ spikes	a fires and spikes
10	ϕ active	ϕ active	ϕ active

neuron 3 fires and spikes (because its spiking rule has 0 as delay). This means that no spike remains in this neuron.

In time units 2 and 3 nothing happens (except the time units are increasing). In time unit 4 neuron 1 spikes (the spikes fired in time unit 1) and again fires rule $a^+/a \to a; 2$. As a consequence of this neuron 1 becomes idle for 2 time units and one spike passes to neuron 2. Notice how the number of spikes is decreased in the spiking neuron when it fires, while it is increased in the receiving neurons when the spiking occurs.

When only one spike is present in neuron 2 the rule present in it cannot be applied. In time units 5 and 6 nothing happens. In time unit 7 neuron 1 spikes again and another spike passes to neuron 2. As a consequence of this in the same time unit neuron 2 spikes.

In time unit 8 nothing happens. In time unit 9 neuron 2 fires, neuron 3 receives the resulting spike and it fires and spikes.

As only the first two spikes of neuron 3 are considered, then the number computed is $9 - 1 = 8$.

After the following time unit the system halts. We can say that $N_2^h(\Pi_1) \supseteq \{3k + 2 | k \in \mathbb{N}_+\}$ and that $N_2^h(\Pi_1) \in \mathbb{N} \, \mathsf{SN}_2^h \mathsf{P}_3(rule_{1+}, cons_k, gen_1, dlay_1, forgo_0,$ $outd_1, k)$. The sequence of configurations of Π_1 for $k = 2$ is:
$(a^2, 0, 0, \phi, 0, 0, a, 0, 0) \Rightarrow (a, 2, 1, \phi, 0, 0, \phi, 0, 0) \Rightarrow (a, 1, 1, \phi, 0, 0, \phi, 0, 0) \Rightarrow$
$(a, 0, 0, a, 0, 0, \phi, 0, 0) \Rightarrow (\phi, 2, 1, a, 0, 0, \phi, 0, 0) \Rightarrow (\phi, 1, 1, a, 0, 0, \phi, 0, 0) \Rightarrow$
$(\phi, 0, 0, a^2, 0, 0, \phi, 0, 0) \Rightarrow (\phi, 0, 0, \phi, 1, 1, \phi, 0, 0) \Rightarrow (\phi, 0, 0, \phi, 0, 0, a, 0, 0) \Rightarrow$
$(\phi, 0, 0, \phi, 0, 0, \phi, 0, 0)$.

Notice that, since in the initial configuration neuron 1 has k spikes, then Π_1 halts and $N_2^h(\Pi_1) = N_2^h(\Pi_1)$. This would not be the case if in the initial configuration neuron 1 had more than k spikes. In this case Π_1 would not halt after $3k + 3$ configuration. $\diamond$

Example 7.2 A non-extended spiking P system Π_2 of degree 3 generating positive even numbers.

We do not give a formal definition of this system; it is depicted in Fig. 7.3.

As Π_2 has only rules of the kind $E/a^r \rightarrow a; d$ and $a^t \rightarrow \epsilon$, then it is non-extended. Moreover, the rules in neuron 3 make Π_2 non-deterministic. The system computes according to the following. Neuron 5, the output neuron, fires in the first time unit, then either one or two spikes can arrive in neuron 5 in the same time units. This number of spikes depends on the non-determinist application of the rules in neuron 3. If one spike arrives into neuron 5, then it fires and spikes (and the system halts after a few more transitions); if two spikes arrive, then the forgetting rule $a^2 \rightarrow \epsilon$ is applied and the system goes on computing. The fact that Π_2 generates positive even numbers depends on the fact that spikes can arrive in neuron 2 only in alternate time units.

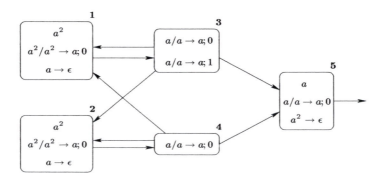

Fig. 7.3 The spiking P system associated with Example 7.2

In the initial configuration only neurons 1, 2 and 5 contain spikes. In this configuration these neurons fire and spike. The spikes generated by neurons 1 and 2 pass to neurons 3 and 4, respectively. Neuron 4 then fires and spikes in the same time unit; neuron 3 fires but it can either spike in the same time unit or after a delay of 1 time unit (this depends on the spiking rule that fired). If the two neurons fire and spike in the same time unit, then:

Two spikes arrive at neuron 5. In the next time unit these spikes are removed by the application of the forgetting rule $a^3 \to \epsilon$.

Two spikes arrive at each of neurons 1 and 2. In the next time unit these three neurons fire and spike.

The configuration is similar to the initial one and, with the exception of the firing of neuron 5, the sequence of transition (cycle) can be repeated. Notice that this cycle requires 2 time units.

If instead neuron 3 remains idle for 1 time unit, then:

One spike arrives at neuron 5. In the next time unit this neuron fires and spikes.

One spike arrives at neurons 1 and 2. In the next time unit these neurons remove the spike, then another spike arrives (from neuron 3) and it is removed, too.

Neuron 5 receives another spike (from neuron 3) and then it fires and spikes.

When this happens the system halts.

It should be clear that the interval between the first and the second spiking of neuron 5 is an even number. We can say that $N_2^h(\Pi_2) \supseteq \{2n \mid n \in \mathbb{N}_+\}$ and $N_2^h(\Pi_2) \in \mathbb{N} \, SN_2^h P_5(rule_2, cons_2, gen_1, dlay_1, for g_2, outd_3, 2)$.

One computation of Π_2 is indicated in Table 7.2 (where 'f & s' means 'fires and spikes'). ◇

Example 7.3 A non-extended spiking P system Π_3 of degree 4 generating all natural numbers.

We do not give a formal definition of this system; it is depicted in Fig. 7.4.

As Π_3 has only rules of the kind $E/a^r \to a; d$ and $a^t \to \epsilon$, then it is non-extended. Moreover, the rules in neuron 3 make Π_3 non-deterministic. The system computes according to the following. Neuron 4, the output neuron, fires in the first time unit, then either two or three spikes can arrive in this neuron in the same time units. This number of spikes depends on the non-determinist application of the rules in neuron 3. If two spikes arrive in neuron 4, then it fires again and then Π_3 halts; if three spikes arrive, then the forgetting rule $a^3 \to \epsilon$ is applied in neuron 4 and the system goes on computing.

The fact that Π_3 generates natural numbers depends on the fact that either two or three spikes can arrive in neuron 4 in any time unit. As long as the rules $a^2/a^2 \to a; 0$ are applied in neurons 1, 2 and 3, then three spikes arrive in

Table 7.2 Table associated with Example 7.2, 'f & s' means 'fires and spikes'

time unit	neuron 1	neuron 2	neuron 3	neuron 4	neuron 5
1	a^2 f & s	a^2 f & s	ϕ active	ϕ active	a f & s
2	ϕ active	ϕ active	a f & s	a f & s	ϕ active
3	a^2 f & s	a^2 f & s	ϕ active	ϕ active	a^2 forgets
4	ϕ active	ϕ active	a fires	a f & s	ϕ active
5	a forgets	a forgets	ϕ idle	ϕ active	a f & s
6	ϕ active	ϕ active	ϕ spikes	ϕ active	ϕ active
7	a forgets	a forgets	ϕ active	ϕ active	a f & s
8	ϕ active	ϕ active	ϕ active	ϕ active	ϕ active

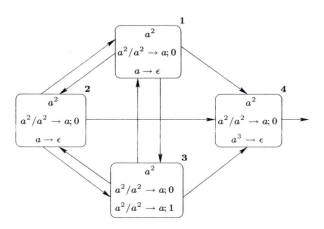

Fig. 7.4 The spiking P system associated with Example 7.3

neuron 4. If, instead, in a configuration neuron 3 fires rule $a^2/a^2 \to a; 1$, then
neuron 4 receives two spikes while neurons 1 and 2 receive one spike. These last
two neurons then apply the rule $a \to \epsilon$ for two consecutive configurations and
the system halts.

We can say that $N_2^h(\Pi_3) \supseteq \mathbb{N}_+$ and $N_2^h(\Pi_3) \in \mathbb{N} \, \mathsf{SN}_2^h\mathsf{P}_4(rule_2, cons_2, gen_1,$ $dlay_1, forg_3, outd_3, 3)$. $\diamond$

Example 7.4 A module Π_4 for a non-extended spiking P system simulating instructions of the kind (s, γ^-, v, w).

These kinds of instructions, studied in Section 4.2, are those used by program machines. We do not give a formal definition for the spiking P system Π_4; it is depicted in Fig. 7.5.

As Π_4 has only rules of the kind $E/a^r \to a; d$ it is non-extended. Neurons n_s, n_v and n_w are associated with the states s, v and w, respectively. Neuron n_γ represents a counter γ in the program machine and we assume that it has either none or an even number of spikes. If n_γ contains an even number of spikes, then no rule is applied in this neuron.

Let us assume that n_γ has no spikes and that n_s contains one spike. In this configuration n_γ cannot fire, while n_s does fire and spike sending one spike to n_γ, n_1 and n_2. All these three neurons fire but only n_1 spikes in the same time unit. In the following time unit n_3 has a spike (coming from n_1) and n_2 and n_γ spike. Neuron n_3 fires and spikes (sending a spike to n_5) and receives another spike (from n_γ), while n_4 receives two spikes (one from n_γ and the other from n_2). In this configuration, n_3 and n_5 both have one spike, while n_4 has two spikes. Neuron n_4 does not spike (and retains the two spikes), similarly for neuron n_5 (and retains one spike). Neuron n_3 fires and spikes so that n_5 can spike in the

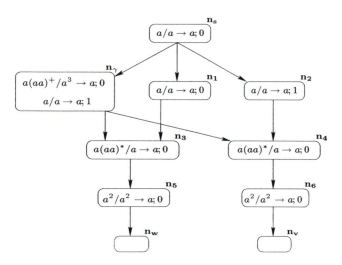

Fig. 7.5 A module for a spiking P system simulating instructions of the kind (s, γ^-, v, w)

next time unit. When this happens a spike is present in n_w. This is the simulation of (s, γ^-, v, w) when $val(\gamma) = 0$.

Let us assume now that n_γ has an even number of spikes and that n_s contains one spike. Similarly as before, in this configuration n_γ cannot fire, while n_s does fire and spike sending one spike to n_γ, n_1 and n_2. All these three neurons fire but only n_1 and n_γ spike in the same time unit. As a consequence of this n_3 receives two spikes while n_4 receives one spike. Neuron n_3 does not fire the two just received spikes (but retains them). Neuron n_4 spikes and fires and receives one spike from n_2. In this configuration n_4 and n_6 both have one spike. Then n_4 fires and spikes and when n_6 receives the second spike fires and spikes, too. When this happens a spike is present in n_v. This is the simulation of (s, γ^-, v, w) when $val(\gamma) > 0$.

It should be clear that, if these sequences of transitions are repeated several times, neurons n_5 and n_6 accumulate an even number of spikes. These spikes do not interfere with the repeated simulation of (s, γ^-, v, w). $\diamond$

Example 7.5 A non-extended spiking P system Π_5 of degree 6 generating $2i + 1$ for $i \in \mathbb{N}_+$.

We do not give a definition of this system; it is depicted in Fig. 7.6.

As Π_2 has only rules of the kind $E/a^r \to a; d$ and $a^t \to \epsilon$, then it is non-extended. Moreover, the rules in neuron 2 make Π_5 non-deterministic. The system computes according to the following. In the initial configuration there are

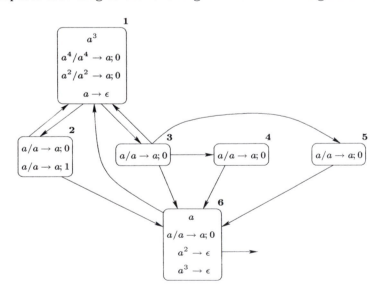

Fig. 7.6 The spiking P system associated with Example 7.5

three spikes in neuron 1 and one spike in neuron 6, the output neuron. In this configuration only the output neuron fires and spikes. This spike passes to neuron 1 so that the rule $a^4/a^4 \rightarrow a; 0$ is applied here. The generated spikes pass to neurons 2 and 3.

Neuron 3 applies the rule present in it, while if the rule $a/a \rightarrow a; 0$ is applied in neuron 2, then in the next time unit neuron 6 applies $a^2 \rightarrow \epsilon$ and neuron 1 fires rule $a^2/a^2 \rightarrow a; 0$ and spikes again. The spike of neuron 3 also passes to neurons 4 and 5 and their subsequent spikes pass to neuron 6 that applies again the rule $a^2 \rightarrow \epsilon$. This sequence of configurations requires an even number of time units.

If instead when neuron 3 applies the rule present in it neuron 4 applies $a/a \rightarrow a; 1$, then neuron 1 receives only one spike and it applies $a \rightarrow \epsilon$. In the same time unit neuron 6 fires and spikes. This spike reaches neuron 1 and 6 in the same time unit in which the spike from neuron 3 reaches neuron 1 and the spikes from neuron 4 and 5 reach neuron 6. While neuron 1 fires and spikes (applying the rule $a^2/a^2 \rightarrow a; 0$), neuron 6 applies the rule $a^3 \rightarrow \epsilon$. Neuron 1 has two spikes as in the configuration following the initial one; the process can be repeated.

It should be clear that Π_5 never halts and that it generates $2i+1$ for $i \in \mathbb{N}_+$. We can than say that $N_\alpha^\beta(\Pi_5) = \mathbb{N} \, \mathsf{SN}_\alpha^\beta \mathsf{P}_6(rule_3, cons_4, gen_1, dlay_1, forg_3, outd_2, 4)$ for $\alpha \in \{\omega, all\} \cup \{k \mid k \geq 2\}$ and for $\beta = a$ (but not for the halting case) or β is omitted. $\diamond$

7.4 Spiking P systems and P/T systems

In this section we show how P/T systems can be simulated by spiking P systems. Our results in this respect are far from being exhaustive. In Section 7.8 we suggest how such investigations can be brought forward (see Suggestion for research 7.3).

Lemma 7.1 *The building blocks* fork *and* join *can be simulated by spiking P systems.*

Proof Given a building block we can define a spiking P system simulating it. In this simulation tokens are represented by spikes, places are represented by neurons and transitions are represented by spiking and forgetting rules.

The simulation of a *fork* is straightforward: it can be performed by the spiking P systems depicted in Fig. 7.7. When a spike is present in neuron 1, then it fires and spikes allowing a spike to pass to neurons 2 and 3.

The simulation of a *join* is slightly more complex. We can consider, for instance, the spiking P system depicted in Fig. 7.8.

Here the neurons 1 and 2 represent $^\bullet t$, where t is the transition present in the *join*, while neuron 3 represents $t^\bullet$. The transition in a *join* fires only if both

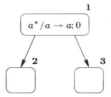

Fig. 7.7 A spiking P system simulating a *fork*

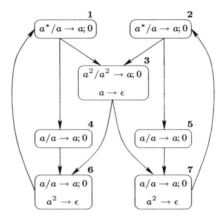

Fig. 7.8 A spiking P system simulating a *join*

places in $\bullet t$ have at least one token. The spiking P system depicted in Fig. 7.8 is such that a spike is generated by neuron 3 only if neurons 1 and 2 have at least one spike in the same configuration. If this happens, then neurons 1 and 2 fire and spike, so that neuron 3 receives two spikes while neuron 4 and 5 receive one spike each. When this happens neurons 3, 4 and 5 spike. In this way neurons 6 and 7 receive two spikes each and they apply the forgetting rule $a^2 \to \epsilon$.

If instead in one configuration neuron 1 (2) spikes, then neuron 3 and 4 (5) receive one spike. Neuron 3 then applies the forgetting rule $a \to \epsilon$ while neuron 4 (5) spikes. This spike passes to neuron 6 (7) and also this neuron spikes and its spike passes to neuron 1 (2).   □

We could now go on as in Section 6.4 showing that spiking P system can simulate the 0-test and then deducing some results on the base of those present in Section 4.4. We do not do this for two reasons. Firstly because we would strictly follow what has already been done in Section 6.4. Secondly because this work would result in spiking P systems more complex than those obtained by direct proofs.

7.5   Finite and semilinear sets of numbers

In this section we consider only non-extended spiking P systems.

Theorem 7.1 $\mathbb{N} \, \mathsf{FIN} = \mathbb{N} \, \mathsf{SN}_2^h\mathsf{P}_1(rule_{*+}, cons_*, gen_1, dlay_*, forg_*, outd_1, *)$
$= \mathbb{N} \, \mathsf{SN}_2^h\mathsf{P}_2(rule_*, cons_*, gen_1, dlay_*, forg_*, outd_1, *)$.

Proof   It is clear that non-extended spiking P systems with only one neuron can generate finite languages. This is due to the fact that the number of spikes present in the neuron can only decrease.

Let $F = \{f_1, \ldots, f_{|F|}\}$ be a finite set of numbers with $f_i \le f_{i+1}$, $1 \le i \le |F| - 1$. Moreover, let Π be a spiking P system and $N_2^h(\Pi) \in \mathbb{N} \, \mathsf{SN}_2^h\mathsf{P}_1(rule_{*+}, cons_1,$ $gen_1, dlay_*, forg_0, outd_1, *)$. The only (output) neuron of Π contains only the rules: $a^2/a \to a; 0$ and $a/a \to a; j$, $j \in \{1, \ldots, |F|\}$. In the initial configuration only two spikes are present in the neuron of Π. In this configuration only the rule $a^2/a \to a; 0$ fires, and then any of the rules $a/a \to a; j$ can fire. In this way the number of time units between the first and the second spike is j, $j \in F$. We can then say that $N_2(\Pi)$ generates F.

Let us consider now spiking P system with two neurons. One of the two neurons is the output one that can spike only twice. This means that the (finite) number of spikes initially present in the other neuron can only be increased by two. So, all computations have a finite number of configurations. $\square$

Knowing that a semilinear set is a finite union of linear sets, in the following we prove

Theorem 7.2 $\mathbb{N} \, \mathsf{SN}_2^h\mathsf{P}_*(rule_3, cons_3, gen_1, dlay_*, forg_3, outd_3, 3) = \mathsf{SLS}_1$.

Lemma 7.2 $\mathbb{N} \, \mathsf{SN}_2^h\mathsf{P}_*(rule_*, cons_*, gen_1, dlay_*, forg_*, outd_*, *) \subseteq \mathsf{SLS}_1$.

Proof   Let Π be a spiking P system with $N_2^h(\Pi) \in \mathbb{N} \, \mathsf{SN}_2^h\mathsf{P}_*(rule_*, cons_*, gen_1,$ $dlay_*, forg_*, outd_*, *)$. The system Π is such that the number of neurons and the maximum number of spikes present in each neuron are bounded. Moreover the number of rules and their delays present in each neuron are known. This means that the set of configurations $\mathbb{C}_\Pi$ of Π is finite.

We then define the right-linear regular grammar $G = (N, \{a\}, (c_0, 0), P)$ with $N = \mathbb{C}_\Pi \times \{0, 1, 2\}$ and P is such that:

$(c, 0) \to (c', 0)$, $(c, 1) \to a(c', 1) \in P$ for $c, c' \in \mathbb{C}_\Pi$, such that $c \Rightarrow c'$ is a transition of Π in which the output neuron does not spike;

$(c, 0) \to (c', 1)$, $(c, 1) \to a(c', 2) \in P$ for $c, c' \in \mathbb{C}_\Pi$, such that $c \Rightarrow c'$ is a transition of Π in which the output neuron spikes;

$(c, 2) \to \epsilon \in P$ for $c \in \mathbb{C}_\Pi$ if there is a halting computation $c \Rightarrow^* c'$ of Π during which the output neuron never spikes, or if there is an infinite computation starting in c during which the output neuron never spikes;

no other production belongs to P.

From Theorem 3.1 and knowing that $\mathbb{N}\,\mathsf{REG} = \mathbb{N}\,\mathsf{CF}$ we have that $N_2^h(\Pi)$ is the length set of the regular language $L(G)$.   □

In the following we prove that spiking P systems generating languages in $\mathbb{N}\,\mathsf{SN}_2^h\mathsf{P}_*(rule_*, cons_*, gen_1, dlay_*, forg_*, outd_*, *)$ can generate finite unions of linear sets, that is, semilinear sets.

We start by saying that, as a consequence of Theorem 7.1, finite sets with only one elements can be generated by systems generating languages in $\mathbb{N}\mathsf{SN}_2^h\mathsf{P}_*(rule_*, cons_*, gen_1, dlay_*, forg_*, outd_1, *)$.

Lemma 7.3 $\{ni\,|\,i \geq 1\} \subseteq \mathbb{N}\,\mathsf{SN}_2^h\mathsf{P}_{n+2}(rule_3, cons_3, gen_1, dlay_1, forg_2, outd_2, 3)$, $n \geq 3$.

Proof Given n we consider the spiking P system depicted in Fig. 7.9.

The output neuron $n+2$ is the only neuron that fires and spikes in the initial configuration. Its spikes pass to neuron 1 and from there to neurons $n-1$ and $n+1$ (through neurons $2, \ldots, n-2$). If neurons $n-1$ and $n+1$ spike in different time units (because neuron $n+1$ fires the rule $a/a \to a; 1$), then neuron $n+2$ does not spike and the cycle through neurons $2, \ldots, n-2$ is repeated. If instead neurons $n-1$ and $n+1$ spike in the same time unit (because neuron $n+1$ fires the rule $a/a \to a; 0$), then neuron $n+2$ spikes again. This can happen only after ni, $i \geq 1$ time units.

The second spike of neuron $n+2$ arrives in neuron 1 together with the spike from neuron n. The application of $a^2 \to \epsilon$ in neuron 1 allows the spiking P system to halt.   □

Lemma 7.4 If $A \in \mathbb{N}\,\mathsf{SN}_2^h\mathsf{P}_m(rule_{p_1}, cons_{p_2}, gen_{p_3}, dlay_{p_4}, forg_{p_5}, outd_{p_6}, p_7)$, then $\{x + r \mid x \in A\} \in \mathbb{N}\,\mathsf{SN}_2^h\mathsf{P}_{m+1}(rule_{p_1'}, cons_{p_2'}, gen_{p_3}, dlay_{p_4'}, forg_{p_5'}, outd_{p_6},$

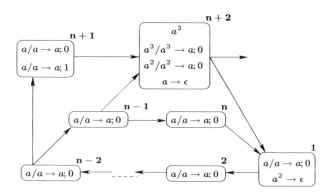

Fig. 7.9 The spiking P system associated with Lemma 7.3

p_7') *for* $r \geq 1$, $p_1' = max(\{p_1, 2\})$, $p_2' = max(\{p_2, 3\})$, $p_4' = max(\{p_4, r\})$, $p_5' = max(\{p_5, 0\})$ *and* $p_7' = max(\{p_7, 3\})$.

Proof Let Π be the spiking P system generating A and let n_{out} be the output neuron of Π. Starting from Π we define another spiking P system Π' having one more neuron n_{out}' having two spikes in the initial configuration and with the rules $a^3/a^3 \rightarrow a; 0$ and $a/a \rightarrow a; r$. The neuron n_{out}' is the output neuron of Π' and there is a synapse from n_{out} to n_{out}'.

If n_{out} in Π spikes in the time units t_1 and t_2 computing the number $x = t_2 - t_1$, then n_{out}' in Π' spikes first in the time units $t_1 + 1$ applying $a^3/a^3 \rightarrow a; 0$ and then in time unit $t_2 + 1 + r$ applying $a/a \rightarrow a; r$. In this way Π' computes $x + r$. □

Lemma 7.5 *For every spiking P system Π there is an equivalent system Π' having at most one spike in its initial configuration and such that $N_2^h(\Pi) = N_2^h(\Pi')$.*

Proof Let us assume that in the initial configuration of a spiking P system $\Pi = (V, \mu, L_1, \ldots, L_m, R_1, \ldots, R_m, comp), \mu = (Q, E)$, of degree m the maximum number of spikes present in a neuron is $z > 1$ and that $n_1, \ldots, n_k$ are the neurons with at least one spike in the initial configuration. We define a spiking P system $\Pi' = (V, \mu', L_1', \ldots, L_m', L_{m+1}', \ldots, L_{m+z+1}', R_1, \ldots, R_m, R_{m+1}', \ldots, R_{m+z+1}', comp), \mu' = (Q \cup Q', E \cup E')$ such that in the initial configuration the only neuron with spikes is $m + 1$ and it has only one spike.

In Π' the neuron $m+1$ has synapses going to the neurons $m+2, \ldots, m+z+1$ and these last neurons have synapses going to the neurons $n_1, \ldots, n_k$. It is not important for this proof to indicate which of the $m + 2, \ldots, m + z + 1$ neurons have synapses with which of the $n_1, \ldots, n_k$ neurons. What is important is that neuron n_i has $L(n_i)$ incoming synapses from $L(n_i)$ of the $m + 2, \ldots, m + z + 1$ neurons, $1 \leq i \leq k$. This is always possible as $L(n_i)$ is at most z.

The neurons $m + 1, \ldots, m + z + 1$ have only the rule $a/a \rightarrow a; 0$. This means that in the initial configuration of Π' neuron $m + 1$ fires so that one spike passes to each neuron $m + 2, \ldots, m + z + 1$. In the second time unit these neurons fire so that spikes pass to the neurons $n_1, \ldots, n_k$. The system Π' goes on computing as Π would. □

Lemma 7.6 *If $A_1, A_2 \in \mathbb{N} \, SN_2^h P_m(rule_{p_1}, cons_{p_2}, gen_{p_3}, dlay_{p_4}, forg_{p_5}, outd_{p_6}, p_7)$, then $A_1 \cup A_2 \in \mathbb{N} \, SN_2^h P_{2m+6}(rule_{p_1'}, cons_{p_2'}, gen_{p_3}, dlay_{p_4'}, forg_{p_5'}, outd_{p_6'}, p_7')$ for $p_1' = max(\{p_1, 2\})$, $p_2' = max(\{p_2, 2\})$, $p_4' = max(\{p_4, 1\})$, $p_5' = max(\{p_5, 0\})$, $p_6' = max(\{p_6, 3\})$ and $p_7' = max(\{p_7, 2\})$.*

Proof Let Π' and Π'' be spiking P systems such that $N_2^h(\Pi'), N_2^h(\Pi'') \in \mathbb{N} \, SN_2^h P_m(rule_{p_1}, cons_{p_2}, gen_{p_3}, dlay_{p_4}, forg_{p_5}, outd_{p_6}, p_7)$. Moreover, Π' and Π''

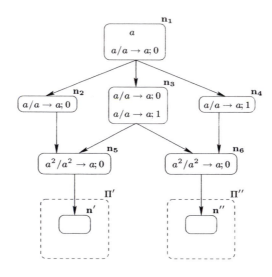

Fig. 7.10 The spiking P system associated with Lemma 7.6

have at most one spike in their initial configurations (see Lemma 7.5). Let n' and n'' be the respective neurons having this one spike. We define a spiking P system Π such as that depicted in Fig. 7.10.

If we consider Fig. 7.10, then in the initial configuration only neuron n_1 has a spike (neurons n' and n'' do not have spikes in the initial configuration of Π). The non-deterministic firing of either $a/a \to a; 0$ or $a/a \to a; 1$ in n_3 determines if either the subsystem Π_1 or Π_2 (suggested by a dashed rectangular in Fig. 7.10) is simulated by Π. $\qquad\square$

Finally, considering the systems in Example 7.2 and Example 7.3 we can say that Theorem 7.2 holds. The closure of $\mathbb{N} \, SN_2^h P_*(rule_3, cons_3, gen_1, dlay_*, forg_3, outd_3, 3)$ under finite union, complementation and intersection and the non-closure under product follow from this theorem.

Following similar proof techniques it is possible to obtain

Theorem 7.3 $\mathbb{N} \, SN_\alpha^\beta P_*(rule_3, cons_1, gen_*, dlay_3, forg_4, outd_*) = SLS_1$ *with* $\alpha \in \{\omega, all\} \cup \{k\underline{k} \mid k \geq 2\}$ *and* $\beta \in \{h, a, ha\}$ *or* β *is omitted.*

7.6 Recursive enumerable sets of numbers

Theorem 7.4 $\mathbb{N} \, SN_2^h P_*(rule_2, cons_3, gen_1, dlay_1, forg_0, outd_*, *) = \mathbb{N} \, RE.$

Proof We define a spiking P system Π simulating a generating program machine $M = (S, I, s_1, s_f)$ having all counters empty in the initial configuration

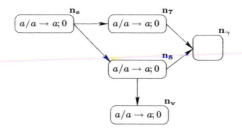

Fig. 7.11 A module for the spiking P system associated with Theorem 7.4 simulating instructions of the kind (s, γ^+, v)

and halting with all counters empty except the output counter γ_{out} and such that this counter is only incremented.

The system Π has *state-neurons* n_i for each $i \in S$ and the presence of a spike in one of these neurons simulates M being in the relative state. Considering how Π is defined, it is not possible for one state-neuron to have more than one spike and for more than one state-neuron to have one spike in the same configuration. Moreover, the system Π has *counter-neurons* n_γ, one for each counter γ of M. The simulation of addition of 1 to counter γ is performed passing two spikes to n_γ; the simulation of subtraction of 1 from counter γ is performed removing two spikes from n_γ.

We consider the program machine M to have instructions of the kind (s, γ^+, v) and (s, γ^-, v, w). The simulation of instructions of the kind (s, γ^+, v) is performed by the module depicted in Fig. 7.11.

When a spike is present in n_s, then it fires and spikes. The resulting spikes are sent to neurons n_7 and n_8. These two neurons fire and spike so that n_γ receives two spikes and n_v receives one spike.

The simulation of instructions of the kind (s, γ^-, v, w) is performed by the module depicted in Fig. 7.5. The dynamics of this module is explained in Example 7.4.

As the counter-neurons are shared between the different modules we have to explain what happens if more than one counter-neuron is shared between different modules simulating instructions of the kind (s, γ^-, v, w). Let us assume, for instance, that the instructions (s, γ^-, v, w) and (s', γ^-, v', w') are in I. Moreover, let as assume that the neurons in the module simulating (s, γ^-, v, w) are $n_s, n_\gamma, n_1, n_2, n_3, n_4, n_5, n_6, n_v$ and n_w (as in Fig. 7.5) while the neurons in the module simulating (s', γ^-, v', w') are $n_{s'}, n_\gamma, n'_1, n'_2, n'_3, n'_4, n'_5, n'_6, n_{v'}$ and $n_{w'}$ (as in Fig. 7.5 but some neurons have a $'$).

If n_s spikes, then after one transitions the neurons n_3, n_4, n'_3 and n'_4 get a spike. This results in either n_v and $n_{w'}$ or n_w and $n_{v'}$ having a spike (not a

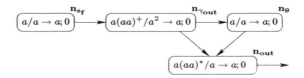

Fig. 7.12 A module for the spiking P system associated with Theorem 7.4 halting the computation

simulation of (s, γ^-, v, w). This can be avoided if there are synapses from n_1 to n_3', from n_1' to n_3, from n_2 to n_4' and from n_2' to n_4.

The system Π renders an output when the neuron n_{s_f}, associated with the final state of M, has a spike. We consider then the module depicted in Fig. 7.12.

Neuron n_{s_f} fires and spikes so that neuron $n_{\gamma_{\text{out}}}$ contains an odd number of spikes. This causes this last neuron to spike in every time unit removing each time two spikes. The spikes generated by neuron $n_{\gamma_{\text{out}}}$ are sent to neurons n_9 and n_{out}. Neuron n_{out} spikes one time unit after neuron $n_{\gamma_{\text{out}}}$ spiked for the first time and it spike a second time one time unit after neuron $n_{\gamma_{\text{out}}}$ spiked for the last time. These are the two time units when neuron n_{out} contains an odd number of spikes.

Table 7.3 indicates what happens in the module depicted in Fig. 7.12 when neuron n_{s_f} has one spike and neuron $n_{\gamma_{\text{out}}}$ has eight spikes in time unit t.

In the example considered in this table neuron n_{out} spikes in time units $t+2$ and $t+6$, so $t+6-t-2 = 4$, half the number of spikes present in time unit t in neuron $n_{\gamma_{\text{out}}}$.

We can say that Π simulated M and that $N(\Pi) \in \mathbb{N} \ \mathsf{SN}_2^h\mathsf{P}_*(rule_2, cons_3, gen_1, dlay_1, forg_0, outd_*, *)$. $\square$

With a few changes in the modules simulating instructions of the kind (s, γ^-, v, w), (s, γ^+, v) and halting the computation it is possible to have computational complete spiking P systems with no delays (but with forgetting rules)

Corollary 7.1 $\mathbb{N} \ \mathsf{SN}_2^h\mathsf{P}_*(rule_2, cons_3, gen_1, dlay_0, forg_2, outd_*, *) = \mathbb{N} \ \mathsf{RE}$.

Proof Similar to the proof of Theorem 7.4 with state-neurons containing two spikes and with the module depicted in Fig. 7.13 for the simulation of instructions of the kind (s, γ^-, v, w), the module depicted in Fig. 7.14 for the simulation of instructions of the kind (s, γ^+, v) and the module depicted in Fig. 7.15 for halting the computation. $\square$

Table 7.3 Table associated with the module depicted in Fig. 7.12. 'f & s' means 'fires and spikes'

time unit	n_{s_f}	$n_{\gamma_{out}}$	n_9	n_{out}
t	1 f & s	8 active	ϕ active	ϕ active
t+1	ϕ active	9 f & s	ϕ active	ϕ active
t+2	ϕ active	7 f & s	1 f & s	1 f & s
t+3	ϕ active	5 f & s	1 f & s	2 active
t+4	ϕ active	3 f & s	1 f & s	4 active
t+5	ϕ active	1 active	1 f & s	6 active
t+6	ϕ active	1 active	ϕ active	7 f & s
t+7	ϕ active	1 active	ϕ active	6 active

It has also been proved that

Theorem 7.5 $\mathbb{N} \, SN_2^\beta P_*(rule_{2+}, cons_2, gen_1, dlay_7, forg_1, outd_2, *) = \mathbb{N} \, RE$ *with* $\beta = h$ *or* β *is omitted.*

The proof of this theorem is rather complex, so we decided not to include it.

Theorem 7.6 $\mathbb{N} \, SN_\omega^a P_*(rule_2, cons_3, gen_1, dlay_1, forg_0, outd_*, *) = \mathbb{N} \, SN_\omega^a P_* (rule_2, cons_3, gen_1, dlay_0, forg_2, outd_*, *) = \mathbb{N} \, RE.$

Proof Let us consider the spiking P system Π described in the proof of Theorem 7.4 and the sketch of a spiking P system Π' depicted in Fig. 7.16.

In this sketch Π is represented by a rectangle with a dashed line in which two neurons n_{s_0} and n_{out} are indicated. We recall that neuron n_{s_0} is the state-neuron associated with the initial state of the simulated program machine M, while n_{out} is the output neuron of Π (and Π'). Every time the output neuron spikes the spike passes to neuron z_1 of Π' so that when a second spike is received by z_1 the rule $a^2/a^2 \to a; 0$ is applied. The generated spike passes to n_{s_0} so that another simulation of the program machine starts.

If instead we start from the spiking P system described in the proof of Corollary 7.1, then it is possible to define a spiking P system Π'' having another

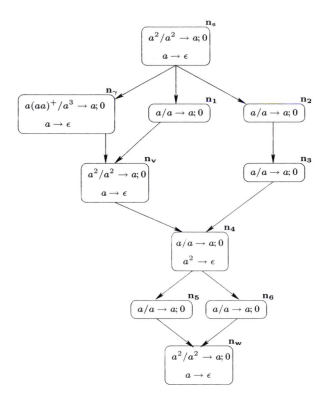

Fig. 7.13 A module for the spiking P system of Corollary 7.1 simulating instructions of the kind (s, γ^-, v, w)

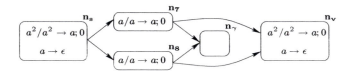

Fig. 7.14 A module for the spiking P system of Corollary 7.1 simulating instructions of the kind (s, γ^+, v)

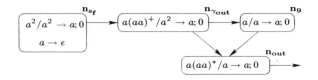

Fig. 7.15 A module for the spiking P system of Corollary 7.1 halting the computation

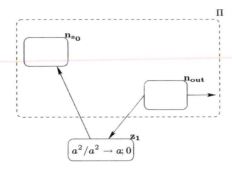

Fig. 7.16 Sketch of one spiking P system associated with Theorem 7.6

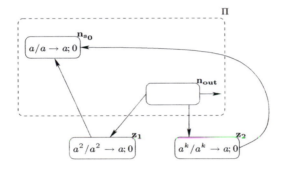

Fig. 7.17 Sketch of one spiking P system associated with Theorem 7.7

neuron as z_1 in Fig. 7.16. In this way two spikes would pass to n_{s_0} so that another simulation of the program machine starts.

It should be clear that the computation of Π' and Π'' is infinite. If $(t_1, t_2, t_3, t_4, \ldots)$ is the spike train of Π' or Π'', then only $t_{2k} - t_{2k-1}$ for $k \in \mathbb{N}_+$ represents the content of the output counter of M. The numbers $t_{2k+1} - t_{2k}$ indicate the number of time units elapsed from one simulation of M to another.

We can then say that the computations of Π' and Π'' are infinite and the alternating intervals between the spikes of n_{out} give $L(M)$. □

Theorem 7.7 $\mathrm{NSN}\alpha^{ha}\mathrm{P}_*(rule_2, cons_{p_2}, gen_1, dlay_1, forg_0, outd_*, *) = \mathrm{NSN}\,\alpha^{ha}$ $\mathrm{P}_*(rule_2, cons_{p_2}, gen_1, dlay_0, forg_2, outd_*, *) = \mathbb{N}\,\mathrm{RE}$ *for* $\alpha \in \{k, \underline{k} \mid k = 2n,\ n \in \mathbb{N}_+\}$ *and* $p_2 = max(\{k, 3\})$.

Proof Let us consider the spiking P system Π described in the proof of Theorem 7.4 and the sketch of a spiking P system Π' depicted in Fig. 7.17.

In this sketch Π is represented by a rectangle with a dashed line in which two neurons n_{s_0} and n_{out} are indicated. We recall that neuron n_{s_0} is the state-neuron associated with the initial state of the simulated program machine M, while n_{out} is the output neuron of Π (and Π'). Every time the output neuron spikes the spike passes to neuron z_1 and z_2 of Π'. The function of neuron z_1 is equal to that used in the proof of Theorem 7.6. Neuron z_2 does not spike until k spikes are present in it. When this happens it spikes and its spikes pass to neuron n_{s_0} together with the spike generated by neuron z_1. As the only rule present in neuron n_{s_0} is $a/a \to a; 0$, then this neuron no longer spikes.

If instead we start from the spiking P system described in the proof of Corollary 7.1, then it is possible to define a spiking P system Π'' having another neuron as z_2 in Fig. 7.17. In this way three spikes would pass to n_{s_0} so that this neuron no longer spikes.

We know that the output neurons of the spiking P systems described in the proofs of Theorem 7.4 and Corollary 7.1 spike twice for each simulation of M. Let us call these two spikes a *spiking pair*. If k was an odd number, then the neuron z_2 would spike in between a spiking pair and its spike would not halt the computation of Π'. This is why k has to be an even number.   □

Lemma 7.7 *Let Π be a spiking P system with $\mathbb{C}_\Pi$ its set of configurations and $c_i \in \mathbb{C}_\Pi, i \geq 0$. For each computation $C = c_0 \Rightarrow c_1 \ldots c_k, k \in \mathbb{N}_+$, computation of Π such that $st(C) = (t, t + n)$, $t, n \in \mathbb{N}_+$, is the spiking train of C there is a spiking P system Π' with $\mathbb{C}'_\Pi$ as set of configurations such that:*

(i) there are $c'_j \in \mathbb{C}'_\Pi$, $j \geq 0$, with $C' = c'_0 \Rightarrow c'_1 \ldots$ a computation of Π' and $st(C') = (t, t + (n + 1), t + 2(n + 1), \ldots)$;

(ii) for each $c'_j \in \mathbb{C}'_\Pi$, $j \geq 0$, such that $C' = c'_0 \Rightarrow c'_1 \Rightarrow \ldots$ is a computation of Π' the vector $st(C')$ is either empty or $st(C') = (t, t + (n+1), t + 2(n+1), \ldots)$ for $st(C) = (t, t + n)$, $t, n \in \mathbb{N}_+$, a spiking train of a computation C of Π.

Proof The system Π', depicted in Fig. 7.18, is an extension of Π. This figure can be logically divided into five regions: the system Π, with the indication of its output neuron *comp*, the modules *Initialise*, *From n_5 to n_{12}* and *From n_{12} to n_5* and *comp'* the final compartment of Π'. The system Π is represented by a dashed rectangle, while the neurons composing the three modules are enclosed in a dotted rectangle. The neurons in the three modules of Π' and neuron *comp'* do not have any spike in the initial configuration.

The module *Initialise* puts $2n + 1$ spikes in neuron n_5 only if neuron *comp* spikes at the time units t and $t + n$ for $t, n \in \mathbb{N}_+$. If neuron *comp* spikes in time unit t, then its spike passes to neurons n_1, n_2 and n_4. The neurons n_1 and n_2 fire and spike and their spikes pass to neuron n_5 and to each other. For this reason the neurons n_1 and n_2 fire and spike in the next time unit. This process, seeing neuron n_5 accumulating an even numbers of spikes, terminates only when neuron *comp* spikes for the second time.

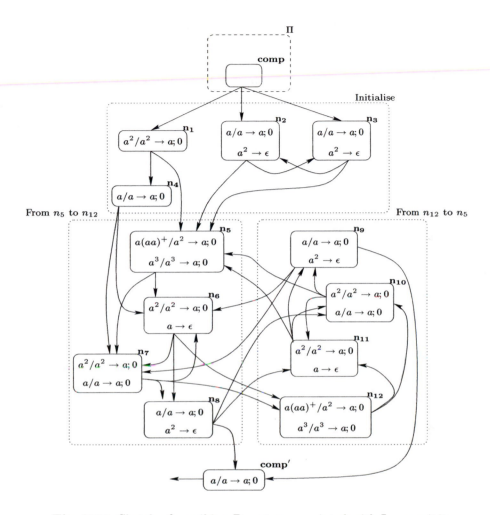

Fig. 7.18 Sketch of a spiking P system associated with Lemma 7.7

When this occurs the neurons n_1, n_2 and n_3 get a token. These last two neurons apply the forgetting rule $a^2 \to \epsilon$, while neuron n_1 fires and spikes and its spike passes to neurons n_4 and n_5. It should be clear that if the second spike of *comp* occurred in the time unit $t+n$ then, when the spike generated by neuron n_1 passes to n_5, neuron n_5 has $2n+1$ spikes.

When neuron n_5 has an odd number of spikes, then the neurons in the module *From n_5 to n_{12}* start to fire. This module allows all the spikes in neuron n_5 to pass to neuron n_{12}. When this is completed, then neuron *comp'*, the output neuron of Π', spikes.

Neurons n_4 and n_5 spike in the same time unit (neuron n_5 fires rule $a(aa)^+ / a^2 \to a; 0$) so that neuron n_6 and n_7 get two spikes each. When this happens both neurons n_6 and n_7 fire and spike so that neurons n_{12} and n_8 get two spikes each. When only two spikes are present in neuron n_8, then the rule $a^2 \to \epsilon$ is applied. This process is repeated $2(n-1)$ times.

When neuron n_5 contains only three spikes, then the rule $a^3/a^3 \to a; 0$ is applied in it. Two more spikes pass to neuron n_{12}, but in the next time unit neuron n_7 fires and spikes applying rule $a/a \to a; 0$ while neuron n_6 applies the forgetting rule $a \to \epsilon$. When this happens neuron n_{12} contains $2n+1$ spikes while n_8 contains one spike. This last neuron fires and spikes and its spike passes to neurons n_{10}, n_{11} and *comp'*.

When *comp'* receives one spike, it fires and spikes.

The module *From n_{12} to n_5* is similar to the module *From n_5 to n_{12}*. This module allows all the spikes in neuron n_{12} to pass to neuron n_5. When this is completed, then neuron *comp'* spikes.

The computation of Π' never halts and $n+1$ time units elapse between two consecutive spikes of *comp'*. The lemma follows.   □

Theorem 7.8 $\mathbb{N} \, \text{SN}_\alpha^\beta \text{P}_*(rule_2, cons_3, gen_1, dlay_1, forg_2, outd_*, *) = \mathbb{N} \, \text{RE}$ *with* $\alpha \in \{\omega\} \cup \{k \mid k \geq 2\}$ *and* $\beta = a$ *or* β *is omitted.*

Proof Let $A \in \mathbb{N} \, \text{RE}$ and let $A' = \{n - 1 \mid n \in A\}$. Clearly $A' \in \mathbb{N} \, \text{RE}$. From Corollary 7.1 we know that there is a spiking P system Π such that $N_2^h(\Pi) = A'$. According to Lemma 7.7 there is a spiking P system Π' such that $N_\omega(\Pi') = N_k(\Pi') = \{n + 1 \mid n \in N_2^h(\Pi)\} = A \setminus \{1\}$ for $k \geq 2$.

If $1 \neq A$, then the theorem holds. Otherwise we consider the spiking P system Π_1 depicted in Fig. 7.19.

This system is such that the output neuron, n_2, spikes in each time unit. So $N_\omega(\Pi_1) = N_k(\Pi_1) = \{1\}$. Moreover, from the proof of Lemma 7.6, we can define a spiking P system Π'', having a delay of 1, given by the union of Π' and Π_1 and such that $N_\omega(\Pi'') = N_k(\Pi'') = A$. The theorem follows.   □

Combining the ideas at the base of the proofs of Theorem 7.6 and Lemma 7.7 we obtain

Theorem 7.9 $\mathbb{N} \, \text{SN}_\alpha^h \text{P}_*(rule_2, cons_{p_2}, gen_1, dlay_1, forg_2, outd_*, *) = \mathbb{N} \, \text{RE}$ *with* $\alpha \in \{k, \underline{k} \mid k \geq 2\}$ *and* $p_2 = max(\{k, 3\})$.

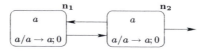

Fig. 7.19 A spiking P system with an infinite spiking train

The following theorem considers *asynchronous spiking P systems*: they do not operate under maximal strategy but in each configuration applicable rules can be applied or not. Moreover, we do not allow spiking rules to have delays.

There are a few consequences because of the lack of maximal strategy:

the concept of *time unit* is absent;

if the system has m neurons, then configurations are m-tuples recording the number of spikes in each neuron;

transitions see the change in the number of spikes in at least one neuron;

if in one configuration a neuron i spikes, then in the next configurations all the neurons j such that (i,j) is a synapse receive the spikes generated by i;

the output of such a system clearly cannot be the time between two spikes of the output neuron. Instead it is given by the number of spikes present in a neuron (called the *output neuron*) when at least one spike is present in a specific neuron (called the *acknowledgement neuron*).

An asynchronous spiking P system is defined by $\Pi = (V, \mu, L_1, \ldots, L_m, R_1, \ldots, R_m, ack, comp)$ where $\mu = (Q, E)$, $V, \mu, L_1, \ldots, L_m, R_1, \ldots, R_m, comp$ are as in Definition 7.1 and $ack \in Q$ is the acknowledgement neuron. The definition of computation follows from that of spiking P systems. In an asynchronous spiking P system a computation *halts* if it reaches a configuration in which at least a spike is present in neuron ack.

With $\mathbb{N} \, ASNP_m(rule_{p_1}, cons_{p_2}, gen_{p_3}, forg_{p_4}, outd_{p_5}, bound_{p_6})$ we denote the family of numbers generated by asynchronous spiking P systems with:

degree at most m;

at most p_1 rules in each neuron;

at most p_2 spikes are removed by rules of the kind $E/a^{p_2} \rightarrow a^s; d$;

at most p_3 spikes are generated by rules of the kind $E/a^r \rightarrow a^{p_3}; d$;

at most p_4 spikes are removed by rules of the kind $a^{p_4} \rightarrow \epsilon$;

at most p_5 synapses outgoing a neuron; formally, $p_5 = max(|E_i|)$, $i \in Q$ where $E_i = \{(i,j) \mid j \in Q\}$;

at most p_6 spikes are present at any time unit in any neuron.

Parameters $m, p_1, p_2, p_3, p_4, p_5$ and p_6 are replaced by $*$ if they are unbounded. If parameter p_6 is specified, then if a system in such a family reaches a configuration where a neuron has more than p_6 spikes, then the system ends without providing any result.

Theorem 7.10 $\mathbb{N} \, ASNP_*(rule_3, cons_3, gen_2, forg_3, outd_*, bound_*) = \mathbb{N} \, RE$.

Proof We define an asynchronous extended spiking P system Π simulating a generating program machine $M = (S, I, s_1, s_f)$ with n counters having all

counters empty in the initial configuration and halting with all counters empty
except the output counter γ_{out} and such that this counter is only incremented.

We consider the program machine M to have instructions of the kind (s,γ^+,v)
and (s,γ^-,v,w). For γ_i, $1 \leq i \leq n$, counter of M, let t_i be the number of
instructions of the kind (s,γ^-,v,w) and let $t = 2\ max(\{t_i \mid 1 \leq n\}) + 1$.
Moreover, let I contain $b \in \mathbb{N}_+$ instructions of the kind (s,γ^-,v,w) and let each
of these instructions be associated with one different number in $\{1, \ldots, b\}$.

The system Π has *state-neurons* n_i for each $i \in S$ and the presence of $3t$ spikes
in one of these neurons simulates M being in the relative state. Considering how
Π is defined, it is not possible for one state-neuron to have more than $3t$ spikes
and for more than one state-neuron to have $3t$ spikes in the same configuration.
Moreover, the system Π has *counter-neurons* n_γ, one for each counter γ of M.
The simulation of the addition of 1 to counter $\gamma \neq \gamma_{\text{out}}$ is performed by passing $3t$
spikes to n_γ, and the simulation of subtraction of 1 from counter γ is performed
by removing $3t$ spikes from n_γ. The simulation of addition of 1 to counter γ_{out}
is performed by passing one spike to $n_{\gamma_{\text{out}}}$.

The simulation of instructions of the kind (s,γ^+,v) when $\gamma \neq \gamma_{\text{out}}$ is performed by the module depicted in Fig. 7.20(a).

The spiking of neuron n_s allows $3t$ spikes to pass to neuron n_γ and neuron
n_v. The simulation of instructions of the kind $(s,\gamma_{\text{out}}^+,v)$ is performed by the
module depicted in Fig. 7.20(b). In this case one spike passes to neuron $n_{\gamma_{\text{out}}}$
and $3t$ spikes pass to neuron n_v.

The simulation of instructions of the kind (s,γ^-,v,w) is performed by the
module depicted in Fig. 7.21.

This module uses the function $\delta : S \to \mathbb{N}_+$ such that:

$$\delta(s) = \begin{cases} 3t & \text{if } (s,\gamma^+,v) \in I; \\ 3t - g & \text{if } g \text{ is the number associated with } (s,\gamma^-,v,w) \in I; \\ 1 & \text{if } s = s_f. \end{cases}$$

Let us assume that neuron n_s in Fig. 7.21 contains $3t$ spikes, simulating M
being in state s. This neuron can spike $3t - g$ spikes, where g is the number
associated with (s,γ^-,v,w), and these spikes pass to neurons n_γ and n_2. In
neuron n_2 one of the rules $a^q \to \epsilon$, $2t < q < 4t$, can be applied. In neuron

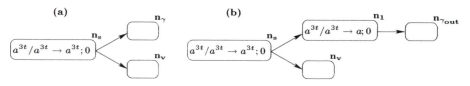

Fig. 7.20 Modules for the spiking P system associated with Theorem 7.10
simulating instructions of the kind (s,γ^+,v)

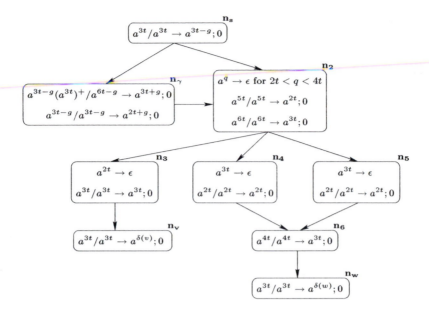

Fig. 7.21 Module for the spiking P system associated with Theorem 7.10 simulating instructions of the kind (s, γ^-, v, w)

n_γ either the rule $a^{3t-g}(a^{3t})/a^{6t-g} \to a^{3t+g}; 4$, simulating the subtraction of 1 from counter γ or $a^{3t-g}/a^{3t-g} \to a^{2t+g}; 4$ can fire. The generated spikes pass to neuron n_2. If one of the rules $a^q \to \epsilon$, $2t < q < 4t$ has been applied in neuron n_2, then no rule can be fired in Π, so the system never halts. Otherwise either $a^{5t}/a^{5t} \to a^{2t}; 0$ or $a^{6t}/a^{6t} \to a^{3t}; 0$ can be applied.

In this last case either $2t$ or $3t$ spikes are generated and pass to neurons n_3, n_4 and n_5. Neuron n_3 spikes only if $3t$ spikes are present in it. When this happens $3t$ spikes pass to neuron n_v. Neurons n_4 and n_5 spike only if $2t$ spikes are present in them; the generated spike passes to neuron n_6. Only when $4t$ spikes are present in this last neuron it spikes $3t$ spikes that pass to neuron n_w.

If a state $s \in S$ is such that there are several instructions of the kind $(s, \gamma^-, v, w), (s, \gamma^+, v) \in I$, then for each such instructions the state-neurons n_s contain one rule of the kind $a^{3t}/a^{3t} \to a^{\delta(s)}; 0$.

Similarly to the proof of Theorem 7.4 we have to explain what happens if more than one counter-neuron is shared between different modules simulating instructions of the kind (s, γ^-, v, w). When a counter-neuron spikes, its spikes pass to all the n_2 neurons. If the n_2 neurons not in the module of the simulated instruction apply a forgetting rule, then the computation can halt. If instead one of these neurons, say n_2', does not apply a forgetting rule, then, if other spikes pass to n'_2, no rule is applicable to it. This means that if n'_2 is present in the module simulating (s', γ'^-, v', w'), then the simulation is not completed and the computation does not halt.

When $3t$ spikes are present in the n_{s_f} state-neuron and acknowledgement neuron of Π, then the generated number is present in the $n_{\gamma_{out}}$ counter-neuron associated with the output counter of M and the final neuron of Π.   □

7.7   Infinite output and astrocytes

In this section we informally define two very recently introduced models of spiking P systems: one generating infinite sequences of 0's and 1's and another introducing *astrocytes*

A spiking P systems can generate infinite binary strings. For this purpose the definition of the output rendered by these systems is changed such that if in one time unit the output neuron fires, then the output is 1; if it does not fire, then the output is 0. We recall that these systems operate in a way synchronised by the ticks (*time units*) of a global clock. It should be clear that the length of the output of such systems is infinite.

The sequences of 0's and 1's generated by these systems are called *binary spike trains*. In the following the superscript ω denotes an infinite repetition. Here are some examples of such spiking P systems.

Example 7.6 A spiking P system generating $1^k 0^\omega$ for $k \in \mathbb{N}_+$.

Two such systems are depicted in Fig. 7.22.

The neuron in the system Π_7 depicted in Fig. 7.22(a) has k spikes in its initial configuration. As this neuron contains the rule $a^*/a \to a; 0$, then it spikes only for the first k time units and then it no longer spikes. As this neuron is the output neuron, then k 1's are rendered for the first time units, then an infinite sequence of 0's is generated.

All the neurons in the systems Π_8 depicted in Fig. 7.22(b) have one spike in their initial configuration. They all fire their rule in this configuration but only neurons 3 and 4 spike in the same time unit. Neuron 2 spikes after $k - 1$ time units.

The spikes generated by neurons 3 and 4 pass to each other so that they keep firing and spiking. Moreover, as neuron 3 is the output neuron, a 1 is rendered as output in each of these time units. When neuron 2 spikes, then its spike passes to neurons 3 and 4 so that they stop spiking, so an infinite sequence of 0's is generated.

It should be clear that Π_7 and Π_8 generate $1^k 0^\omega$.   ◇

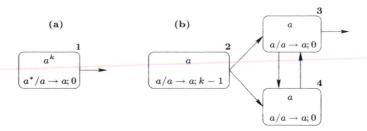

Fig. 7.22 Spiking P systems associated with Example 7.6

Example 7.7 A spiking P system Π_9 generating $1^k 0^\omega$ for $k \geq 2$.

We do not give a formal definition of this system; it is depicted in Fig. 7.23.

This system can be logically divided into two parts (which we could call modules). Neurons 4 (being the output neuron) and 5 behave in the same way as neurons 3 and 4, respectively, in Fig. 7.22(b). They spike from the initial configuration and they stop spiking only when another spike passes to them.

This can only happen when neuron 3 spikes and this depends on the non-deterministic application of the rules in neuron 2. Let us look at this closely.

In each time unit, neuron 3 can either receive only one spike (from neuron 1) or two spikes (one from neuron 1 and another from neuron 2). If in a time unit neuron 3 receives only one spike, then it fires; if it receives two spikes, then the forgetting rule $a^2 \to \epsilon$ is applied in this neuron.

In the initial configuration neurons 1 and 2 have one spike each. Neuron 1 fires and spikes in the same time unit, neuron 2 can fire and spike in the same time unit or it can fire and spike after one time unit. As long as rule $a/a \to a; 0$ is applied in neuron 2, then neuron 3 receives two spikes. If instead the rule $a/a \to a; 1$ is applied in neuron 2, then neuron 3 receives one spike.

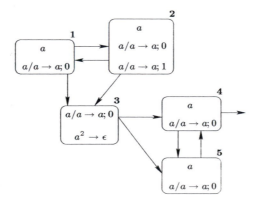

Fig. 7.23 Spiking P systems associated with Example 7.7

The shortest sequence of 1's that can be generated by Π_9 is 2: in the initial configuration neurons 1, 4 and 5 fire and spike while neuron 2 only fires; in the second configuration neuron 3, 4 and 5 fire and spike and then neuron 3 no longer spikes.

It should be clear that Π_9 generates $1^k 0^\omega$ for $k \geq 2$.   ◇

Spiking P systems can also behave as *transducers* of infinite binary strings. To this aim an input is provided to the system through *input neurons* receiving spikes from outside the system itself. The only restriction imposed on this kind of transducer is that the input neurons can never be idle. In the figures representing transducer spiking P systems, the input neurons are the only ones having an incoming arrow not originating from a neuron.

Astrocytes are cells different from neurons present in the central nervous system. Recent biochemical literature suggests that astrocytes have an important role in influencing the concentration of chemicals relevant to neural activity and neuronal blood flow. Astrocytes are, in turn, influenced by neural activity.

This led to the formalisation of astrocytes in the framework of spiking P systems. *Spiking P systems with astrocytes* add to Definition 7.1 a set of astrocytes, connections between astrocytes and synapses and sets of rules for each astrocyte. These rules are of the kind $(h, h'; f, f', f'')$ with:

$h, h' \in \mathbb{N}_+,\ h \leq h'$, called *thresholds*;

f, f' and f'' functions from $\mathbb{N}_+$ to $\mathbb{N}_+$ changing the number of spikes sent through the synapses to which the astrocyte is connected.

These rules work in the following way. Let us assume that there is a synapse (i, j) from neuron i to neuron j and that there is an astrocyte a connected to (i, j). If in a time unit an astrocyte receives w spikes and neuron i passes v spikes to neuron j, then:

if $w \leq h$, then $v = f(w)$;

if $h \leq w \leq h'$, then $v = f'(w)$;

if $w \geq h'$, then $v = f''(w)$.

Example 7.8 A spiking P system with astrocytes simulating the logic AND.

We do not give a formal definition of this system; it is depicted in Fig. 7.24.

In this system there are five neurons and one astrocyte. The astrocyte is represented by a parallelogram and it is connected to the synapse from neuron 2 to neuron 3. The astrocyte receives spikes from two neurons that can emit at most one spike in each time unit.

Neurons 1 and 2 fire and spike in each time unit but one spike passes from neuron 2 to neuron 3 only if the astrocyte receives at least two spikes as input, otherwise no spike reaches neuron 3. There is only one rule associated with the

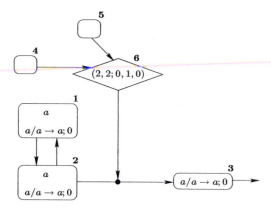

Fig. 7.24 Spiking P systems with astrocytes associated with Example 7.8

astrocyte: $(2, 2; 0, 1, 0)$. This rule indicates that if in a time unit the number of spikes received by the astrocyte is 2 (that is, ≥ 2 and ≤ 2), then the synapses from neuron 2 to neuron 3 can allow one spike to pass. In this case neuron 3 spikes, otherwise not.

So, if we regard the presence of a spike as the Boolean value TRUE and the absence of a spike as the Boolean value FALSE, then in one time unit neuron 3 spikes (that is, emits TRUE) only if in the previous time unit neurons 4 and 5 spike (that is, the astrocyte received two TRUE as input).   ◇

7.8   Final remarks and research topics

There are some differences between what is known of the activity of real neurons (described in Section 7.1) and the definition of neurons in spiking P systems (Section 7.2). Real neurons first spike and then they have a refractory period; neurons in a spiking P system first fire a rule, then they (eventually) become idle and then they spike. So, in nature the refractory period is after the spiking, while in spiking P systems the refractory period (that is, a neuron being idle) is before the spiking.

Suggestion for research 7.1 *Study models of spiking P systems in which neurons first fire, spike and then become idle.*

Several of the examples in this chapter, and in the present monograph in general, read: 'It should be clear' when it is about defining the set generated by the given P systems. A formal verification of the generative capabilities of the considered P system would be appropriate in these cases, mainly for spiking P systems as these devices can have a rather complex behaviour originating from a simple definition.

Suggestion for research 7.2 *Introduce and study formal verification methodologies for (spiking) P systems.*

The links between spiking P systems and P/T systems definitely are a very interesting research topic. The simulation of a *fork* given in Lemma 7.1 is just one of the many possible. How can a *fork* be simulated by spiking P systems without forgetting rules? What about P/T systems simulating spiking P systems?

In addition to being interesting on their own, such investigations could lead to finding lower and upper bounds of the parameters defining spiking P systems (see Suggestion for research 6.2).

Suggestion for research 7.3 *Investigate further the links between spiking P systems and P/T systems.*

Moreover, results related to the previous suggestion for research could be proved useful in the understanding of Theorem 7.10 in the light of Theorem 4.1. By this we mean the following: Theorem 7.10 says that a maximal strategy is not needed for spiking P systems as long as extended rules are present. Theorem 4.1 says that $SCG(N)$ can weakly simulate instructions of the kind (s, γ^-, v, w) only if N is a P/T system with an infinite number of places and transitions or if N has inhibitor arcs. Where are these infinities in Theorem 7.10? Or do we have to regard forgetting rules (present in Theorem 7.10) as inhibitor arcs?

On the same line of thought we have:

Suggestion for research 7.4 *Study the computational power of asynchronous non-extended spiking P systems and asynchronous extended spiking P systems without forgetting rules.*

In relation to the previous suggestion for research we state

Theorem 7.11 *A set $A \subseteq \mathbb{N}^k$ is generated by a k-output partially blind program machine if and only if it can be generated by an asynchronous non-extended k-output unbounded spiking P system using spiking rules without delays.*

The proof of this theorem is not simple and it is rather long (seven pages in the original paper) and for these reasons we decided not to include it. The next section indicates where interested readers can find this proof.

7.9 Bibliographical notes

Spiking P systems were introduced in [132]. Some of the examples presented in the present chapter appeared previously: Examples 7.1, 7.2 and 7.3 are from [132], Example 7.4 is from [214], Example 7.5 is from [216], Examples 7.6 and 7.7 are from [215] while Example 7.8 is from [31].

Section 7.5 is mainly based on results presented in [132]; Theorem 7.3 was originally published in [216].

Section 7.6 is mainly based on results presented in [132, 216]; the proof of Theorem 7.10 was originally published in [44].

Spiking P systems generating infinite binary strings and behaving as transducers were introduced in [215], see also [217]. Spiking P system with astrocytes were introduced in [31] and further studied in [212].

The proof of Theorem 7.11 can be found in [121]; also [123] has results on the same line of research.

Spiking P systems were also studied as generators of strings. Interested readers can refer to [7, 47–50, 131].

A software tool for spiking P systems was reported in [226].

A few papers show how spiking P systems can solve instances of **NP** problems. Interested readers can refer to [159–161].

Some formal verification for P systems has been introduced and studied. The interested reader can refer to [96, 142, 162, 192, 193]. On this topic the interested readers can refer to the simulation and verification software tools available from [238].

8 Conformons

Differently from the rest of the models of membrane systems presented in this monograph, conformon P system are not based on a specific intra- or intercell process. They have been inspired by a theoretical model of the living cell based on the concept of *conformon*. Conformons are sequence-specific mechanical strains embedded in the (macro)molecules present in cells. DNA supercoils and proteins are examples of conformational deformations that provide both the information and the energy needed for biopolymers to drive molecular processes.

Conformons are at the base of the *Bhopalator*: a theoretical model of the living cell in which the processes present in a cell are described using conformons arranged in space and time with appropriate force vectors. In this model conformons interact exchanging all or part of their free-energy.

The mathematical abstraction of a biological conformon is an ordered pair of a symbol in an alphabet and a number. The symbol represents the information while the number represents the energy of the conformon. The interaction between two such mathematical conformons becomes then the passage of part of the number from one pair to another.

Compartments define the locality of these interactions. The passage of conformons is possible between different compartments. In the model we present, the passage is solely based on the number present in a conformon: a conformon can pass from one compartment to another only if its number is in between certain boundaries.

Following closely what happens in nature, in the model of membrane systems considered in the present chapter nothing is created or destroyed but everything is conserved.

8.1 Basic definitions

Let V be an alphabet, $\mathbb{N}_+$ the set of natural numbers and $\mathbb{N} = \mathbb{N}_+ \cup \{0\}$. A *conformon* is an element of $V \times \mathbb{N}$, denoted by $[\Phi, a]$. We refer to Φ as the *name* (that is, the information) of the conformon $[\Phi, a]$ and to a as its *value* (that is, the energy). If, for instance, $V = \{A, B, C, \ldots, Z\}$, then $[A, 5], [C, 0], [Z, 14]$ are conformons, while $[AB, 21]$ and $[D, 0.5]$ are not. Provided there is an unambiguous context, the sole name identifies the conformon itself.

Two conformons can interact according to an *interaction rule*. An interaction rule is of the form $\Phi \xrightarrow{e} \Upsilon, \Phi, \Upsilon \in V$, and $e \in \mathbb{N}_+$, and it says that a conformon with name Φ can give e from its value to the value of a conformon having name Υ. If, for instance, there are conformons $[G, 5]$ and $[R, 9]$ and the rule $G \xrightarrow{3} R$, one application of this rule leads to $[G, 2]$ and $[R, 12]$. Since for the moment we consider only conformons whose value cannot be a negative number, then the rule $G \xrightarrow{3} R$ cannot be applied to $[G, 2]$.

The compartments present in a conformon P system have a label, every label being different. Membrane compartments present in a conformon P system can be unidirectionally connected to each other and for each connection there is a *predicate*. A predicate is an element of the set $\{\geq n, \leq n \mid n \in \mathbb{N}\}$. Examples of predicates are: $\geq 5, \leq 2$, etc. If, for instance, there are two compartments (with labels) m_1 and m_2 and there is a connection from m_1 to m_2 having predicate ≥ 4, then conformons having value greater than or equal to 4 can pass from m_1 to m_2. These are *unidirectional connections*: m_1 connected to m_2 does not imply that m_2 is connected to m_1. Moreover, each connection has its own predicate. If, for instance, m_1 is connected to m_2 and m_2 is connected to m_1, the two connections can have different predicates.

The interaction with another conformon and the passage to another compartment are the only *operations* that can be performed by a conformon.

We are going to study a few models of conformon P systems. We introduce one of these models here.

Definition 8.1 *A (generating/accepting) conformon P system is a construct*

$$\Pi = (V, \mu, \omega_z, ack, L_1, \ldots, L_m, R_1, \ldots, R_m)$$

where:

V is an alphabet.

$\mu = (Q, E, lr)$ *is an edge-labelled directed multigraph (a cell-graph) underlying* Π *where:*

   $Q \subset \mathbb{N}_+$ *contains vertices. For simplicity we define* $Q = \{1, \ldots, m\}$. *Each vertex in Q defines a* compartment *of* Π.

   $E \subseteq Q \times Q$ *defines directed labelled edges between vertices, denoted by* (i, j), $i, j \in Q$, $i \neq j$.

   $lr : Q \times Q \rightarrow pred(\mathbb{N})$ *is the labelling relation where for each* $n \in \mathbb{N}$ *we consider* $pred(n) \in \{\geq n, \leq n\}$ *set of predicates.*

ω_z *with* $\omega \in \{input, output\}$ *and* $z \in Q$ *denote whether* Π *is an accepting* $(\omega = input)$ *or generating* $(\omega = output)$ *device. The compartment z contains the input or output, respectively.*

$ack \in Q$ *denotes the* acknowledge *compartment.*

$L_i : (V \times \mathbb{N}) \to \mathbb{N} \cup \{+\infty\}$, $i \in Q$, *are multisets of conformons initially associated with the vertices in* Q.

R_i, $i \in Q$, *are finite sets of interaction rules associated with the vertices in* Q.

If $(i, j) \in E$ and $lr(i, j) = pred(n)$, then we write $(i, j, pred(n))$, a *labelled edge*. Let M_i and R_i be the multiset of conformons and the set of interaction rules, respectively, associated with the compartment $i, i \in Q$. Two conformons present in i can interact according to a rule also present in i such that the multiset of conformons M_i changes into M_i'. If, for instance, $[\Phi, a], [\Upsilon, b] \in M_i$, $\Phi \xrightarrow{e} \Upsilon \in R_i$, and $a \geq e$, then $M_i' = M_i \backslash \{[\Phi, a], [\Upsilon, b]\} \cup \{[\Phi, a - e], [\Upsilon, b + e]\}$.

A conformon $[\Phi, a]$ present in i can *pass* to compartment j if $(i, j, pred(n)) \in E$ and $pred(a)$ holds on a. That is, if $pred(n)$ is $\leq n$, then $a \leq n$; if $pred(n)$ is $\geq n$, then $a \geq n$. This passage changes the multisets of conformons M_i and M_j into M_i' and M_j', respectively. In this case $M_i' = M_i \backslash \{[\Phi, a]\}$ and $M_j' = M_j \cup \{[\Phi, a]\}$.

At the moment we do not assume any requirement (as maximal parallelism, priorities, etc.) on the application of operations. That is, conformon P systems operate in an asynchronous way. If a conformon can pass into another compartment or interact with another conformon according to a rule, then one of the two operations or none of them is non-deterministically chosen.

The possibility of carrying out one of the two allowed operations in the same compartment or none of them allows conformon P systems to be non-deterministic. Non-determinism can also arise from the configurations of a conformon P system if in a compartment a conformon can interact with more than one conformon and also from the cell-graph underlying Π if a compartment has edges with the same predicate going to different compartments.

A *configuration* of Π is an m-tuple $(M_1, \ldots, M_m)$ of multisets over $V \times \mathbb{N}$. The m-tuple $(L_1, \ldots, L_m)$, $supp(L_{ack}) = \emptyset$, is called the *initial configuration* (so in the initial configuration the acknowledge compartment does not contain any conformon) while any configuration having $supp(M_{ack}) \neq \emptyset$ is called the *final configuration*. In a final configuration no operation is performed even if it could be. If in a configuration with no conformon in M_{ack} no operation can be performed, then we say that the system *stops*.

For two configurations $(M_1, \ldots, M_m), (M_1', \ldots, M_m')$ of Π we write $(M_1, \ldots, M_m) \Rightarrow (M_1', \ldots, M_m')$ to denote a *transition* from $(M_1, \ldots, M_m)$ to $(M_1', \ldots, M_m')$, that is, the application of one operation to at least one conformon. In other words, in any configuration in which $supp(M_{ack}) \neq \emptyset$ any conformon present in a compartment can either interact with another conformon present in the same compartment or pass into another compartment or remain in the same compartment unchanged. If no operation is applied to a multiset M_i, then $M_i' = M_i$. The reflexive and transitive closure of $\Rightarrow$ is denoted by $\Rightarrow^*$.

A *computation* is a sequence of transitions between configurations of a system Π starting from $(L_1, \ldots, L_m)$. If a computation is finite, then the last configuration is called *final*.

Let $\Pi = (V, \mu, \omega_z, ack, L_1, \ldots, L_m, R_1, \ldots, R_m)$ be a conformon P system. If Π is an accepting device ($\omega = input$), then the input is given by the number of conformons (counted with their multiplicity) present in L_z. The input is accepted by Π if it reaches a final configuration in which a (any) conformon is present in ack. When this happens the computation *halts*, that is, no transition takes place even if it could.

Formally:

$$N(\Pi) = \{|L_z| \mid (L_1, \cdots, L_m) \Rightarrow^* (M'_1, \cdots, M'_m) \Rightarrow^* (M_1, \cdots, M_m),$$
$$supp(M'_{ack}) = \emptyset, supp(M_{ack}) \neq \emptyset\}.$$

If Π is a generating device ($\omega = output$), then $supp(L_z) = \emptyset$. Such a system renders a result if it reaches a configuration in which a (any) conformon is present in ack. When this happens the computation *halts*, that is, no transition takes place even if it could. The result of a computation is given by the number of conformons (counted with their multiplicity) present in M_z in the final configuration of a halting transition. This number of conformons defines the *number generated* by Π, denoted by $N(\Pi)$.

Formally:

$$N(\Pi) = \{|M_z| \mid (L_1, \cdots, L_m) \Rightarrow^* (M'_1, \cdots, M'_m) \Rightarrow^* (M_1, \cdots, M_m),$$
$$supp(M'_{ack}) = \emptyset, supp(M_{ack}) \neq \emptyset\}.$$

We also consider *conformon P systems where interaction rules have priority on passage rules*. In these systems if in a configuration a conformon can be subject to an interaction rule and to a passage rule, then the interaction rule is applied. In other words, in each compartments a passage rule is applied to a conformon only if no interaction rule can be applied to that conformon.

We do not provide formal definitions for most of the conformon P systems considered in this chapter. Figures depicting them are provided instead.

8.2 Examples

Figures representing conformon P systems have compartments represented by rectangles having their label written in **bold** on their right-upper corner. Ovals with a label in them refer to the compartment having that label. Conformons and interaction rules related to a compartment are written inside a rectangle.

Conformons present in the initial configuration of a system are written in **bold** inside a rectangle while those written in normal font are present in that compartment in one of the possible configurations of the system. A slash (/) between values in a conformon denotes that a conformon can have any of the indicated values. For instance, $[A, 3/5/10]$ denotes that there are configurations in which $[A, 3], [A, 5]$ or $[A, 10]$ can be present in a compartment. If m conformons $[\Phi, a]$ can be present in a compartment, then $([\Phi, a], m)$ is indicated. If an infinite number of conformons $[\Phi, a]$ is present in a compartment, then $([\Phi, a], +\infty)$ is indicated.

If in a compartment the interaction rules $\Phi \xrightarrow{e} \Upsilon$ and $\Upsilon \xrightarrow{e} \Phi$ are present, then $\Phi \xleftrightarrow{e} \Upsilon$ is indicated. The interaction rules $\Phi \xrightarrow{e} \Upsilon$ and $\Upsilon \xrightarrow{e} \Phi$ are each the *reverse* of the other.

Directed edges between compartments are represented as arrows with their predicate indicated close to them. Several edges connecting two compartments are depicted as just one edge with different predicates separated by a slash (/). For instance, an edge with predicate $\leq 2/ \geq 5$ denotes two edges, one with predicate ≤ 2 and the other with predicate ≥ 5.

Further explanations related to the figures in this chapter are present in the next section.

Example 8.1 A conformon P system generating any set $\{q, \ldots, q + p\} \subset \mathbb{N}$.

The conformon P system related to this example is depicted in Fig. 8.1. Its formal definition is:

$$\Pi_1 = (\{A, B\}, (\{1, 2, 3\}, \{(1, 2, \geq 0), (2, 3, \geq 1)\}, output_2, 3,$$
$$\{([A, 0], p)\}, \{[B, 1], ([A, 0], q)\}, \phi, \phi, \phi, \phi).$$

In the initial configuration $p \in \mathbb{N}$ occurrences of $[A, 0]$ are present in compartment 1, compartment 2 has $q \in \mathbb{N}$ occurrences of $[A, 0]$ and one copy of $[B, 0]$. This initial configuration is denoted by the conformons in **bold** in Fig. 8.1. Compartment 3 is initially empty. In Fig. 8.1 $[B, 1]$ present in compartment 3 indicates that that conformon can pass into this compartment.

This conformon P system is a generating device whose output compartment is compartment 2. This is indicated by $output_2$ in the formal definition of Π_1. The acknowledge compartment is 3 and it is initially empty (as it has to be for a generating conformon P system). There are no interactions rules.

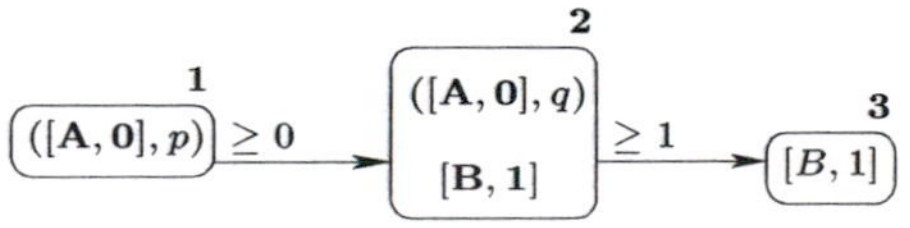

Fig. 8.1 The conformon P system associated with Example 8.1

From the initial configuration a few things can happen:

Some of the $[A, 0]$ conformons present in compartment 1 can pass into compartment 2. Once in this compartment they remain here.

The $[B, 1]$ conformon can pass into the acknowledgement compartment. When this happens the computation halts.

It should be clear that the $[A, 0]$ conformons in compartment 1 behave independently of each other: they can all remain in that compartment, they can all pass at the same time into compartment 2, some of them can pass and then other ones can pass, etc. Similarly for the $[B, 1]$ conformon: it can stay in compartment 2 while some $[A, 0]$ conformons can pass from compartment 1 to compartment 2. When $[B, 1]$ passes to the acknowledgement compartment any number between q and $q + p$ of $[A, 0]$ conformons can be in compartment 2. This number is the output of the system.

If when $[B, 1]$ passes to the acknowledgement compartment some $[A, 0]$ are in compartment 1, then they cannot pass into compartment 2 (because the system halted). We can say that Π_1 generates $\{q, \dots, q + p\}$. $\diamond$

Example 8.2 A conformon P system accepting any positive even number

The conformon P system related to this example is depicted in Fig. 8.2. In order to have a simple system we slightly change the condition of acceptance into: the input is accepted if the system reaches a configuration in which a (any) conformon is present in the acknowledge compartment and the input compartment is empty.

We only give an informal definition of this system, a formal definition would be tedious. This system is called Π_2, its input compartment is 1 and its acknowledgement compartment is 11.

In the initial configuration $p \in \mathbb{N}$ occurrences of $[A, 0]$ (the input) are present in compartment 1, $[B, 3]$ and $[C, 11]$ are present in compartment 2. The system works as follows: pairs of occurrences of $[A, 0]$ are taken out from compartment 1. This is performed with the help of conformons C and B. If when a (any) conformon is present in the acknowledgement compartment the input compartment is empty, then there was a positive even number of $[A, 0]$ in this compartment. The system is non-deterministic.

Before going on with the explanation of this system we remind readers that a conformon can be the subject of an operation or not.

The conformons $[B, 3]$ and $[C, 11]$, initially present in compartment 2, can pass into compartment 1. Here:

(i) the B or C conformon interacts with two different occurrences of $[A, 0]$;

(ii) the B or C conformon interacts with the same occurrences of $[A, 0]$;

(iii) the B or C conformon passes into compartment 3.

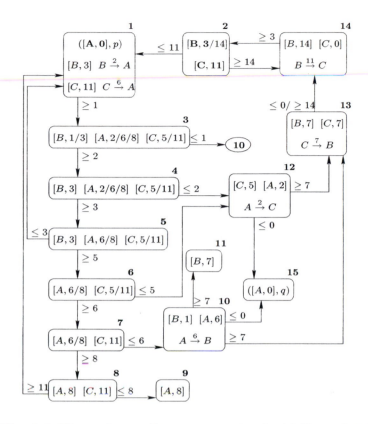

Fig. 8.2 The conformon P system associated with Example 8.2

If item (i) in the previous list takes place, then $[B, 1], [A, 2], [C, 5]$ and $[A, 6]$ are generated and then they can pass into compartment 3. The compartments from 3 to 8 can be regarded as filtering conformons depending on their value (see Definition 8.2). From these compartments conformons with value 1 and 6 (B and A, respectively) pass into compartment 10, while those with value 5 and 2 (C and A, respectively) pass into compartment 12. This sequence of compartments allows conformons with value 3 and 11 to pass back to compartment 1 (this is related to item (iii) in the previous list).

If in compartment 10 $[B, 1]$ and $[A, 6]$ are present, then two occurrences of $[A, 0]$ have certainly been removed from compartment 1. One of these occurrences became $[A, 6]$ while the other interacted with the B conformon so that $[B, 1]$ is generated.

In compartment 10 the B and A conformon can interact so that $[A, 0]$ and $[B, 7]$ are generated. The $[A, 0]$ conformon can pass into compartment 15, while the $[B, 7]$ conformon can either pass into compartment 11 (the acknowledgement compartment) or into compartment 13. What happens in compartment 12 between $[C, 5]$ and $[A, 2]$ is similar to what was just described for compartment 10.

If $[B, 7]$ passes to the acknowledgement compartment the computation halts and the input is accepted only if compartment 1 is now empty. If instead $[B, 7]$ passes to compartment 13, then it stays here until $[C, 7]$ (generated in compartment 12) arrives. When $[B, 7]$ $[C, 7]$ are both in compartment 13 they can interact so that $[B, 14]$ and $[C, 0]$ are generated. These two last conformons can pass into compartment 14 and interact so that $[B, 3]$ and $[C, 11]$ are generated and they can pass into compartment 2.

A configuration similar to the initial one is recreated: two occurrences of $[A, 0]$ have been removed from compartment 1 and passed to compartment 10, 12 or 15.

Item (ii) in the previous list sees $[A, 8]$ being generated. This conformon can only pass into compartment 9. In this case a halting configuration is never reached.

Item (iii) in the previous list has been discussed in the above.

If in the initial configuration there is an odd number of $[A, 0]$ conformons in compartment 1, then at a certain point either $[B, 1]$ or $[A, 6]$ are present in compartment 10. In this case the system stops.

In case in the initial configuration there are no occurrences of $[A, 0]$ conformons in compartment 1, then $[B, 3]$ and $[C, 11]$ keep looping through compartments 1, 3, ..., 8.

We can say that Π_2 accepts $\{p \mid p = 2n, n \in \mathbb{N}_+\}$. $\diamond$

8.3   Modules of conformon P systems

In the following we use *modules*: groups of compartments with conformons and interaction rules in a conformon P system able to perform a specific task.

As a module is supposed to be part of a bigger system it has edges coming from or going to compartments not defined in the module itself. These compartments are depicted with a dashed line in the figures representing detailed modules. Some modules are depicted with a rectangle with a thicker line. Such modules have a *label* written in **bold** on its right upper corner denoting the kind of module. A subscript is added to differentiate labels referring to the same kind of module present in one system.

Definition 8.2 *A* splitter *is a module that can select conformons depending on their value. This module is such that when a conformon* $[\Phi, a_i]$, $1 \leq i \leq h$ *and* $a_j < a_{j+1}$ *for* $1 \leq j \leq h - 1$, *is present in a specific compartment of it, then this conformon can pass into another specific compartment depending on* a_i.

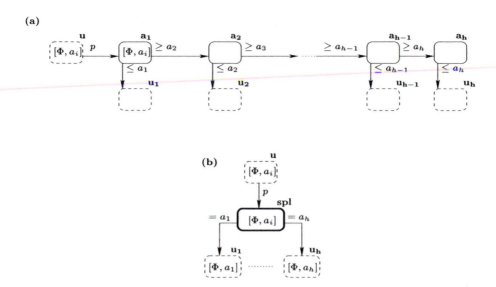

Fig. 8.3 (a) A detailed splitter and (b) its module representation, $1 \leq i \leq h$

Figure 8.3(a) represents a detailed splitter.

No conformon and no rule is present in the initial configuration of this module; h compartments with labels a_i, $1 \leq i \leq h$, are present. From compartment u, external to the module, a conformon $[\Phi, a_i]$, $1 \leq i \leq h$, can reach compartment a_1 (there can be any predicate for this edge, in Fig. 8.3 we indicated p as a generic predicate). Each of the a_i compartments has edges $(a_i, a_{i+1}, \geq a_{i+1})$ for $1 \leq i \leq h - 1$, and $(a_i, u_i, \leq a_i)$ for $1 \leq i \leq h$, where u_i are compartments external to the module. If a conformon $[\Phi, a_i]$ is present in compartment a_j, $1 \leq j \leq h$, then (considering what has just been described, that $1 \leq i \leq h$, and that $a_j < a_{j+1}$ for $1 \leq j \leq h - 1$) it can pass into compartment u_k only if $a_i = a_k$. Otherwise $a_i > a_k$ and $[\Phi, a_i]$ passes into compartment a_{k+1}.

Considering what has just been stated, the operation performed by a splitter can be indicated in a more convenient way by more specific predicates on the edges leaving the module. Figure 8.3(b) illustrates the module representation of a splitter having **spl** as label and edges with predicates $= a_i$, $1 \leq i \leq h$, indicating that a conformon $[\Phi, a]$ can pass from compartment u to compartment u_i only if $a = a_i$. It is important to notice that the module representations of splitters can miss the edges having labels $= a_1$ and $= a_h$ and having one outgoing edge with label $\leq a_1$ and one outgoing edge with label $\geq a_h$.

The compartments from 3 to 8 in Fig. 8.2 define a splitter. This means that the conformon P system Π_2 can be depicted was as done in Fig. 8.4

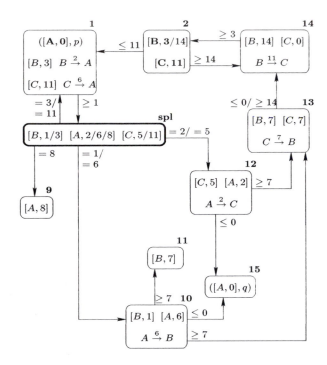

Fig. 8.4 The conformon P system with a splitter associated with Example 8.2

Definition 8.3 *A separator is a module that can select conformons depending on their name. This module is such that when conformons of type* $[\Phi_i, a]$, $1 \leq i \leq h$, *are present in a specific compartment of it, then they can pass into specific different compartments depending on* Φ_i.

Figure 8.5(a) represents a detailed separator.

The number of compartments in this module and the conformons present in them depend on h and a. The dotted lines in Fig. 8.5 suggest the presence of more compartments, modules and edges than those depicted in that figure. From compartment u, external to the module, a conformon $[\Phi_i, a]$ can reach compartment 1. There can be any predicate for this edge; in Fig. 8.5(a) we indicated p as a generic predicate. In the initial configuration compartment 1 contains the conformons $[C_i, 2a + h + 4i - 1]$, $C_i \neq \Phi_j$, $1 \leq i, j \leq h$. In order to differentiate the conformons $[\Phi_i, a]$, $a + h + 2i$ is added to the value of them and subtracted from the value of $[C_i, 2a + h + 4i - 1]$. In this way $[\Phi_i, 2a + h + 2i]$ and $[C_i, a + 2i - 1]$ are generated and they can pass to spl_1. The pairs of conformons that interacted in compartment 1 can now be selected on their value. Conformons with values $2a + h + 2i$ and $a + 2i - 1$ pass into specific compartments where the

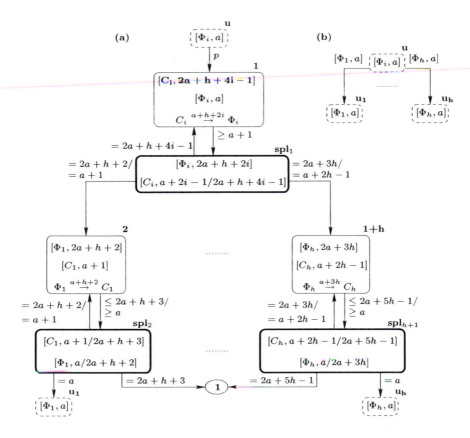

Fig. 8.5 (a) A detailed separator and (b) its shorthand representation, $1 \le i \le h$

rule $\Phi_i \stackrel{a+h+2i}{\rightarrow} C_i$ is present. In these compartments $[\Phi_i, a]$ and $[C_i, 2a+h+4i-1]$ are generated and they can pass into compartment u_i and 1, respectively.

The operation performed by a separator can be denoted in a more convenient way by a label on an edge. In Fig. 8.5(b) a conformon P system having edges with labels $[\Phi_i, a], 1 \le i \le h$, is depicted. This indicates that a conformon $[\Phi_j, a]$ can pass from compartment u to compartment u_i only if $j = i$.

Example 8.3 A separator for $[A_1, 5], [A_2, 5]$ and $[A_3, 5]$.

The module related to this example is depicted in Fig. 8.6.

If we consider the definition of separator, then we have $a = 5$ and $h = 3$. The C_i conformons initially present in compartment 1 are: $[C_1, 16], [C_2, 20]$ and $[C_3, 24]$. The interaction rules present in compartment 1 are: $C_1 \stackrel{10}{\rightarrow} A_1, C_2 \stackrel{12}{\rightarrow} A_2$, and $C_3 \stackrel{14}{\rightarrow} A_3$. These rules allow the conformons $[A_1, 15], [C_1, 6], [A_2, 17]$,

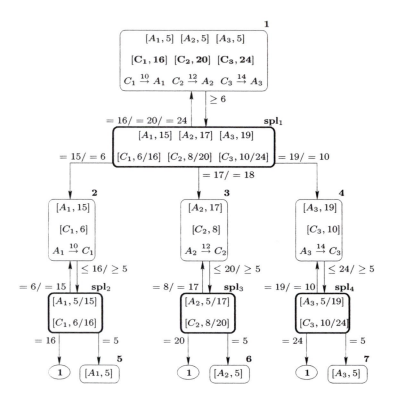

Fig. 8.6 The separator associated with Example 8.3

$[C_2, 8]$, $[A_3, 19]$ and $[C_3, 10]$ to be generated. These conformons can pass (through **spl$_1$**) to compartments 2, 3 and 4. In compartment 2, for instance, only the conformons $[A_1, 15]$ and $[C_1, 6]$ can pass. Here they interact such that $[A_1, 5]$ and $[C_1, 16]$ are generated. These two conformons can then pass into compartment 5 and 1, respectively.   ◇

In the pictorial representations of conformon P systems several conformons separated by a slash (/) close to an edge denote several shorthand representations of separators.

Definition 8.4 *An* increaser *is a modules that can increase the value of conformons to a specific amount. This module is such that when a conformon $[\Phi, a]$ is present in a specific compartment of it, then its value can increase until q, $q \geq 1$, so that $[\Phi, q]$ can pass into another specific compartment.*

In Fig. 8.7(a) a detailed increaser is depicted. From compartment u, external to the module, a conformon $[\Phi, a]$ can reach compartment 1. There can be

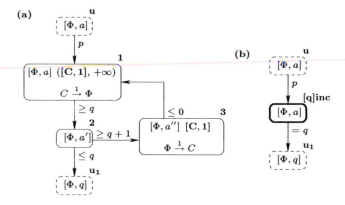

Fig. 8.7 **(a)** A detailed increaser and **(b)** its module representation

any predicate for this edge; in Fig. 8.7(a) we indicated p as a generic predicate. In the initial configuration an infinite number of $[C, 1]$ conformons is present in compartment 1 and one occurrence of $[C, 1]$ is present in compartment 3. In compartment 1 the value of Φ can be increased to any value but only when the value of this conformon is $\geq q$, can it then pass into compartment 2.

In this compartment the Φ conformon can pass into compartment u_1, external to the module, only if its value is $\leq q$, that is, if it is equal to q. Otherwise, this conformon can pass into compartment 3 where its value can be decreased (and added to that of the C conformon present in this compartment). When the value of Φ is equal to zero, then it can pass back to compartment 1.

In Fig. 8.7(b) the module representation of an increaser is depicted. The label of this module is **[q]inc**.

Definition 8.5 *A decreaser is a modules that can decrease the value of conformons to a specific amount. This module is such that when a conformon $[\Phi, a]$ is present in a specific compartment of it, then its value can decrease until $q, q \geq 1$, so that $[\Phi, q]$ can pass into another specific compartment.*

This module can be easily obtained by an increaser. If we consider Fig. 8.7, then it is enough to remove the edge from compartment u to compartment 1 and to add an edge from compartment u to compartment 3. The module representation of such a module is similar to that of an increaser with the difference that the label is **[q]dec**.

The combination of separators, decreasers and increasers allows us to define a *strict interaction* rule: $\Phi^{(a)} \xrightarrow{e} \Upsilon_{(b)}$ where $a, b, e \in \mathbb{N}$, meaning that a conformon with name Φ can interact with Υ passing just e only if the value of Φ and Υ before the interaction is a and b, respectively. The detailed module for strict

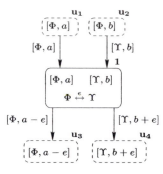

Fig. 8.8 A detailed strict interaction

interaction is depicted in Fig. 8.8. Notice that in a strict interaction just e is passed even if the value of Φ could be decreased by any multiple of e.

Interactions of the kind $\Phi \xrightarrow{e} \Upsilon_{(b)}$ (before the interaction Φ can have any value while Υ must have b as value) and $\Phi^{(a)} \xrightarrow{e} \Upsilon$ (before the interaction Φ must have a as value while Υ can have any value) can be defined, too.

8.4   Conformon P systems and P/T systems

Let us call the *total value* the sum of all the values of the conformons present in a conformon P system.

A conformon P system can be simulated by a P/T system in the following way. A place in the P/T system identifies a conformon present in a compartment. If, for instance, the conformon $[A, 5]$ is present in compartment 1, then the relative place is $[A, 5]_{[1]}$. The number of tokens present in a place identifies the number of occurrences of that conformon in that compartment and the transitions identify passage and interaction rules.

For instance, the passage of $[A, 5]$ from compartment 1 to compartment 2 can be simulated by the net depicted in Fig. 8.9(a). The interaction of $[A, 5]$ and $[B, 0]$ according to the rule $A \xrightarrow{3} B$ in compartment 1 can be simulated by the net depicted in Fig. 8.9(b).

We know from Chapter 4 that the building blocks can be arranged so as to obtain the nets depicted in Fig. 8.9 (see Fig. 4.18).

The simulation of conformon P systems with finite total value and finite number of conformons can then be easily done with a P/T system. Theorems 8.1, 8.2 and 8.3 consider conformon P systems with an infinite number of conformons. This kind of conformon P system can be simulated by P/T systems as described in Section 4.6.

In the following sections we return to the links between conformon P systems and P/T systems.

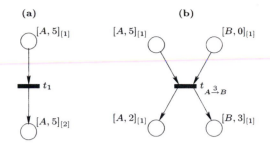

Fig. 8.9 (a) Nets simulating the passage and **(b)** the interaction of conformons

8.5 One and two infinite elements

In the following theorem we consider conformon P systems such that all the conformons present in an infinite number of occurrences have 0 as value. In this way the total value is finite.

Theorem 8.1 *The family of sets numbers generated by conformon P systems coincides with that generated by partially blind register machines.*

Proof *Part I: (Such conformon P systems can simulate partially blind register machines)* The proof is based on the conformon P system depicted in Fig. 8.10.

In the initial configuration of the conformon P system for each register γ of the simulated partially blind register machine there is an infinite number of occurrences of $[\gamma, 0]$ in compartment 5. Compartment 4 is the input compartment and initially it contains $val(\gamma)$ occurrences of $[\gamma, 0]$, where $val(\gamma)$ is the value of γ in the initial configuration of the simulated machine. The addition of 1 to γ is simulated moving one occurrence of $[\gamma, 0]$ from compartment 5 to compartment 4; the subtraction of 1 is simulated with the passage of one occurrence of $[\gamma, 0]$ from compartment 4 to compartment 5. The acknowledgement compartment is compartment 2. The system stops if there is no $[\gamma, 0]$ in compartment 4 while the system tried to simulate a subtraction from γ.

Let $S = \{s_1, \ldots, s_f\}$ be the set of states of the register machine $M = (S, I, s_1, s_f)$. For each state $s \in S$, the initial configuration of the conformon P systems has $[s_j, 0]$ in compartment 6. For each instruction of the kind $(s, \gamma^+, v) \in I$, $s, v \in S$, the initial configuration of the conformon P system has $[s'_{j,\gamma}, 0]$ in compartment 1; for each instruction of the kind $(s, \gamma^-, v) \in I$ the initial configuration of the conformon P system has $[s''_{j,\gamma}, 0]$ in compartment 1.

At most one conformon of the kind $[s_i, 7], 1 \leq i \leq f$, can be present in compartment 1 ($[s_i, 0]$ is not present in compartment 6, then). Initially it is the one related to the initial state of the simulated machine.

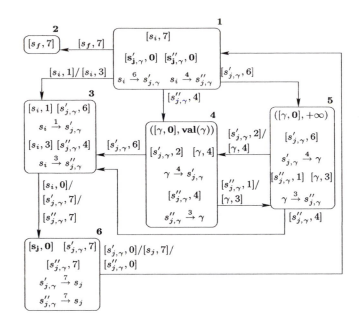

Fig. 8.10 The conformon P system related to Theorem 8.1

If the instruction $(s_i, \gamma^+, s_j) \in I$, then the interaction rule $s_i \xrightarrow{6} s'_{j,\gamma} \in R_1$. The conformons $[s'_{j,\gamma}, 6]$ and $[s_i, 1]$ are generated when this rule is applied, then they can pass into compartment 5 and compartment 3, respectively. In compartment 5 $[s'_{j,\gamma}, 6]$ is needed to let an occurrence of $[\gamma, 0]$ pass from this compartment to the final one. This is performed by first increasing to 4 the value of an occurrence $[\gamma, 0]$. When this happens $[s'_{j,\gamma}, 2]$ and $[\gamma, 4]$ can pass into compartment 4. Here these two conformons can interact by $\gamma \xrightarrow{4} s'_{j,\gamma}$ so that $[\gamma, 0]$ and $[s'_{j,\gamma}, 6]$ are generated. Conformon $[\gamma, 0]$ remains in this compartment, while $[s'_{j,\gamma}, 6]$ can pass into compartment 3. Here the conformons $[s'_{j,\gamma}, 6]$ and $[s_i, 1]$ can interact so as to generate $[s'_{j,\gamma}, 7]$ and $[s_i, 0]$. These last two conformons can pass into compartment 6 where $[s'_{j,\gamma}, 7]$ interacts with $[s_j, 0]$. The result of this interaction generates $[s'_{j,\gamma}, 0]$ and $[s_j, 7]$ and both these conformons can pass into compartment 1.

If the instruction $(s_i, \gamma^-, s_j) \in I$, then the rule $s_i \xrightarrow{4} s''_{j,\gamma} \in R_1$. The passage of one occurrence of $[\gamma, 0]$ from compartment 4 to compartment 5 is performed in a way similar to what was just described. In case no conformon $[\gamma, 0]$ is present in compartment 4, then $[s'_{j,\gamma}, 4]$ can only remain in compartment 4.

When $[s_f, 7]$ (s_f is the final state of the register machine) is present in compartment 1, then it can only pass into the acknowledge compartment, compartment 2 in the figure, in this way halting the computation.

Part II: (Partially blind register machines can simulate such conform P systems) It is a consequence of:

the number of compartments, the different conformons and the total value are finite quantities in such conformon P systems;

operations are applied in an asynchronous way;

the operations in such conformon P systems can be simulated by the nets depicted in Fig. 8.9;

the nets depicted in Fig. 8.9 can be obtained by compositions of building blocks (see Fig. 4.18);

Theorem 4.3. □

If we consider Section 8.4, then the simulation of instructions of the kind (s_i, γ^-, s_j) can be simulated by the P/T system depicted in Fig. 8.11(a). This net is composed of the sub-nets depicted in Fig. 8.9.

The net in Fig. 8.11(a) can be simulated by that depicted in Fig. 8.11(b). Here the conformons $[\gamma, 0]$ in compartment 5 are disregarded as they are present in an infinite amount (that is, they are invariant). Notice that what was depicted in Fig. 8.11(b) is just a *join* building block.

The remaining nets present in this chapter are not composed of the sub-nets depicted in Fig. 8.9; for simplicity only rewritings such as that in Fig. 8.11(b) are depicted.

If one of the features discussed in Section 4.2 is added to the conformon P systems considered in the previous theorem, then computationally complete systems are obtained.

Theorem 8.2 *The family of sets of numbers generated by conformon P systems where interactions rules have priority on passage rules coincides with that generated by register machines.*

Proof The conformon P system, depicted in Fig. 8.12, is very similar to that related to Theorem 8.1 and depicted in Fig. 8.10. We assume that readers are familiar with Part I of the proof of Theorem 8.1; here we only describe the differences related to this proof.

Let $M = (S, I, s_1, s_f)$ be a register machine. For each instruction $(s_i, \gamma^-, s_j, s_k) \in I$ the rule $s_i \xrightarrow{4} s''_{j,\gamma} \in R_1$. If the conformon $[s_i, 7]$ is present in this compartment, then the conformons $[s_i, 3]$ and $[s''_{j,\gamma}, 4]$ are generated and they can pass into compartment 3 and 4, respectively.

If in compartment 4 there is an occurrence of $[\gamma, 0]$, then the priorities force the application of $s''_{j,\gamma} \xrightarrow{3} \gamma$, otherwise $[s''_{j,\gamma}, 4]$ can pass into compartment 7. This is the only application of priorities in the conformon P system.

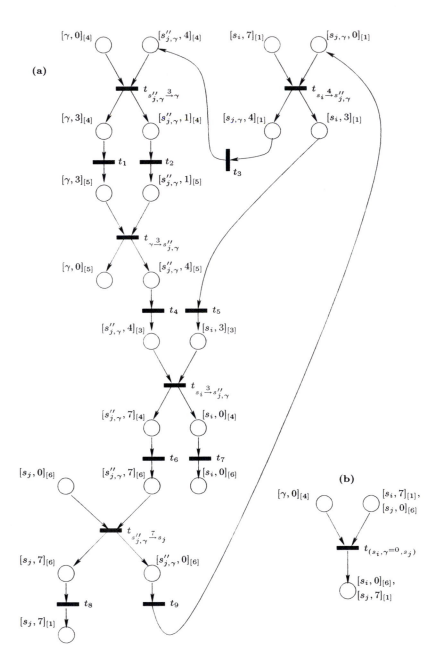

Fig. 8.11 (a) Net for the simulation of (s_i, γ^-, s_j) related to Theorem 8.1 and (b) its rewriting

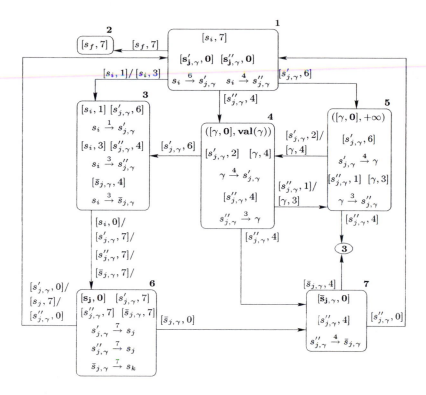

Fig. 8.12 The conformon P system with priorities related to Theorem 8.2

In case $[s''_{j,\gamma}, 4]$ passes to compartment 7 it can interact with $[\bar{s}_{j,\gamma}, 0]$ initially present in this compartment. The conformons $[s''_{j,\gamma}, 0]$ and $[\bar{s}_{j,\gamma}, 4]$ are generated and they can pass into compartments 1 and 3, respectively. In compartment 3 $[\bar{s}_{j,\gamma}, 4]$ can interact with $[s_j, 3]$. This generates $[\bar{s}_{j,\gamma}, 7]$ and $[s_j, 0]$ and they can both pass into compartment 6. Here $[\bar{s}_{j,\gamma}, 7]$ and $[s_k, 0]$ can interact so as to generate $[\bar{s}_{j,\gamma}, 0]$ and $[s_k, 7]$. These two conformons can pass into compartment 1. □

Similarly to what was done for Theorem 8.1 the simulation of instructions of the kind $(s_i, \gamma^-, s_j, s_k)$ can be performed by a P/T system having as underlying net that depicted in Fig. 8.13. This net is similar to that depicted in Fig. 4.8.

We now consider conformon P systems with an infinite number of conformons and these conformons can have a positive integer as value. In this way the total value becomes infinite, too.

Theorem 8.3 *The family of sets of numbers generated by conformon P systems with infinite value coincides with that generated by register machines.*

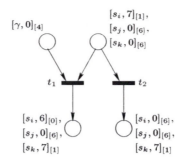

Fig. 8.13 Net related to Theorem 8.2 and underlying a P/T system simulating $(s_i, \gamma^-, s_j, s_k)$

Proof Figure 8.14 represents such a conformon P system with infinite value simulating a register machine $M = (S, I, s_1, s_f), S = \{s_1, \ldots, s_f\}$. During this proof we refer to this figure.

For each state $s_j \in S, 1 \leq j \leq f$, the initial configuration of the conformon P systems has $[s_j, 0]$ in compartment 5. For each instruction $(s_i, \gamma^+, s_j) \in R$ the initial configuration of the conformon P system has $[s'_{j,\gamma}, 0]$ in compartment 1;

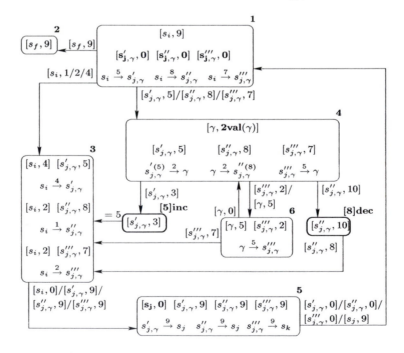

Fig. 8.14 The conformon P system with infinite value related to Theorem 8.3

for each instruction $(s_i, \gamma^-, s_j, s_k) \in I$ the initial configuration of the conformon P system has $[s''_{j,\gamma}, 0]$ and $[s'''_{j,\gamma}, 0]$ in compartment 1. Moreover, if $val(\gamma)$ is the initial value of the register γ in M, then the conformon $[\gamma, 2val(\gamma)]$ is present in compartment 4 in the initial configuration. The simulation of the addition of 1 to a register corresponds to the addition of 2 to the value of the related conformon; the simulation of subtraction of 1 from a register corresponds to the subtraction of 2 from the value of the related conformon. Considering that initially empty registers are represented by conformons having 0 as value we have that the value of any γ conformon is an even number.

At most one conformon of the kind $[s_i, 9]$ can be present in compartment 1. Initially it is that related to the initial state of the register machine.

If the instruction $(s_i, \gamma^+, s_j) \in I$, then the interaction rule $s_i \xrightarrow{5} s'_{j,\gamma}$ is in R_1. The conformons $[s_i, 4]$ and $[s'_{j,\gamma}, 5]$ are generated when this rule is applied and then they can pass into compartments 3 and 4, respectively.

In compartment 4 the conformons $[s'_{j,\gamma}, 5]$ and $[\gamma, 2k_\gamma]$ can strictly interact: only when the value of the former conformon is 5 can it pass two units of its value to the latter conformon. When this happens $[s'_{j,\gamma}, 3]$ can pass into an increaser changing the value of this conformon into 5. The resulting conformon can then pass into compartment 3 and interact with $[s_i, 4]$ so that $[s'_{j,\gamma}, 9]$ and $[s_i, 0]$ are generated. These last two conformons can pass into compartment 5. Here $[s'_{j,\gamma}, 9]$ can interact with conformon $[s_j, 0]$. The outcome of this interaction, conformons $[s'_{j,\gamma}, 0]$ and $[s_j, 9]$, can pass into compartment 1.

The simulation of the instruction $(s_i, \gamma^-, s_j, s_k) \in I$ is performed by 'gambling' (see Section 4.3). For each of these instructions two interaction rules $s_i \xrightarrow{8} s''_{j,\gamma}$ and $s_i \xrightarrow{7} s'''_{j,\gamma}$ are in R_1. In case the former interaction rule is applied, then the conformon P system 'gambles' that the register is not empty. This is performed by the strict interaction between the γ and the $s''_{j,\gamma}$ conformons: the former can pass two units of its value to the latter only if the value of the latter is 8. When this happens $[s''_{j,\gamma}, 10]$ can pass to a decreaser that brings its value to 8. From here $[s''_{j,\gamma}, 8]$ can pass into compartment 3.

In case the conformon P system 'gambles' that the register is empty, then $[s'''_{j,\gamma}, 7]$ is generated in compartment 1 and it can pass into compartment 4. Here it can interact with $[\gamma, 2k_\gamma]$ so that $[s'''_{j,\gamma}, 2]$ and $[\gamma, 2k_\gamma + 5]$ are generated. The former conformon can pass into compartment 6 while the latter one can pass into this compartment only if $k_\gamma = 0$. If this does not happen, then the conformon P system never reaches a configuration with a conformon in the acknowledge compartment. It is relevant to notice that the value of the γ conformon can be 5 (an odd number) only because of this sequence of configurations. If k_γ was 0, then $[\gamma, 5]$ can pass into compartment 6 and here interact with $[s'''_{j,\gamma}, 2]$ such that $[\gamma, 0]$ and $[s'''_{j,\gamma}, 7]$ are generated. The former conformon can pass back to compartment 4, while the latter is passed to compartment 3.

When $[s_f, 7]$ (s_f is the final state of the register machine) is present in compartment 1, then it can only pass into the acknowledge compartment, compartment 2 in the figure, in this way halting the computation.   □

As indicated in the proof of this theorem the simulation of instructions of the kind $(s_i, \gamma^-, s_j, s_k)$ is performed by 'gambling' (see Section 4.3). A net performing such a simulation is depicted in Fig. 8.15, where the dotted lines suggest the presence of an infinite number of places, transitions and elements of the flow relation. We have not detailed the name of the places in this net. Instead of indicating the full configuration for each place in this net we only indicate the conformon and compartment that is essential in the configuration to perform some operations. We are also not precise in the indication of the transitions: we indicate only one transition instead of several.

The net underlying a P/T system depicted in Fig. 8.15 recalls the one depicted in Fig. 4.13 if p_γ is replaced by $p_{\gamma=0}$, $p_{\gamma=1}$, etc. In Fig. 8.15 the place $[s_i, 9]_{[1]}$

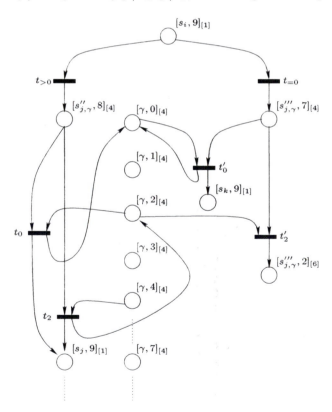

Fig. 8.15 Net related to Theorem 8.3 and underlying a P/T system simulating $(s_i, \gamma^-, s_j, s_k)$

refers to all the configurations in which conformon $[s_i, 9]$ is in compartment 1. As we know from the proof of Theorem 8.3, when this happens, then either $s_i \xrightarrow{8} s''_{j,\gamma}$ or $s_i \xrightarrow{7} s'''_{j,\gamma}$ can be applied. The conformon $[s''_{j,\gamma}, 8]$ can then be present in compartment 4. This is indicated in Fig. 8.15 by the place $[s''_{j,\gamma}, 8]_{[4]}$. The transition $t_{>0}$ denotes all the operations applied to reach this configuration. In a similar way the remaining net depicted in Fig. 8.15 can be understood. In this net the place $[s''''_{j,\gamma}, 2]_{[6]}$ is equivalent to the place p_{wrong} present in the net depicted in Fig. 4.13.

8.5.1 Negative values

What if the value of a conformon can become negative? In this case we have *conformon P systems with negative values*. If, for instance, we consider the conformons $[G, 5]$ and $[R, 4]$ and the rule $G \xrightarrow{3} R$, then the rule can be applied forever leading to $[G, 5 + 3n]$ and $[R, 4 - 3n], n \geq 1$. Despite the fact that the previous proofs do not consider this possibility, they can easily be adapted to this new feature.

As splitters do not contain interaction rules they are not directly affected by possible negative values. Separators, on the other hand, are strongly based on the fact that some interactions occur only once. A separator working also with negative values is depicted in Fig. 8.16. Also in this figure the dotted lines suggest the presence of more compartments, modules and edges.

It works essentially in the same way as the separator depicted in Fig. 8.5 with the difference that conformons which interacted in compartment 1 more than once pass (through $\mathbf{spl}_1$) to compartment $1'$ where the reverse interactions can take place. Considering what we said in Section 8.3 in relation to splitters, the edge from $\mathbf{spl}_1$ to compartment $1'$ having label $\geq 3a + 2h + 4i$ in Fig. 8.16 is allowed.

The definition of separators working with negative values simplifies the description of conformon P systems with negative values. Increasers and decreasers work also for negative values.

Theorem 8.4 *Conformon P systems with negative values can simulate partially blind register machines.*

Proof Such a conformon P system can be obtained by the method described in Part I of the proof of Theorem 8.1 where for each interaction rule its reverse is present. $\qquad \square$

Theorem 8.5 *The class of numbers generated by conformon P systems with negative values and infinite value coincides with that generated by register machines.*

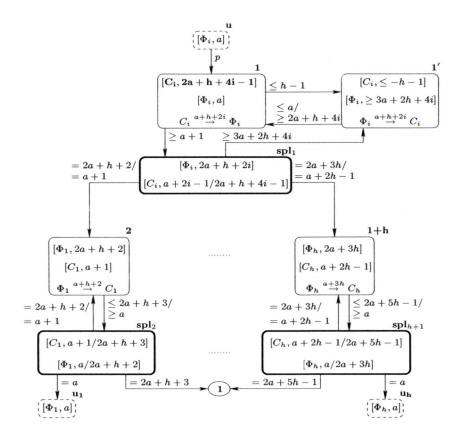

Fig. 8.16 A detailed separator working with negative values

Proof Such a conformon P system can be obtained by the method described in the proof of Theorem 8.3 where for each interaction rule its reverse is present.

□

8.6   Infinite hierarchies

Some time ago the search for a non-computationally complete model of membrane system for which the number of membranes induces an infinite hierarchy on the computation that can be performed by such a system was begun.

The answer to this quest was found by Oscar H. Ibarra. His answer used a *communicating membrane system* working under maximal parallelism, a model of a membrane system that we do not treat in this monograph, simulating restricted register machines. From Theorem 4.4 and Theorem 4.1 we know that restricted register machines can be weakly simulated by SCG related to P/T systems.

In this section we describe how restricted models of conformon P systems can induce infinite hierarchies on the computation they can perform.

Conformon-restricted conformon P systems can have more than one input compartment and they have only one conformon with a distinguished name, let us say l, encoding the input. The formal definition of such conformon P systems then changes into:

$$\Pi = (V, \mu, input_{z_1}, \dots, input_{z_n}, L_1, \dots, L_m, R_1, \dots, R_m)$$

where $z_1, \dots, z_n \in Q$ denote the input compartments. In the initial configuration the input compartments contain only l conformons and no other compartment contains l conformons. The definitions of configuration, transition, computation, halt, stop and set of numbers accepted follow from those given in Section 8.1.

Lemma 8.1 *Conformon-restricted conformon P systems can simulate restricted register machines with two registers.*

Proof The conformon P system related to this proof is depicted in Fig. 8.17. It is very similar to that depicted in Fig. 8.14. The only difference between the two figures is that the increaser and decreaser modules present in Fig. 8.14 have been substituted by compartment 7 in Fig. 8.17. This proof is then similar to that of Theorem 8.2.

Let $M = (S, I, s_1, s_f)$ be a restricted register machine with two registers: γ_1 and γ_2. The only instructions present in this register machine are of the kind $(s_i, \alpha^-, \beta^+, s_j, s_k)$ with $\alpha, \beta \in \{\gamma_1, \gamma_2\}, \alpha \neq \beta$. If the initial values of these registers is $val(\gamma_1)$ and $val(\gamma_2)$, respectively, then the value of the γ conformons initially present in compartments 4 and 7, the input compartments of the conformon P system, is $2val(\gamma_1)$ and $2val(\gamma_2)$, respectively. We refer by $\gamma_{[4]}$ to the conformon $[\gamma, 2val(\gamma_1)]$ initially present in compartment 4 and with $\gamma_{[7]}$ to the conformon $[\gamma, 2val(\gamma_2)]$ initially present in compartment 7. These conformons are the only ones present in compartments 4 and 7 in the initial configuration, so γ is the distinguished conformon name encoding the input. This is similar to what happens to $\gamma_{[4]}$ and $\gamma_{[7]}$.

The conformons with name $s'_{j,\gamma}$ are used to increase the value of $\gamma_{[4]}$ and then to decrease, if possible, that of $\gamma_{[7]}$. The conformons with name $s''_{j,\gamma}$ are use to decrease, if possible, the value of $\gamma_{[4]}$ and then to increase that of $\gamma_{[7]}$. The conformons with name $s'''_{j,\gamma}$ and $s''''_{j,\gamma}$ test if the value of respectively $\gamma_{[4]}$ and $\gamma_{[7]}$ is 0. If any of these operations cannot be performed, then the system never reaches a configuration with a conformon in the acknowledgement compartment.

The system halts when the conformon $[s_f, 9]$ passes to compartment 2, the acknowledgement compartment. $\square$

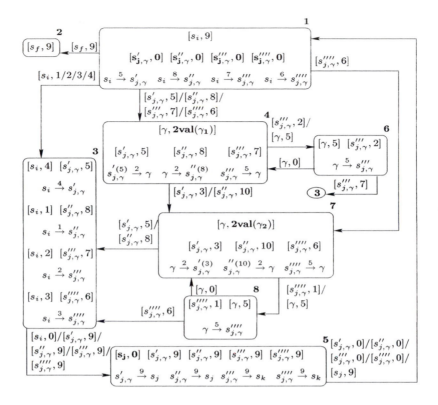

Fig. 8.17 The conformon-restricted conformon P system related to Lemma 8.1

This constructive proof can be used to define conformon-restricted conformon P systems simulating restricted register machines with any number of registers.

Let us assume that a specific restricted register machine has n registers $\{\gamma_1, \ldots, \gamma_n\}$ each with an initial value. Then it is possible to build a conformon-restricted conformon P system Π' having conformons with name γ initially present in n different input compartments. Considering the proof of Lemma 8.1 this seems to be a must as collecting more than one conformon with name γ in the same compartment would not allow the system Π' to perform a simulation on the restricted register machine (we discuss this in Section 8.7). The system Π' would be such that every time the value of a γ conformon is increased (decreased), then the value of its connected register (for the particular simulated instruction) is decreased (increased, respectively) by the same amount. The subscripts or the value of the s', s'', s''' and s'''' conformons can be used to pass these conformons to specific compartments in order to interact with the appropriate γ conformons.

So we can state:

Corollary 8.1 *Conformon-restricted conformon P systems can simulate restricted register machines.*

Here is the reverse of this inclusion:

Lemma 8.2 *Restricted register machines with* n *registers can simulate conformon-restricted conformon P systems having* n *input compartments.*

Proof This is a consequence of:

the number of compartments, the different conformons and the total value are finite quantities in such conformon P systems;

operations are applied in an asynchronous way;

the operations in such conformon P system can be simulated by the nets depicted in Fig. 8.9(b);

the nets depicted in Fig. 8.9(b) can be obtained by compositions of building blocks (see Fig. 4.18);

Theorem 4.4. □

Knowing from Theorem 3.3 that restricted register machines induce an infinite hierarchy on the number of registers, Theorem 3.3, then we can state:

Theorem 8.6 *Conformon-restricted conformon P systems induce an infinite hierarchy on the number of compartments.*

Compartment-restricted conformon P systems have only one input compartment but only conformons with name from a finite set of names can be used as input. Conformons having there names from this set are called *input conformons*.

An initial configuration is such that some input conformons are present in the input compartment, no input conformon is present in the remaining compartments, and the acknowledgement compartment is empty. The definition of configuration, transition, computation, halt, stop and set of numbers accepted follow from those given in Section 8.1.

Lemma 8.3 *Compartment-restricted conformon P systems with two input conformons can simulate a restricted register machine with two registers.*

Proof The conformon P system related to this proof is depicted in Fig. 8.18. It is very similar to that depicted in Fig. 8.17. This proof is then similar to that of Lemma 8.1 (and, in turn, to that of Theorem 8.2).

Let $M = (S, I, s_1, s_f)$ be a restricted register machine with two registers: γ_1 and γ_2. The only instructions present in this register machine are of the kind $(s, \alpha^-, \beta^+, v, w)$ with $\alpha, \beta \in \{\gamma_1, \gamma_2\}, \alpha \neq \beta$. Let compartment 4 be the input compartment of the conformon P system and let γ_1 and γ_2 be the two input

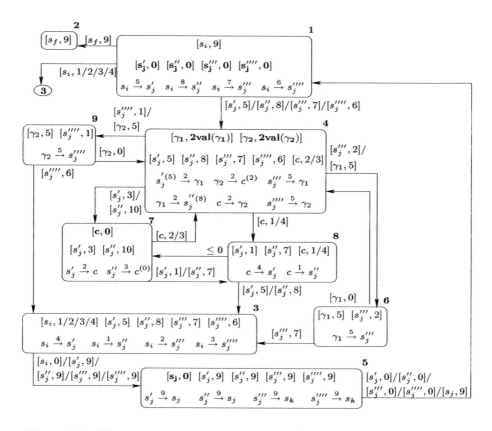

Fig. 8.18 The compartment-restricted conformon P system related to Lemma 8.3

conformons. If the initial values of the registers of M is $val(\gamma_1)$ and $val(\gamma_2)$, then the conformons $[\gamma_1, 2val(\gamma_1)]$ and $[\gamma_2, 2val(\gamma_2)]$ are present in compartment 4 in the initial configuration.

Let us assume that the instruction $(s_i, \gamma_1^-, \gamma_2^+, s_j, s_k) \in I$ and that the conformon P system 'gambles' that the γ_1 conformon has a value bigger than 0 (so, the system tries to decrease the value of the conformon with name γ_1). The conformon $[s_j'', 8]$ is generated in compartment 1 and it can pass into compartment 4. Here it can strictly interact with the γ_1 conformon decreasing its value by 2 (the reason why 2 and not 1 is explained in the proof of Theorem 8.2). Then $[s_j'', 10]$ can pass into compartment 7 and strictly interact with $[c, 0]$ (present here from the initial configuration) so that $[s_j'', 7]$ and $[c, 3]$ are generated. These last two conformons can pass into compartments 8 and 4, respectively. In compartment 8 $[c, 3]$ can pass two units of its value to the γ_2 conformon, then $[c, 1]$ can pass into compartment 8. Here $[c, 1]$ and $[s_j'', 7]$ interact so that $[c, 0]$ and $[s_j'', 8]$ are generated. They can then pass into compartment 7 and 3, respectively. What happens in compartment 3 is described in the proof of Theorem 8.2.

The conformons with name $s'_{j,\gamma}$ are used to increase the value of $\gamma_{[4]}$ and then to decrease, if possible, that of $\gamma_{[7]}$. The conformons with name $s'''_{j,\gamma}$ and $s''''_{j,\gamma}$ test if the value of respectively $\gamma_{[4]}$ and $\gamma_{[7]}$ is 0. If any of these operations cannot be performed, then the system never halts.

The system halts when the conformon $[s_f, 9]$ passes into compartment 2, the acknowledgement compartment. $\qquad\square$

Let us assume that a specific restricted register machine has n registers $\{\gamma_1, \ldots, \gamma_m\}$. Then it is possible to build a compartment-restricted conformon P system having input conformons with name $\gamma_1, \ldots, \gamma_m$ in the input compartment. The equivalence between the number of registers and the number of different names of input conformons seems to be a must (we discuss this in Section 8.7). The compartment-restricted conformon P system would be such that every time the value of an input conformon is increased (decreased), then one of its connected registers (for the particular simulated instruction) is decreased (increased, respectively) by the same amount. The information on the connected register can be present in the name or in the value of the state conformons (that is, conformons which are not input conformons).

So we can say:

Corollary 8.2 *Compartment-restricted conformon P systems can simulate restricted register machines.*

Here is the reverse of this inclusion whose proof is similar to that of Lemma 8.2:

Lemma 8.4 *Restricted register machines with n registers can simulate compartment-restricted conformon P systems having n input conformons.*

Knowing from Theorem 3.3 that restricted register machines induce an infinite hierarchy on the number of registers, Theorem 3.3, then we have:

Theorem 8.7 *Compartment-restricted conformon P systems induce an infinite hierarchy on the number of input conformons.*

8.7 Final remarks and research topics

Differently from other models of membrane systems several conformon P systems do not contains pairs of rules of the kind $A \rightarrow B$ and $B \rightarrow A$. Such a pair is called a *local cycle* and these rules have to be regarded as pseudo-rules, that is, any kind of rule that can occur in a compartment in a model of a membrane system. So, in a conformon P system a local cycle is given by an interaction rule and its reverse present in the same compartment. Notably, the conformon P system defined in the proof of Theorem 8.1 does not have local cycles.

The absence of local cycles becomes important the moment in which one tries to implement membrane systems using (bio)molecules: if a local cycle is present, then the solution never goes to equilibrium.

The concept of *cycle* can be extended to the entirety of a system and refers to the possible flow of information in it. If, for instance, one considers the conformon P system depicted in Fig. 8.10, then the cycles present in it are: (1, 3, 6), (1, 4, 3, 6), (1, 5, 3, 6), (1, 4, 5, 3, 6), etc. where each number refers to a compartment.

Suggestion for research 8.1 *Study how the topology of these cycles is related to the computational power of the systems. Are there different topologies 'equivalent' from a computational point of view?*

Is it possible to simulate with models of conformon P systems other models of register machines? *Blind register machines*, for instance, are register machines that cannot sense (so neither halt or check) if the value of a register is zero. It is known that these machines are strictly less powerful from a computational point of view than partially blind register machines.

Suggestion for research 8.2 *Find a model of conformon P systems equivalent to blind register machines and try to deduce more general characterisation laws (such as those discussed in Chapter 4).*

Theorem 8.4 gives only a lower limit on the computational power of conformon P systems with negative values. It would be very interesting to know their upper limit.

Suggestion for research 8.3 *Can partially blind register machine simulate conformon P systems with negative values?*

We conclude this section posing a suggestion for research related to models of membrane systems inducing an infinite hierarchy. The proofs in the presence of such hierarchies are based on simulations from the set of configurations of the simulated system (for instance, a restricted register machine) to the considered membrane system model. In this simulation the content of the registers of the restricted register machine is represented, for instance, by the number of symbols in the P system.

Suggestion for research 8.4 *Is it possible to have another simulation such that the infinite hierarchy is not induced? If not, why not?*

In the previous section we wrote: 'considering the proof of Lemma 8.1 this seems to be a must as collecting more than one conformon with name γ in the same compartment would not allow the system Π' to perform a simulation on the restricted register machine' and also 'The equivalence between the number of registers and the number of different names of input conformons seems to be a must'. These 'must' have no proof even if our common sense suggests that it is not possible to do otherwise.

8.8 Bibliographical notes

The concept of *conformon* was introduced independently in [94] and [250]. Following the definition given in [94] conformons and conformon-like entities were classified into 10 families according to their biological functions [136]. More about the Bhopalator can be found in [135, 137]. The term *conformon* was adopted in [143, 144] where the authors started to develop a quantum mechanical theory based on this concept. It is fair to say that conformons and models based on them do not play a significant role in biology.

Conformon P systems were introduced in [78]. This chapter is mainly based on [78, 80, 82].

Proofs of Theorem 8.1 and Lemma 8.2 not based on P/T systems can be found in [78] and [80], respectively.

The search indicated in Section 8.6 was begun in [210] and an award related to it was advertised in [238].

Candidate solutions to this problem were reported in [64, 148], but they were based on definitions that were considered too restrictive, so they were not accepted as solutions.

In [117] several models of membrane systems answering the problem were proposed and accepted as solutions to it. The research presented in [117], based on communicating membrane system [233, 236], was followed by [120, 122] where infinite hierarchies on the number of symbols, the number of membranes or on the computational power of (restricted) variants of membrane systems were presented.

The problems related to a possible implementation of membrane systems are described in [93], and [90] indicates how some of these problems can be solved with the help of conformon P systems.

Conformon P systems were successfully used as a platform to model biological processes. Interested readers can refer to [52, 85, 86].

9 Splicing

The model of a membrane system presented in this chapter has been inspired by a complex biological process involving DNA molecules and enzymes. In order to fully understand it several biological concepts and processes have to be introduced. This is done in the following section.

9.1 Biological background

9.1.1 DNA structure

DNA, or *deoxyribonucleic acid*, is a molecule that carries the genetic information in living cells and it can be synthesised in a biological laboratory. A *single stranded* DNA molecule is in fact a sequence of *nucleotides*. Each nucleotide consists of three parts: a *sugar* (deoxyribose sugar), a *phosphate group* and a *base*. While in DNA there is only one kind of sugar and one kind of phosphate group in all nucleotides, there are four types of possible bases: *adenine, thymine, cytosine* and *guanine*, abbreviated by A, T, C and G, respectively. Accordingly, there are four types of nucleotides also denoted by A, T, C and G, respectively.

The structure of a single stranded DNA molecule is sketched in Fig. 9.1. In this figure each group composed of $C_{5'}, \ldots, C_{1'}$ and four short segments between them, represents the sugar component of the nucleotide while the phosphate group is indicated with P. One sugar component and the phosphate group belonging to one nucleotide are highlighted by the grey shadow in Fig. 9.1. The sugar component has five attachment points: the carbon atoms indicated with C in the figure and numbered $1'$–$5'$. The phosphate group is attached to $5'$ and the base is attached to $1'$. The bond between two consecutive nucleotides is established between the $3'$ attachment point of one nucleotide and the (phosphate group at the) $5'$ attachment point of the next nucleotide. Figure 9.2 gives a much more detailed view of a single stranded DNA molecule that includes many chemical details.

Note that the nucleotide at one ('leftmost') end of the molecule has the $5'$-*end* available for binding with yet another nucleotide, while the nucleotide at the other ('rightmost') end of the molecule has the $3'$-*end* available for binding. Therefore the single stranded molecule has an *orientation* (polarity): it is possible to distinguish one end of the molecule from the other one. Thus one can read

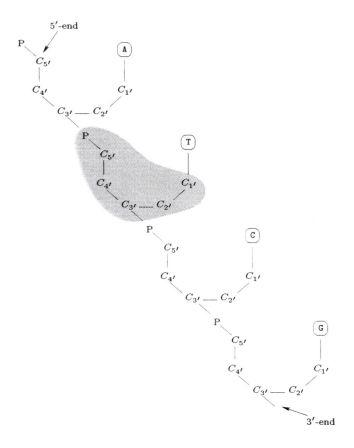

Fig. 9.1 Sketch of a single stranded DNA molecule

the sequence of nucleotides forming a single stranded DNA molecule either from its 5′-end to the 3′-end or from the 3′-end to the 5′-end. It turns out that nature 'favours' the 5′ to 3′ direction for some processing of DNA.

Two single strands of DNA molecules can bind together to form a *double stranded* DNA molecule. The two strands hold together by (hydrogen) bonds between their bases. These bonds hold between complementary bases, that is, between A in one strand and T in the other, and between C in one strand and G in the other. This complementarity between A and T, and C and G is referred to as the *Watson–Crick complementarity*. The other necessary condition for forming a double stranded DNA molecule is that the two single stranded components align with opposite orientations, that is, the first nucleotide at the 5′-end of one strand binds (by complementarity) with the first nucleotide at the 3′-end of the other strand, the second nucleotide at the 5′-end of one strand binds with the second

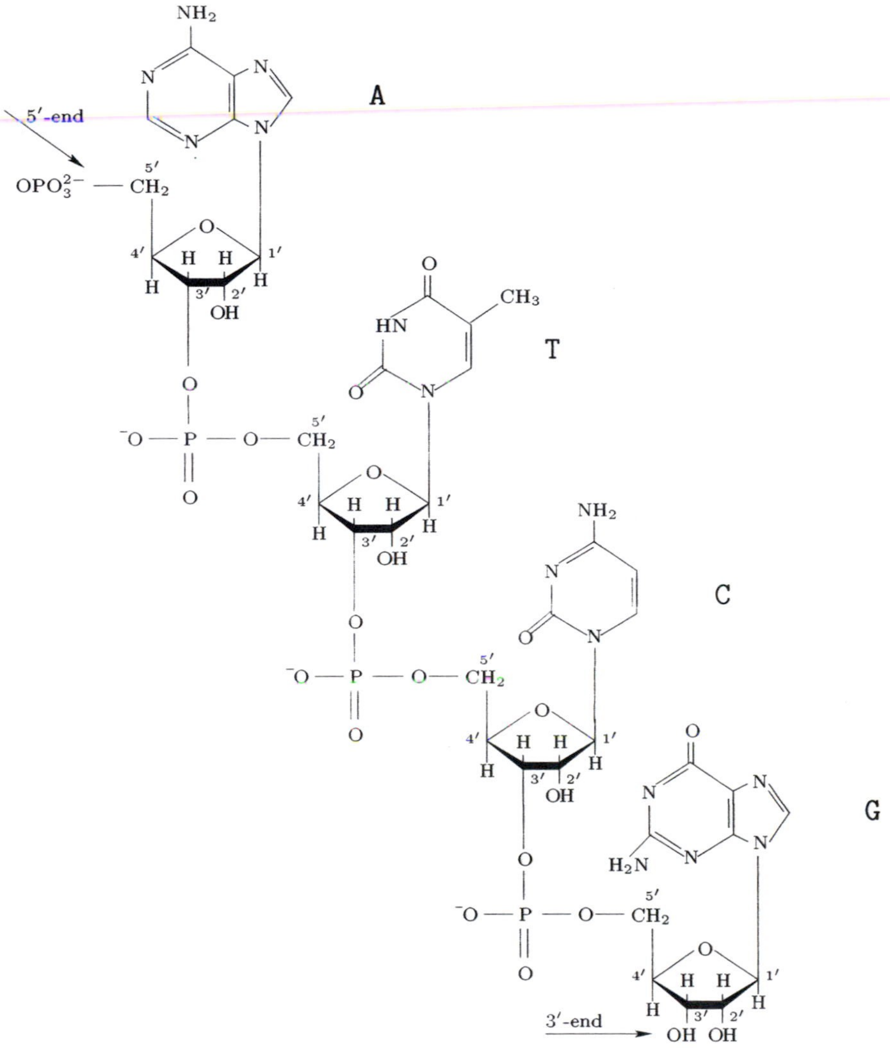

Fig. 9.2 The chemical structure of a single stranded DNA molecule

nucleotide at the 3'-end of the other strand, etc. This situation is illustrated in Fig. 9.3 where the horizontal thick lines indicate the sequences of sugars and phosphate groups.

Two single stranded DNA molecules can also bind together in such a way that the resulting double stranded DNA molecule is not 'perfect', that is, not every nucleotide has a partner (the nucleotide to which it binds) in the other strand. In particular one can get molecules such as that depicted in Fig. 9.4,

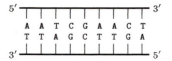

Fig. 9.3 Double stranded DNA

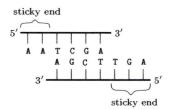

Fig. 9.4 A DNA molecule with sticky ends

where nucleotides without partners are positioned at one or both ends of the molecule, forming so-called *sticky ends*. DNA molecules of this kind are also called *partially double*.

9.1.2   Restriction enzymes, ligases and polymerases

Enzymes are molecules that control chemical reactions in cells by acting as catalysts (see Section 6.1), that is, speeding up the rate of chemical reactions without being consumed themselves in the process. Three groups of enzymes are relevant for the model of membrane systems considered in this chapter: restrictions enzymes, ligases and polymerases.

Restriction enzymes recognise specific and continuous sub-parts, called *restriction sites*, of a double stranded DNA molecule and cut it (that is, they break the bonds between the sugar and the phosphate group and the bonds between complementary bases) either within the restriction site or outside it. These enzymes help bacteria to defend themselves from invasion by viral DNA by chopping the invading DNA molecules into pieces ('restricting' it in this way). Figure 9.5 illustrates the action of an enzyme called *TaqI* (having restriction site $\frac{TCGA}{AGCT}$) on a double stranded DNA molecule. This action results in two DNA molecules with sticky ends.

The cut performed by some restriction enzymes is such that the two resulting molecules have *blunt ends* (that is, they are two double stranded DNA molecules). The cut performed by a restriction enzyme is made in such a way that the phosphate group (P) is preserved at the 5'-end and the hydroxyl (OH) group is preserved at the 3'-end of the two created DNA molecules.

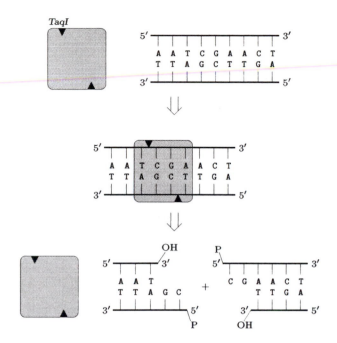

Fig. 9.5 Action of *TaqI* on a DNA molecule

Two DNA molecules with complementary sticky ends can form a new DNA molecule through the binding of the sticky ends. This is illustrated in Fig. 9.6 where the sticky end GC-5′ of molecule **a** binds to the sticky end 5′-CG of molecule **b** forming in this way a new molecule **c** (the underlying reason for the term 'sticky end' should be clear now). Note that **c** has two *nicks* at the junction points of **a** and **b** (the term 'nick' refers to a missing bond between two consecutive nucleotides).

Ligases are enzymes that 'repair' nicks, that is, they create the missing bond between a sugar and a phosphate group, providing the hydroxyl and the phosphate groups are present at the corresponding 3′- and 5′- ends. This is illustrated in Fig. 9.7. Therefore, if the sticky ends used for binding two molecules (like **a** and **b** in Fig. 9.6) were created by restriction enzymes, then the resulting nicks can be repaired by ligase. It is less likely but still possible that two double stranded DNA molecules (that is, without sticky ends) can come close to each other and form a double stranded DNA molecule if ligase is also there.

Polymerases are enzymes that can make double stranded DNA molecules from partially double ones. If polymerase, partially double DNA molecules and nucleotides are put in a solution, then polymerase binds the free nucleotides to the sticky ends of the DNA molecules extending them so that double stranded DNA molecules are formed. Polymerases can extend only following the 5′ − 3′

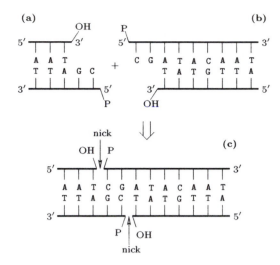

Fig. 9.6 Formation of a DNA molecule with nicks

orientation. This process is depicted in Fig. 9.8. The activity of polymerases is essential for the biological protocol that we describe in the next section.

It should be clear that the activity of enzymes does not stop after one application. If, for instance, in a test tube there are many DNA molecules with nicks and a few ligases, then, under the appropriate conditions, the ligases repair the nicks of the vast majority of DNA molecules.

9.1.3   Multiplying DNA

The models of membrane systems that we describe in this chapter require infinite numbers of strings. Since, in these models, strings represent DNA molecules, the readers may wonder how it is possible to create many copies of one DNA molecule. A biological protocol allows us to copy in a fast and efficient way small amounts of DNA molecules. This protocol is called a *polymerase chain reaction (PCR)*. It requires a solution composed of (single/double/partially double) DNA molecule with known flanking sequences, single stranded DNA molecules (*primers*) complementary to the flanking sequences, polymerase, nucleotides and the repetition of the following steps:

denaturation: the solution is heated so that the two strands of the DNA molecules come apart;

priming: the solution is cooled down so that the primers can bind to the flanking sequences;

extension: the polymerase extends the partially double DNA molecules created in the previous step.

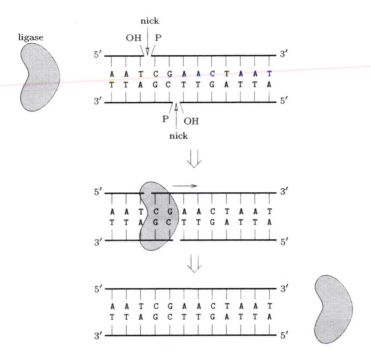

Fig. 9.7 The action of ligase on a DNA molecule with nicks

Each repetition of these steps can double the number of DNA molecules present at the beginning of the procedure.

9.2   From biological to formal splicing

As informally described in Section 9.1.2, a restriction enzyme is able to cut, with a high specificity, a double stranded DNA molecule. Because of their high specificity, restriction enzymes have been widely used in biology. If DNA molecules and restriction enzymes are present in a test tube, then these enzymes can cut the DNA. If also ligase is present, then 'new' DNA molecules (meaning molecules not initially present in the test tube) can be formed by *recombination* (of complementary sticky ends), as explained in Section 9.1.2. Of course, also DNA molecules initially present in the tube can be formed (restored).

In 1987 Tom Head translated the biological phenomenon described above into the framework of formal languages. In particular this has led to the definition of *splicing systems* which were later renamed 'H systems', where 'H' stays for 'Head'.

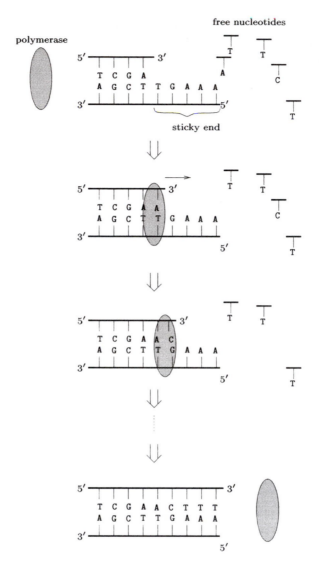

Fig. 9.8 Process performed by polymerase

The effect of restriction enzymes on double stranded DNA molecules, and the recombination of resulting molecules with sticky ends, can actually be formalised as operations on (usually) strings. It is straightforward to represent a single stranded DNA molecules with strings. In case of a double stranded DNA molecule we can consider representing only one of the two (complementary) strands. For instance, the double stranded DNA molecule depicted at the bottom of Fig. 9.8 can be represented by the string $TCGAACTTT$. It is possible to extend the alphabet of these strings grouping bases. If, for instance, we indicate that TCG is rewritten as x_1, AAC is rewritten as x_2 and TTT is rewritten as x_3, then the DNA molecule depicted in Fig. 9.8 can be represented by $x_1x_2x_3$.

The recognition site for a restriction enzyme together with the cut (performed within the site) can be represented by a triple of strings (u_1, u, u_2), where u gives the cut, u_1 is the part of the restriction site to the left of u, and u_2 is the part of the restriction site to the right of u. For example, for the restriction enzyme *TaqI* considered in Section 9.1.2, $u_1 = T$, $u = CG$, and $u_2 = A$. Continuing in this fashion, the cutting of two strings $x = x_1u_1uu_2x_2$ and $y = y_1u_3uu_4y_2$ (with the restriction sites u_1uu_2 and u_3uu_4, respectively) by the triples (u_1, u, u_2) and (u_3, u, u_4), and the subsequent recombination of the resulting string (jointly referred to as splicing of x and y) produces the strings $z = x_1u_1uu_4y_2$ and $w = y_1u_3uu_2x_2$. This is called the *splicing* of x and y. Similarly, one can formalise the cut-and-recombine action by using ordered pairs rather than triples, that is, using either (u_1u, u_2) or (u_1, uu_2).

We are now ready to introduce three main formulations of splicing systems (all using an arbitrary alphabet V rather than the four letter alphabet of nucleotides). We introduce them to show different ways to formalise the same process.

A *Head splicing system* is defined by the tuple $S_H = (V, A, R_1, R_2)$, where V is an alphabet, $A \subseteq V^*$ a finite set of strings called *axioms*, and R_1 and R_2 define sets of triples of the form (u_1, u, u_2) called *patterns*, with $u_1, u, u_2 \in V^*$. R_1 and R_2 represent sets of restriction enzymes producing DNA molecules with sticky and blunt ends, respectively. Given two strings $x = x_1u_1uu_2x_2$ and $y = y_1u_3uu_4y_2$ and two patterns (u_1, u, u_2) and (u_3, u, u_4) that are both in R_1 or both in R_2, the splicing produces $z = x_1u_1uu_4y_2$ and $w = y_1u_3uu_2x_2$ with $x_1, x_2, y_1, y_2 \in V^*$.

A *Pixton splicing system* is defined by the tuple $S_{Pi} = (V, A, R)$, where V and A are defined in the above while R is a finite set of triples of the form (u_1, u, u_2), with $u_1, u, u_2 \in V^*$. Given two strings $x = x_1u_1x_2$ and $y = y_1u_2y_2$, the splicing operation produces $z = x_1uy_2$ and $w = y_1ux_2$ with $x_1, x_2, y_1, y_2 \in V^*$. This formalisation allows us to substitute part of a string involved in a splicing.

A *Păun splicing system* is defined by the tuple $S_{Pă} = (V, A, R)$, where V and A are defined in the above, while R is a finite set of *splicing rules*.

Definition 9.1 *A* splicing rule *over an alphabet V is $r \subseteq V^*\#V^*\$V^*\#V^*$ with $\#, \$ \notin V$. If $r = u_1\#u_2\$u_3\#u_4$ and $x = x_1u_1u_2x_2$, $y = y_1u_3u_4y_2$ with $x, y \in V^*$ we write:*

$$(x_1u_1|u_2x_2, y_1u_3|u_4y_2) \vdash_r (x_1u_1u_4y_2, y_1u_3u_2x_2)$$

for some $x_1, x_2, y_1, y_2 \in V^*$, *or*

$$(x, y) \vdash_r (z, w)$$

with $z = x_1 u_1 u_4 y_2$ *and* $w = y_1 u_3 u_2 x_2$.

The application of the splicing rule can also be indicated with:

$$x_1 u_1 | u_2 x_2 \qquad x_1 u_1 u_4 y_2$$
$$\vdash_r$$
$$y_1 u_3 | u_4 y_2 \qquad y_1 u_3 u_2 x_2$$

The bars (|) in the previous definitions are only present to clearly show where the splicing takes place.

What we have just defined is called *2-splicing* as two strings, z and w, are obtained as output. For a 2-splicing we denote by z and w the first and the second output string, respectively. If instead only z is considered as output, then Definition 9.1 identifies *1-splicing*. It is known that 1-splicing can simulate 2-splicing. If, for instance, $r = u_1 \# u_2 \$ u_3 \# u_4$ is a 2-splicing rule, then the two 1-splicing rules $r' = u_1 \# u_2 \$ u_3 \# u_4$ and $r'' = u_3 \# u_4 \$ u_1 \# u_2$ simulate r.

In the rest of this chapter we use only splicing rules as in Definition 9.1.

9.3   Basic definitions

Splicing P systems, combining splicing rules in a topological compartment structure, were defined on both cell-trees and (asymmetric) cell-graphs.

Definition 9.2 *A* (generating) splicing P system *of degree m is a construct*

$$\Pi = (V, T, \mu, L_1, \ldots, L_m, R_1, \ldots, R_m) \tag{9.1}$$

where:

V *is an alphabet with* $\#, \$ \notin V$.

$T \subseteq V$ *is the terminal alphabet.*

$\mu = (Q, E)$ *is a cell-tree underlying* Π *where:*

   $Q \subset \mathbb{N}$ *contains vertices. For simplicity we define* $Q = \{0, 1, \ldots, m\}$. *Each vertex in* Q *defines a compartment of* Π.

   $E \subseteq Q \times Q$ *defines arcs between vertices, denoted by* (i, j), $i, j \in Q, i \neq j$.

$L_i \subseteq V^*$, $i \in \{1, \ldots, m\}$, *are finite languages associated with the vertices of* Q *such that every string in the language, called an* axiom, *is present in an infinite number of occurrences.*

R_i for $i \in \{1, \ldots, m\}$, are sets of evolution rules of the kind $(r; tar_1, tar_2)$ associated with the vertices in Q. These rules are such that:

$r \subseteq V^* \# V^* \$ V^* \# V^*$ is a splicing rule.

$tar_1, tar_2 \in \{here, in, out\}$ are target indicators.

The use of target indicators is an abstraction of the passage of molecules across membranes (see Section 5.1). We remind readers that a vertex in μ and a compartment in Π are used as synonymous. In Π there is one vertex of its underlying topological structure without languages and evolution rules associated with it. This vertex is the *environment*, that is, the root of μ. A *configuration* of Π is an m-tuple $(M_1, \ldots, M_m)$ of languages over V. The m-tuple $(L_1, \ldots, L_m)$ is called the *initial configuration*.

For two configurations $(M_1, \ldots, M_m), (M'_1, \ldots, M'_m)$ of Π we denote by $(M_1, \ldots, M_m) \Rightarrow (M'_1, \ldots, M'_m)$ a *transition* from $(M_1, \ldots, M_m)$ to $(M'_1, \ldots, M'_m)$, that is, the application of splicing rules in Π in a maximally parallel way. For $i \in Q$ if $x = x_1 u_1 u_2 x_2$, $y = y_1 u_3 u_4 y_2 \in M_i$ and $(r = u_1 \# u_2 \$ u_3 \# u_4; tar_1, tar_2) \in R_i$, $x_1, x_2, y_1, y_2, u_1, u_2, u_3, u_4 \in V^*$, the splicing $(x, y) \vdash_r (z, w)$, $z, w \in V^*$ can take place in i. The strings z and w pass into the vertices indicated by tar_1 and tar_2, respectively, as explained in the following.

For $j = 1, 2$:

if $tar_j = here$, then the string remains in vertex i;

if $tar_j = in$, then the string passes to a child of vertex i, that is, to one of the vertices k only if $(i, k) \in E$;

if $tar_j = out$, then the string passes to the parent of vertex i, that is, to the vertex k only if $(k, i) \in E$.

It is important to recall that the strings in the languages M_i, $i \in Q$, have infinite multiplicity. Moreover, one evolution rule can be applied to many pairs of strings and generate many pairs of output strings. Let us assume that $x, y \in M_i$ and $r \in R_i$ for $1 \le i \le m$ and that r can be applied to x and y. If the application of r to x and y generates z and w (with z and w different from x and y), then x and y still belong to M_i. If z or w pass into another compartment, then no copy of z or w, respectively, remains in that compartment.

A *computation* is a (possibly infinite) sequence of transitions between configurations of a system Π starting from the initial configuration. The result of a computation is given by all the different strings over T that passed into the environment during a computation of Π. All strings of this kind define the *language generated* by Π denoted by $L(\Pi)$.

If a string $x \notin T^*$ passes to the environment, then it is not included in $L(\Pi)$. Strings over T present in a vertex different from the environment are not included in $L(\Pi)$.

The *diameter* of a splicing P system Π, denoted by $dia(\Pi) = (p_1, p_2, p_3, p_4)$, is defined by

$$p_i = \max(\{|u_i| \mid (u_1 \# u_2 \$ u_3 \# u_4; tar_1, tar_2) \in R_1 \cup \cdots \cup R_m,$$
$$tar_1, tar_2 \in \{here, in, out\}\}),\ 1 \le i \le 4.$$

If any of the p_i is infinite, then it is replaced by $*$.

We denote by $\mathsf{SPL}_m(d, p_1, p_2, p_3, p_4)$ the family of languages $\mathsf{L}(\Pi)$ generated by splicing P systems having a cell-tree as underlying structure of degree at most m, depth at most d and diameter at most (p_1, p_2, p_3, p_4).

Splicing P systems on asymmetric cell-graphs have also been studied. In this case the underlying topological structure $\mu = (Q, E)$ is a cell-graph in which for $i, j \in Q$, either $(i, j) \in E$ or $(j, i) \in E$. Also in this case there is one vertex to which no language or evolution rule is associated. This vertex is called the *environment*. The degree of a splicing P system on an asymmetric graph is $m = |Q| - 1$. The target indicators for such splicing P systems range over $\{here, out, go\}$. If $(r; tar_1, tar_2) \in R_i$, then for $j = 1, 2$:

if $tar_j = here$, then the string remains in vertex i;

if $tar_j = out$, then the string passes to vertex k only if $(i, k) \in E$ and k is the environment;

it $tar_j = go$, then the string passes to one of the vertices k only if $(i, k) \in E$.

The definitions of *configuration, computation, language generated* and *diameter* are similar to those for splicing P systems.

We denote by $\mathsf{tSPL}_m(p_1, p_2, p_3, p_4)$ the family of languages $\mathsf{L}(\Pi)$ generated by splicing P systems on asymmetric cell-graphs of degree at most m and diameter at most (p_1, p_2, p_3, p_4).

From Definition 9.1 and Definition 9.2 it follows that the evolution rule $(u_1 \# u_2 \$ u_3 \# u_4; tar_1, tar_2)$ is equivalent to $(u_3 \# u_4 \$ u_1 \# u_2; tar_2, tar_1)$. That is: $\mathsf{SPL}_m(d, p_1, p_2, p_3, p_4) = \mathsf{SPL}_m(d, p_3, p_4, p_1, p_2)$ and $\mathsf{tSPL}_m(p_1, p_2, p_3, p_4) = \mathsf{tSPL}_m(p_3, p_4, p_1, p_2)$.

In both kinds of splicing P system, evolution rules can be applied only if the splicing rule can occur and the resulting strings can pass into the vertices denoted by both target indicators.

9.4   Examples

Example 9.1 A splicing P system generating $\{aa, aaa, aaaa\}$.

The formal definition of the system associated with this example is:

$$\Pi_1 = (\{A, B, a\}, \{a\}, (\{0, 1\}, \{(0, 1)\}), \{aA, Ba, Baa, Baaa\},$$
$$\{(a \# A \$ B \# a; out, here)\}):$$

There are two vertices in the cell-tree underlying Π_1, so the degree of Π_1 is 1. The environment (vertex 0) has no language or evolution rule associated with it (as specified by Definition 9.2). The external compartment (vertex 1) initially contains the strings aA, Ba, Baa and $Baaa$ in an infinite number of occurrences. The only evolution rule present in the system is associated with the external compartment and it is $(a\#A\$B\#a; out, here)$.

In the initial configuration the evolution rule can be applied in the following ways:

$(a|A, B|a) \vdash (aa, BA);$
$(a|A, B|aa) \vdash (aaa, BA);$
$(a|A, B|aaa) \vdash (aaaa, BA).$

The strings aa, aaa and $aaaa$ pass to the environment, while the string BA remains in the external compartment. After any of these application the external compartment contains an infinite number of $aA, Ba, Baa, Baaa$ and BA. The system goes on applying the evolution rule and passing to the environment aa, aaa and $aaaa$.

As the terminal alphabet is $\{a\}$ it should be clear that Π_1 generates $\{aa, aaa, aaaa\}$. The diameter of Π_1 is $(1, 1, 1, 1)$. ◇

Example 9.2 A splicing P system generating $\{a^{2n} \mid n \in \mathbb{N}_+\}$.

The formal definition of the system associated with this example is:

$$\Pi_2 = (\{A, B, Z, a\}, \{a\}, (\{0, 1\}, \{(0, 1)\}), \{Aa^2B, a^2B, Z\},$$
$$\{1 : (\#B\$A\#; here, here), 2 : (\#B\$Z\#; out, here)\}).$$

In order to facilitate the explanation, evolution rules have been numbered. The following sequences of applications of splicing rules

$$
\begin{array}{ccc}
a^2|B & a^4|B & a^4 \\
A|a^2B \;\; \vdash_1 & AB \;\; \vdash_2 & ZB \\
& Z| &
\end{array}
$$

allow the string a^4 to pass to the environment. As a^4 is over $\{a\}$, it belongs to the language generated by Π_2. The string a^2 is generated by the splicing of the axioms a^2B and Z according to rule 2. The application of rule 1 on strings of the kind $a^{2k}B$, $k \in \mathbb{N}_+$, and Aa^2B leads to $a^{2(k+1)}B$ and AB. If then rule 2 is applied to $a^{2(k+1)}B$ and Z, then $a^{2(k+1)}$ passes to the environment.

Strings not in $\{a\}^*$ can pass to the environment, too. For instance, let us consider the following sequence of applications of splicing rules:

$$
\begin{array}{ccc}
Aa^2|B & Aa^4|B & Aa^4 \\
A|a^2B \;\; \vdash_1 & AB \;\; \vdash_2 & ZB \\
& Z| &
\end{array}
$$

The string Aa^4 passes to the environment but, as it does not belong to $\{a\}^*$. it is not part of the language generated by Π_2. Other computations are possible but the generated strings do not belong to $\{a\}^*$. We can say that Π_2, having diameter $(0, 1, 1, 0)$, generates $\{a^{2n} \mid n \in \mathbb{N}_+\}$.   ◇

9.5 Rotate-and-simulate

One of the techniques employed to prove the computational power of systems using splicing rules is known as *rotate-and-simulate.*

Let x be a string over an alphabet V with $x = x_1 \ldots x_n$, $x_1, \ldots, x_n \in V$. To *rotate* a string means to move one symbol from one end to another end of the string, that is, to perform a *circular permutation* of x. For instance, one clockwise rotation of one symbol in x leads to $x_n x_1 \ldots x_{n-1}$, another similar rotation leads to $x_{n-1} x_n x_1 \ldots x_{n-2}$. In general a string x cannot be distinguished by its circular permutations. One possibility to overcome this is to use a symbol $B \notin V$ and to put it before x_1, the leftmost symbol of x. In this way if $Bx = Bx_1 \ldots x_n$, then one clockwise circular permutation of Bx leads to $x_n Bx_1 \ldots x_{n-1}$ and another such rotation leads to $x_{n-1} x_n Bx_1 \ldots x_{n-2}$. A system able to detect the presence of B as the leftmost symbol 'knows' that what comes after B is x (and not a circular permutation of x).

In the present chapter we also consider rotation of substrings. If, for instance, $Xx_1 \ldots x_n Yx_{n+1} \ldots x_m Z$ is a string, then one clockwise rotation of the substring $x_1 \ldots x_n$ leads to $Xx_n x_1 \ldots x_{n-1} Yx_{n+1} \ldots x_m Z$.

The *simulate* part of the technique refers to the simulation of productions of a grammar. From Section 3.2 we know that a production can be applied anywhere in a sentential form.

Let us assume that N and T are the alphabets of non-terminal and terminal symbols, respectively, of a grammar G generating an infinite language and that $\alpha \to \beta$, $\alpha, \beta \in (N \cup T)^+$, is a production in G. Moreover, let $y = y_1 \alpha y_2$, $y' = y'_1 \alpha y'_2$, $y_1, y_2, y'_1, y'_2 \in (N \cup T)^+$, $y_1 \neq y'_1$, $y_2 \neq y'_2$, be sentential forms of G such that $y \Rightarrow z$ and $y' \Rightarrow z'$ in G with $z = y_1 \beta y_2$ and $z' = y'_1 \beta y'_2$. We want to create a splicing P system Π having a finite number of splicing rules and be able to simulate in parallel all derivations of G. We assume that all sentential forms of G can be present in each configuration of Π.

Moreover, we assume that Π simulates the application of a production $\alpha \to \beta$ of G splicing strings as y and y' into $y_1\alpha$, y_2, $y'_1\alpha$ and y'_2. When this is done α can be replaced by β by one splicing rule so that the strings $y_1\beta$, y_2, $y'_1\beta$ and y'_2 are present in Π. At this point Π could try to recreate z and z' concatenating $y_1\beta$, y_2, $y'_1\beta$ and y'_2. Unfortunately, as Π has a finite number of splicing rules and as Π tries to simulate in parallel all the derivations of the grammar G (generating an infinite language), then the strings $w = y_1 \beta y'_2$ and $w' = y'_1 \beta y_2$ can result in Π

(obviously, also z and z' can result). It should be clear that the strings w and w' cannot be distinguished from sentential forms of G. As it is possible that w and w' are not sentential forms of G, then Π would not have performed a simulation of G.

Such a splicing P system Π can simulate productions of G if, after indicating the leftmost symbol of a string, the string is rotated and productions are simulated on one end of the string. So, for instance, the simulation of $\alpha \to \beta$ in $y = y_1 \alpha y_2$ can start indicating the leftmost symbol of y so as to get $By = By_1 \alpha y_2$, rotating By so as to get $y_2 By_1 \alpha$, simulating the production so as to get $y_2 By_1 \beta$, and then rotating the string again until B is again the leftmost symbol.

Example 9.3 A splicing P system generating circular permutations of $\beta \in \{a, b, c\}^*$, $|\beta| \geq 2$.

The formal definition of the system associated with this example is

$$\Pi_3 = (V, T, \mu, L_1, L_2, R_1, R_2)$$

with:

$$
\begin{aligned}
V &= \{X, Y, W, Z\} \cup \{Y_i, X_i, Y'_{i-1}, X'_{i-1} \mid 0 \leq i \leq 3\} \cup T; \\
T &= \{a, b, c\}; \\
\mu &= (\{0, 1, 2\}, \{(1, 0), (1, 2)\})); \\
L_1 &= \{X\beta Y\} \cup \{ZY_i, X_i Z \mid 1 \leq i \leq 3\} \cup \\
&\quad \{X_1 a Z, X_2 b Z, X_3 c Z, ZY, XZ, W\}; \\
L_2 &= \{ZY'_i, X'_i Z' \mid 0 \leq i \leq 2\}; \\
R_1 &= \{1 : (\#aY\$Z\#Y_1; here, out), 2 : (\#bY\$Z\#Y_2; here, out), \\
&\quad 3 : (\#cY\$Z\#Y_3; here, out)\} \cup \\
&\quad \{4 : (X_1 a \# Z\$X \#\alpha; in, out), 5 : (X_2 b \# Z\$X \#\alpha; in, out), \\
&\quad 6 : (X_3 c \# Z\$X \#\alpha; in, out), 7 : (\alpha \# Y'_0 \$ Z \# Y; here, out), \\
&\quad 8 : (X \# Z\$X'_0 \#\alpha; here, out), 9 : (\alpha \# Y\$W\#; here, out), \\
&\quad 10 : (\#W\$X\#\alpha; out, out) \mid \alpha \in \{a, b, c\}\} \cup \\
&\quad \{11 : (\#Y'_i \$ Z \# Y_i; here, out), 12 : (X_i \# Z\$X'_i \#\alpha; in, out) \mid \\
&\qquad\qquad\qquad\qquad\qquad 1 \leq i \leq 2, \ \alpha \in \{a, b, c\}\} \\
R_2 &= \{13 : (\alpha \# Y_{i+1} \$ Z \# Y'_i; out, here), 14 : (X'_i \# Z\$X_{i+1} \#\alpha; here, here) \mid \\
&\qquad\qquad\qquad\qquad\qquad 0 \leq i \leq 2, \ \alpha \in \{a, b, c\}\}.
\end{aligned}
$$

The system Π_3 works in the following way. There is a mapping between symbols in $\{a, b, c\}$ and the subscript i present in X_i and Y_i, $1 \leq i \leq 3$. The symbol a is associated with the subscript 1, the symbol b is associated with the subscript 2 and the symbol c is associated with the subscript 3. Starting from stings of the kind $X\beta Y = X\beta_1 \ldots \beta_k Y$, $\beta_1, \ldots, \beta_k \in \{a, b, c\}$, the system Π_3

replaces $\beta_k Y$ with Y_j, j being the subscript associated with β_k. Moreover, X is replaced with $X_j \gamma$, $\gamma \in \{a, b, c\}$, and j being the subscript associated with γ. The subscripts of the symbols X and Y in the resulting string $X_j \gamma \beta_1 \ldots \beta_{k-1} Y_j$ are decreased. If a string of the kind $X_0' \gamma \beta_1 \ldots \beta_{k-1} Y_0'$ is obtained in compartment 1, then $\beta_k = \gamma$ and the system performed a rotation of β. When this happens other rotations can be performed or any cyclic permutation of β can pass to the environment.

Now a step-by-step explanation. Let $X abc Y \in L_1$. The evolution rules $3, 4, 5,$ $6, 9$ and 10 can then be applied in compartment 1. If rule 10 is applied, then $(|W, X| abc Y) \vdash_{10} (abc Y, XW)$. Both output strings pass to the environment but, as they are not over T, they do not contribute to $\mathsf{L}(\Pi_3)$. Rules 9 and 10 can be applied one after the other: $(X abc | Y, W|) \vdash_9 (X abc, WY)$ followed by $(|W, X| abc) \vdash_{10} (abc, XW)$. The string $abc \in T^*$ passes to the environment and contributes to $\mathsf{L}(\Pi_3)$. If one of the rules $4, 5$ or 6 is applied, then one string passes to the environment (but it is not in $\mathsf{L}(\Pi_3)$ as it is not over T), while the other passes into compartment 2. Here this string can be spliced according to rule 14 but the resulting strings remain in this compartment where they cannot be subject to any splicing rule.

If in compartment 1 one of the rules $4, 5$ or 6 is applied after rule 3 then: $(X ab | cY, Z | Y_3) \vdash_3 (X abY_3, ZcY)$ and then (for instance) $(X_3 c | Z, X | abY_3) \vdash_6 (X_3 cabY_3, XZ)$. The strings $ZcY, XZ \notin T^*$ pass to the environment, the string $X_3 cabY_3$ passes into compartment 2. If after the application of rule 3 either 4 or 5 is applied, then the strings passing to compartment 2 are $X_1 aabY_3$ and $X_2 babY_3$, respectively. In general we can say that a string of the kind $X_j \gamma abY_3$ can pass into compartment 2 (j being the subscript associated with γ).

In compartment 2 the flanking symbols X and Y are replaced by primed ones with the subscript decreased by 1. This is performed by the application of rules 13 and 14. If a rule 13 is applied first, then only one of the resulting strings passes into compartment 1 where it is no longer spliced. If instead rule 13 is applied after rule 14, then the string passing to compartment 1 can be spliced further. In the current example the string passing to compartment 1 is $X_2' cabY_2'$. If the subscript of the flanking symbols X' and Y' is different from 0, then in compartment 1 these symbols are replaced by non-primed ones. This is performed by the application of rules 11 and 12. Similarly to what described in the above this happens only if rule 12 is applied after rule 11. If instead rule 12 is applied first, then the resulting string, passing to compartment 2, is no longer spliced.

If a string flanked by X_0' and Y_0' is in compartment 1, then these symbols are replaced by X and Y, respectively. This occurs by the application of rules 7 and 8. In the above we described how a string flanked by X and Y can be spliced.

Several other splicings are possible but none of them lead to strings which are circular permutations of β. The resulting strings, not over T, do not belong to $\mathsf{L}(\Pi_3)$. For instance, we consider the following example:

$$
\begin{array}{ccccccc}
Xab|cY & & X|abY_3 & & X_1|aabY_3 & & X_0'aab|Y_3 \\
Z|Y_3 & \vdash_3 & ZcY & \vdash_4 & XZ & \vdash_{14} & X_1Z & \vdash_{13} \\
& & X_1a|Z & & X_0'|Z & & Z|Y_2'
\end{array}
$$

$$
\begin{array}{ccccccc}
X_0'|aabY_2' & & Xaab|Y_2' & & X|aabY_2 & & aabY_2 \\
ZY_3 & \vdash_8 & X_0'Z & \vdash_{11} & ZY_2' & \vdash_{10} & XW \\
X|Z & & Z|Y_2 & & |W & &
\end{array}
$$

The string $aabY_2 \notin T^*$ passes to the environment but it does not belong to $\mathsf{L}(\Pi_3)$. The diameter of Π_3 is $(2, 2, 1, 1)$. $\diamond$

The technique used to rotate strings in this example is similar to that used in the proof of Theorem 9.3.

9.6   Computational power

In this section we recall results concerning the computational power of splicing P systems with either a cell-tree or an asymmetrical cell-graph as underlying structure.

Theorem 9.1 $\mathsf{SPL}_1\,(1, 0, 1, 1, 1) = \mathsf{SPL}_1\,(1, 1, 1, 0, 1) = \mathsf{tSPL}_1\,(0, 1, 1, 1) = \mathsf{tSPL}_1(1, 1, 0, 1) = \mathsf{REG}$.

Proof   Here we prove that $\mathsf{SPL}_1(1, 0, 1, 1, 1) = \mathsf{REG}$.

Part I: (Such a splicing P system can simulate regular grammars) Let $G = (N, T, S, P)$ be a right-linear regular grammar and $W, Z \notin N \cup T$. The splicing P system $\Pi = (V, T, \mu, L_1, R_1)$ of degree 1, depth 1 and diameter $(0, 1, 1, 1)$ simulating G is defined in the following.

$$
\begin{aligned}
V &= N \cup T \cup \{W, Z\}; \\
\mu &= (\{0,1\}, \{(0,1)\}); \\
L_1 &= \{S\} \cup \{ZaY \mid X \to aY \in P,\ X, Y \in N,\ a \in T\} \cup \\
&\quad \{Wa \mid X \to a \in P,\ X \in N,\ a \in T\}; \\
R_1 &= \{1 : (\#X\$Z\#a; here, here) \mid X \to aY \in P,\ X, Y \in N,\ a \in T\} \cup \\
&\quad \{2 : (\#X\$W\#a; out, here) \mid X \to a \in P,\ X \in N,\ a \in T\}.
\end{aligned}
$$

From the initial configuration pairs of axioms of the kind ZaY and Wa can be subject to evolution rules but the resulting strings do not contribute to $\mathsf{L}(\Pi)$ as they contain Z. Only sequences of splicings starting from S and an axiom of the kind ZaY or Wa can lead to string over T. These sequences see the application of evolution rules of type 1 and they end with the application of an evolution rule of type 2.

Part II: (The language generated by such splicing P systems is regular) This proof requires concepts that are outside the scope of this monograph. For this reason we do not present it but in Section 9.10 we indicate where it can be found. □

Theorem 9.2 $\text{SPL}_2(2,1,2,2,1) = \text{SPL}_2(2,2,1,1,2) = \text{RE}$.

Proof Here we prove that $\text{RE} \subseteq \text{SPL}_2(2,1,2,2,1)$. Let $G = (N,T,S,P)$ be a type-0 grammar in Kuroda normal form with $|P| = m$ and $B \notin N \cup T$. The symbols in $N \cup T \cup \{B\}$ are numbered in a one-to-one manner so that $N \cup T \cup \{B\} = \{\alpha_1, \ldots, \alpha_n\}$. As G is in Kuroda normal form the productions in P can be divided into two sets: $P_1 = \{u_i \to v_i \mid u_i \to v_i \in P, |u_i| = 1\}$ and $P_2 = \{u_i \to v_i \mid u_i \to v_i \in P, |u_i| = 2\}$ with $P_1 \cup P_2 = P$, $P_1 \cap P_2 = \emptyset$, $|P_1| = m_1$, $|P_2| = m_2$ and $m_1 + m_2 = m$. We consider also $P' = \{u \to u \mid u \in \{\alpha_1, \cdots, \alpha_n\}\}$. Moreover, $\{o, X, X_1, X_2, X_3, Y, Y_1, Y_2, Y_3, Z_X, Z_{X_1^+}, Z_{X_2^+}, Z_{X_3^+}, Z_Y, Z_{Y_1^+}, Z_{Y_2^+}, Z_{Y_3^+}, Z_\lambda, Z_\lambda'\} \cup \{Z_{X_i}, Z_{Y_i} \mid 1 \le i \le n+m\} \cup \{Y_i', Z_{Y_i'} \mid n+m_1 \le i \le n+m, u_i \to v_i \in P_2\}$ are symbols not in $N \cup T$.

The splicing P system $\Pi = (V, T, \mu, L_1, L_2, R_1, R_2)$ of degree 2, depth 2 and diameter $(1, 2, 2, 1)$ simulating this grammar is defined in the following.

$$
\begin{aligned}
V \ = \ & N \cup T \cup \{Z_{X_i}, Z_{Y_i} \mid 1 \le i \le n+m\} \cup \\
& \{Y_i', Z_{Y_i'} \mid n+m+1 \le i \le n+m, u_i \to v_i \in P_2\} \cup \\
& \{o, B, X, X_1, X_2, X_3, Y, Y_1, Y_2, Y_3, Z_X, Z_{X_1^+}, Z_{X_2^+}, Z_{X_3^+}, Z_Y, Z_{Y_1^+}, Z_{Y_2^+}, \\
& \qquad\qquad\qquad\qquad\qquad\qquad\qquad\qquad\qquad Z_{Y_3^+}, Z_\lambda, Z_\lambda'\}, \\
\mu \ = \ & (\{0,1,2\}, \{(0,1),(1,2)\}); \\
L_1 \ = \ & \{XBSY, X_2 Z_{X_2^+}, Z_{Y_1^+} Y_1, X Z_X, Z_\lambda, Z_\lambda'\} \cup \\
& \{Z_{Y_i} o^i Y_1 \mid 1 \le i \le n+m\} \cup \\
& \{Z_{Y_i'} Y_i' \mid n+m_1+1 \le i \le n+m, u_i \to v_i \in P_2\}; \\
L_2 \ = \ & \{Z_{Y_2^+} Y_2, X_1 Z_{X_1^+}, Z_Y Y\} \cup \{X_1 o^i v_i Z_{X_i} \mid 1 \le i \le n+m\}; \\
R_1 \ = \ & \{1 : (\#u_i Y \$ Z_{Y_i} \#; in, out) \mid 1 \le i \le n+m\} \cup \\
& \{2 : (\#u_2 Y \$ Z_{Y_i'} \#; here, out), 3 : (\#u_1 Y_i' \$ Z_{Y_i} \#; in, out) \mid u_i = u_1 u_2, \\
& \qquad\qquad n+m_1+1 \le i \le n+m, u_1, u_2 \in N, u_i \to v_i \in P_2\} \cup \\
& \{4 : (\# Z_{X_2^+} \$ X_1 \# o; here, out), 5 : (o\# o Y_2 \$ Z_{Y_1^+} \#; in, out), \\
& \quad 6 : (\alpha \# o Y_2 \$ Z_{Y_1^+} \#; in, out), 7 : (\# Z_X \$ X_1 \# \alpha; in, out), \\
& \quad 8 : (\# Z_X \$ X_3 \# \alpha; in, out), 9 : (\# BY \$ Z_\lambda \#; here, out), \\
& \quad 10 : (\# Z_\lambda' \$ X \#; out, out) \mid \alpha \in N \cup T \cup \{B\}\}; \\
R_2 \ = \ & \{11 : (\# Y_1 \$ Z_{Y_2^+} \# Y_2; here, out), 12 : (\# Z_{X_i} \$ X \#; out, out), \\
& \quad 13 : (X_1 \# Z_{X_1^+} \$ X_2 o \# o; out, out), 14 : (X_3 \# Z_{X_3^+} \$ X_2 o \# \alpha; out, out), \\
& \quad 15 : (\alpha \# Y_2 \$ Z_Y \# Y; out, out) \mid 1 \le i \le n+m, \alpha \in N \cup T \cup \{B\}\}.
\end{aligned}
$$

The proof is based on rotate-and-simulate. Some of the strings present in Π are of the kind Xw_2Bw_1Y, $w_1, w_2 \in \{N \cup T\}^*$ for different X and Y. These strings are present in Π if and only if w_1w_2 is a sentential form of G. It is possible to remove the non-terminal symbol Y only together with B from strings of the kind Xw_1w_2BY. In this way the desired circular permutation of w_1w_2 is ensured.

If we consider that $w_1w_2 = wu_i$, $w, u_1 \in \{N \cup T \cup \{B\}\}^*$, with either $|u_i| = 1$ or $|u_i| = 2$, then the substring wu_i between X and Y can be rotated. This is performed by replacing u_i with o^i, $o \notin N \cup T \cup \{B\}$. Subsequently $o^j u_j$ is concatenated to the left-hand side of w. The o's on the two sides of the resulting substring $o^j u_j wo_i$ are removed one by one. If the o's on the two sides of the string terminate at the same moment, then $i = j$ and a rotation of the string wu_i was performed. If $i \neq j$, then the resulting string does not contribute to $\mathsf{L}(\Pi)$. The simulation of productions in P is simulated in a similar way.

Now the formal description. Let us assume that in compartment 1 there is a string of the kind Xwu_iY with w and u_i as defined before (initially there is $XBSY$). If a production in $P_1 \cup P'$ is simulated $|u_i| = 1$, then $(Xw|u_iY, Z_{Y_i}|o^iY_1)$ $\vdash_1 (Xwo^iY_1, Z_{Y_i}u_iY)$, $1 \leq i \leq n+m_1$. If a production in P_2 is simulated $|u_i| = 2$, then $(Xwu_1|u_2Y, Z_{Y'_i}|Y'_i) \vdash_2 (Xwu_1Y'_i, Z_{Y'_i}u_2Y)$ followed by $(Xw|u_1Y'_i, Z_{Y_i}|o^iY_1) \vdash_3 (Xwo^iY_1, Z_{Y_i}u_1Y'_i), n+m_1+1 \leq i \leq n+m$. In both cases the suffix u_iY is replaced by $o^iY_1, 1 \leq i \leq n+m$ and the resulting strings pass into compartment 2. The strings passing to the environment do not belong to T^* so they do not contribute to the language generated by Π.

In compartment 2 a string of the kind Xwo^iY_1 can be spliced: $(Xwo^i|Y_1, Z_{Y_2^+}|Y_2) \vdash_{11} (Xwo^iY_2, Z_{Y_2^+}Y_1)$. The second output string passes into compartment 1 and no evolution rule can be applied on it. The string Xwo^iY_2 remains in compartment 2 and it is subject to $(X_1o^jv_j|Z_{X_j}, X|wo^iY_2)\vdash_{12}(X_1o^jv_jwo^iY_2, XZ_{X_j})$, $1 \leq j \leq n+m$. Both output strings pass into compartment 1 but only the first one is involved in splicing operations. In compartment 2 a string of the kind Xwo^iY_1 can also be spliced by rule 12 so as to have $(X_1o^jv_j|Z_{X_j}, X|wo^iY_1) \vdash_{12}$ $(X_1o^jv_jwo^iY_1, XZ_{X_j}), 1 \leq j \leq n+m$. Both output strings pass into compartment 1. The second one cannot be involved in any splicing, while the first one is subject to $(X_2|Z_{X_2}^+, X_1|o^jv_jwo^iY_1) \vdash_4 (X_2o^jv_jwo^iY_1, X_1Z_{X_2}^+)$. Both output strings remain in compartment 1 and they are no longer spliced.

A string of the kind $X_1o^jv_jwo^iY_2$ is spliced in compartment 1 so that X_1 is replaced with X_2 and oY_2 is replaced with Y_1. This happens by $(X_2|Z_{X_2^+}, X_1|o^jv_j wo^iY_2) \vdash_4 (X_2o^jv_jwo^iY_2, X_1Z_{X_2^+})$ and $(X_2o^jv_jwo^{i-1}|oY_2, Z_{Y_1}^+|Y_1) \vdash_5 (X_2o^jv_j wo^{i-1}Y_1, Z_{Y_1}^+oY_2)$. In compartment 1 $(X_1o^jv_jwo^{i-1}|oY_2, Z_{Y_1}^+|Y_1) \vdash_5 (X_1o^jv_jw o^{i-1}Y_1, Z_{Y_1}^+oY_2)$ it is also possible, while $(X_1o^jv_jwo^{i-1}|Y_1, Z_{Y_2^+}|Y_2) \vdash_{11} (X_1o^jv_j wo^{i-1}Y_2, Z_{Y_2^+}Y_1)$ is possible in compartment 2. In these last two cases both output strings are no longer spliced. The strings passing to compartment 0 do not belong to T^* so they do not contribute to the language generated by Π.

In compartment 2 a string of the kind $X_2o^jv_jwo^{i-1}Y_1$ is spliced so that Y_1 is replaced with Y_2 and X_2o is replaced with X_1. This is obtained by $(X_2o^jv_jwo^{i-1}|Y_1, Z_{Y_2^+}|Y_2) \vdash_{11} (X_2o^jv_jwo^{i-1}Y_2, Z_{Y_2^+}Y_1)$ and $(X_1|Z_{X_1^+}, X_2o|o^{j-1}v_jwo^{i-1}Y_2) \vdash_{13} (X_1o^{j-1}v_jwo^{i-1}Y_2, X_2oZ_{X_1^+})$ if $j > 1$. In compartment 2 it is also possible to have $(X_1|Z_{X_1^+}, X_2o|o^{j-1}v_jwo^{i-1}Y_1) \vdash_{13} (X_1o^{j-1}v_jwo^{i-1}Y_1, X_2oZ_{X_1^+})$ if $j > 1$. We describe later on what happens if $j = 1$. In this last case both output strings pass into compartment 1 but only the first one is spliced by rule 4 together with $X_2Z_{X_2}$ so as to obtain $X_2o^{j-1}v_jwo^{i-1}Y_1$, remaining in compartment 1 and no longer spliced, and $X_1Z_{X_2} \notin T^*$ passing to the environment.

The process of removing o's on the left and on the right goes on between compartments 1 and 2 until three kinds of strings can be present: $X_1v_jwY_2$, $X_1o^kv_jw \, Y_2$ in compartment 1 and $X_2v_jwo^kY_1$ in compartment 2, $1 \leq k \leq n + m - 1$.

As described before, a string of the kind $X_1o^kv_jwY_2$ is spliced with $X_2Z_{X_2^+}$ by rule 4 so as to obtain $X_2o^kv_jwY_2$, remaining in compartment 1 and no longer spliced, and $X_1Z_{X_2^+} \notin T^*$ passing to the environment.

In compartment 2 a string of the kind $X_2v_jwo^kY_1$ is spliced so that Y_1 is replaced with Y_2. The output strings $X_2v_jwo^kY_2$, remaining in compartment 2, and $Z_{Y_2^+}Y_1$, passing to compartment 1, are no longer spliced.

Strings of the kind $X_1v_jwY_2$ are spliced in compartment 1 by $(X|Z_X, X_1|v_jw \, Y_2) \vdash_7 (Xv_jwY_2, X_1Z_X)$. The first output string passes into compartment 2 while the second (not in T^*) passes to the environment. In compartment 2 it is possible to have $(Xv_jw|Y_2, Z_Y|Y) \vdash_{15} (Xv_jwY, Z_YY_2)$. Both output strings pass into compartment 1 but only the first one is subject to splicing.

If a string of the kind $X_2ov_jwo^{i-1}Y_2$ is present in compartment 2, then it cannot be subject to rule 13. Instead it can be spliced by rule 14: $(X_3|Z_{X_3^+}, X_2o|v_jw \, o^{i-1}Y_2) \vdash_{14} (X_3v_jwo^{i-1}Y_2, X_2oZ_{X_3^+})$. Both strings pass into compartment 1 but only the first one can be spliced. If rule 5 is applied to $X_3v_jwo^{i-1}Y_2$, then the resulting string, $X_3v_jwo^{i-2}Y_1$, passes into compartment 2 where it is spliced by rule 11. The resulting strings are no longer spliced. Similarly, if rule 8 is applied to $X_3v_jwo^{i-1}Y_2$, then the generated string, $Xv_jwo^{i-1}Y_2$, passes into compartment 2 where, after the application of rule 11 it is no longer spliced. A similar process occurs if strings of the kind $X_1o^{j-1}v_jwoY_2$ are present in compartment 1.

This process indicates how in the splicing P system a string of the kind Xwu_iY can be changed into Xv_jwY.

Strings of the kind XyY, $y \in (N \cup T \cup \{B\})$ can be spliced in compartment 1 also by evolution rules 7 and 8. If $(|Z'_\lambda, X|yY) \vdash_8 (yY, XZ'_\lambda)$ is performed, the first output string passes to the environment but, as $Y \notin T$, it does not contribute to the language generated by Π; the second output string remains in compartment 1 and cannot be involved in any splicing.

If $y = xB$ with $x \in \{N \cup T\}^*$, then $(Xx|BY, Z_\lambda|) \vdash_7 (Xx, Z_\lambda BY)$ and $(|Z'_\lambda, X|x) \vdash_8 (x, XZ'_\lambda)$ is performed. The strings x, $Z_\lambda BY$ and XZ'_λ pass to the environment but only x can contribute to the language generated by Π. If $x \in T^*$ the system Π has simulated a derivation of G. Other computations are possible but none of them results in strings over T passing to the environment.

All derivations in G can be simulated in Π. As only strings over T^* passing to the environment contribute to the language generated by Π, then we can say that $\mathsf{L}(G) = \mathsf{L}(\Pi)$. □

Theorem 9.2 can be used to prove that $\mathsf{tSPL}_3(1, 2, 2, 1) = \mathsf{tSPL}_3(2, 1, 1, 2) = \mathsf{RE}$. If we consider the asymmetric graph depicted in Fig. 9.9 we can imagine that compartment 1 has the the same language and evolution rules defined in the proof of Theorem 9.2 but with all the *in* target indicators changed into *out*.

The alphabet V of the splicing P system described in the proof of Theorem 9.2 is changed into $V' = V \cup \{K\}$, $K \notin T$, while the language and evolution rules associated with compartment 2 are unchanged. Compartment 3 has $L_3 = \{K\}$ and $R_3 = \{13 : (\alpha\#\$K\#; go, here) \mid \alpha \in \{Y, Y_1, Y_2\} \cup \{Z_{X_i} \mid 1 \leq i \leq n + m\}\}$. Compartment 3 passes strings from compartment 2 into compartment 1. No splicing is possible in the initial configuration of compartment 3.

Anyhow, it is possible to define a splicing P system on an asymmetric graph generating RE with diameter $(0, 2, 1, 0)$.

Theorem 9.3 $\mathsf{tSPL}_3(0, 2, 1, 0) = \mathsf{tSPL}_3(1, 0, 0, 2) = \mathsf{RE}$.

Proof Here we prove that $\mathsf{RE} \subseteq \mathsf{tSPL}_3(0, 2, 1, 0)$.

Let $G = (N, T, S, P)$ be a type-0 grammar in Kuroda normal form with $|P| = m$ and $B \notin N \cup T$. The symbols in $N \cup T \cup \{B\}$ are numbered in a one-to-one manner so that $N \cup T \cup \{B\} = \{\alpha_1, \ldots, \alpha_n\}$. As G is in Kuroda normal form the productions in P can be divided in two sets: $P_1 = \{u_i \rightarrow v_i \mid u_i \rightarrow v_i \in P, |u_i| = 1\}$ and $P_2 = \{u_i \rightarrow v_i \mid u_i \rightarrow v_i \in P, |u_i| = 2\}$ with $P_1 \cup P_2 = P$, $P_1 \cap P_2 = \emptyset$, $|P_1| = m_1$, $|P_2| = m_2$ and $m_1 + m_2 = m$. We consider also $P' = \{u \rightarrow u \mid u \in \{\alpha_1, \cdots, \alpha_n\}\}$. Moreover, $\{X, X', Y, Y', Z_X, Z_{X'}, Z_Y, Z_{Y'}, Z_\lambda, Z'_\lambda\} \cup \{X_i, Y_i, Z_i, Z_{X_i}, Z_{Y_i} \mid 1 \leq i \leq n + m\} \cup \{Y'_i, Z_{Y'_i} \mid n + m_1 + 1 \leq i \leq n + m, u_i \rightarrow v_i \in P_2\}$ are symbols not in $N \cup T \cup \{B\}$.

The splicing P system $\Pi = (V, T, \mu, L_1, L_2, L_3, R_1, R_2, R_3)$ of degree 3, and diameter (0, 2, 1, 0) simulating this grammar is defined in the following.

$$\Pi = \{V, T, g, L_1, L_2, L_3, R_1, R_2, R_3\};$$
$$V = N \cup T \cup \{B, X, X', Y, Y', Z_X, Z_{X'}, Z_Y, Z_{Y'}, Z_\lambda, Z'_\lambda\} \cup$$
$$\{X_i, Y_i, Z_i, Z_{X_i}, Z_{Y_i} \mid 1 \leq i \leq n + m\} \cup$$
$$\{Y'_i, Z_{Y'_i} \mid n + m_1 + 1 \leq i \leq n + m, \ u_i \to v_i \in P_2\};$$
$$\mu = (\{0, 1, 2, 3\}, \{(1, 0), (1, 2), (2, 3), (3, 1)\});$$
$$L_1 = \{XBSY, X'Z_{X'}, Z_\lambda, Z'_\lambda\} \cup \{Z_{Y_i} Y_i \mid 1 \leq i \leq n + m\} \cup$$
$$\{X_i Z_{X_i} \mid 1 \leq i \leq n + m - 1\} \cup$$
$$\{Z_{Y'_i} Y'_i \mid n + m_1 + 1 \leq i \leq n + m, \ u_i \to v_i \in P_2\};$$
$$L_2 = \{Z_{Y'} Y'\} \cup \{X_i v_i Z_i \mid 1 \leq i \leq n + m\} \cup \{Z_{Y_i} Y_i \mid 1 \leq i \leq n + m - 1\};$$
$$L_3 = \{XZ_X, Z_Y Y\} \cup \{X_i Z_{X_i} \mid 2 \leq i \leq n + m\};$$
$$R_1 = \{1 : (\#u_i Y \$ Z_{Y_i} \#; go, out) \mid 1 \leq i \leq n + m\} \cup$$
$$\quad \{2 : (\#u_2 Y \$ Z_{Y'_i} \#; here, out), 3 : (\#u_1 Y'_i \$ Z_{Y_i} \#; go, out) \mid u_i = u_1 u_2,$$
$$\quad\quad n + m_1 + 1 \leq i \leq n + m, \ u_1, u_2 \in N, \ u_i \to v_i \in P_2\} \cup$$
$$\quad \{4 : (\#Z_{X_{i-1}} \$ X_i \#; go, out) \mid 2 \leq i \leq n + m\} \cup$$
$$\quad \{5 : (\#BY \$ Z_\lambda \#; here, out), 6 : (\#Z'_\lambda \$ X \#; out, out)\};$$
$$R_2 = \{7 : (\#Z_i \$ X \#; go, go) \mid 1 \leq i \leq n + m\} \cup$$
$$\quad \{8 : (\#Y_i \$ Z_{Y_{i-1}} \#; go, go) \mid 2 \leq i \leq n + m\} \cup$$
$$\quad \{9 : (\#Y_1 \$ Z_{Y'} \#; here, go), 10 : (\#Z_{X'} \$ X_1 \#; go, go)\};$$
$$R_3 = \{11 : (\#Z_{X_i} \$ X_i \#; go, here) \mid 2 \leq i \leq n + m\} \cup$$
$$\quad \{12 : (\#Z_X \$ X' \#; here, go), 13 : (\#Y' \$ Z_Y \#; go, go\}.$$

The asymmetric graph underlying Π is depicted in Fig. 9.9 where the environment is vertex 0. The proof is based on rotate-and-simulate.

Assume that in compartment 1 there is a string of the form $Xwu_i Y$ with $w, u_i \in \{N \cup T \cup \{B\}\}^*$ with either $|u_i| = 1$ or $|u_i| = 2$ (initially there is $XBSY$).

A production in $P_1 \cup P'$ is simulated starting from $(Xw|u_iY, Z_{Y_i}|Y_i) \vdash_1 (XwY_i, Z_{Y_i}u_iY)$, $1 \leq i \leq n + m_1$. The first output string passes into compartment 2 while the second output string passes to the environment.

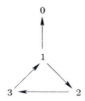

Fig. 9.9 Asymmetric graph related to Theorem 9.3

A production in P_2 is simulated starting from $(Xwu_1|u_2Y, Z_{Y_i'}|Y_i') \vdash_2 (Xwu_1 Y_i', Z_{Y_i'}u_2Y)$, $n + m_1 + 1 \leq i \leq n + m$. The first output string remains in compartment 1 while the second output string passes to the environment. In compartment 1 $(Xw|u_1Y_i', Z_{Y_i}|Y_i) \vdash_3 (XwY_i, Z_{Y_i}u_1Y_i')$, $n+m_1+1 \leq i \leq n+m$, can take place. The first output string passes into compartment 2 while the second output string passes to the environment.

In both simulations the suffix u_iY is changed with $Y_i, 1 \leq i \leq n + m$. The strings passing to the environment do not belong to T^* so they do not contribute to the language generated by Π.

In compartment 2 a string of the kind XwY_i can be spliced by $(X_jv_j|Z_j, X| wY_i) \vdash_7 (X_jv_jwY_i, XZ_j)$, for some $1 \leq j \leq n+m$. Both output strings pass into compartment 3, where only the first output string can be subject to evolution rules.

Strings of the kind $X_jv_jwY_i$ are spliced so that the value of the subscripts of X and Y is decreased until specific strings are present. The subscript of X is decreased in compartment 1, that of Y is decreased in compartment 2. Compartment 3 is used to pass strings between compartments 1 and 2 during this process.

When a string of the kind $X_jv_jwY_i, 2 \leq j \leq n + m$, is present in compartment 3 it can be spliced by $(X_j|Z_{X_j}, X_j|v_jwY_i) \vdash_{11} (X_jv_jwY_i, X_jZ_{X_j})$. The first output string passes into compartment 1 while the second output string remains in compartment 3 where it cannot be spliced.

In compartment 1 it is possible to have $(X_{j-1}|Z_{X_{j-1}}, X_j|v_jwY_i) \vdash_4 (X_{j-1}v_jw Y_i, X_jZ_{X_{j-1}})$. The first output string passes into compartment 2, the second passes to the environment but it does not contributes to the language generated by Π as it is not in T^*.

A string of the kind $X_{j-1}v_jwY_i$ can be spliced in compartment 2 so as to have $(X_{j-i}v_jw|Y_i, Z_{i-1}|Y_{i-1}) \vdash_8 (X_{j-i}v_jwY_{i-1}, Z_{i-1}Y_i)$, both output strings pass into compartment 3 but the second one cannot be involved in any splicing.

Strings that can be generated in the system while decreasing the indexes of X and Y are: $X_1v_jwY_k$ in compartment 1 and $X_kv_jwY_1$ in compartment 2, where $2 \leq k \leq n + m$. As X_1 and Y_1 can be replaced by X' and Y' by rules 10 and 9, respectively, then it is also possible to have $X'v_jwY'$ in compartment 3.

In the first case $(X'|Z_{X'}, X_1|v_jwY_k) \vdash_5 (X'v_jwY_k, X_1Z_{X'})$ takes place. The string $X_1Z_{X'}$ passes to the environment and, as it is not in T^*, it does not contribute to the language generated by Π. The first output string passes into compartment 2 where it is subject to evolution rules replacing the Y_k with Y_{k-1}. The string $X'v_jwY_{k-1}$ is then generated and passes into compartment 3. Here X' is replaced with X by $(X|Z_X, X'|v_jwY_{k-1}) \vdash_{12} (Xv_jwY_{k-1}, X'Z_X)$. Both output strings, one remaining in compartment 3 and the other passing into compartment 1, are no longer spliced.

In the second case the Y_1 in $X_k v_j w Y_1$ is replaced in compartment 2 with Y' by $(X_k v_j w | Y_1, Z_{Y'} | Y') \vdash_9 (X_k v_j w Y', Z_{Y'} Y_1)$. Also this time both output strings are no longer spliced.

In the third case both evolution rules 12 and 13 can be applied. If $(X' v_j w | Y', Z_Y | Y) \vdash_{13} (X' v_j w Y, Z_Y Y')$ takes place, then the first output string passes into compartment 1, while the second remains in compartment 3 and it is no longer spliced. As we already saw, in compartment 1 strings of the kind $X' v_j w Y$ with $w = w' v_k$, $w' \in V^*$, $v_k \in V$, can be spliced by rule 1 (or rules 2 and 3) so that a string of the kind $X' v_j w' Y_k$ passes into compartment 2. Here such a string can be spliced by rule 7, but the resulting strings, passing into compartment 3, are no longer spliced. If instead rule 13 is applied after rule 12 on a string of the kind $X' v_j w Y'$, then the string $X v_j w Y$ passes into compartment 1 (the other output strings are no longer spliced).

What we have just described is the process to pass from $X w u_i Y$ to $X v_j w Y$ simulating a production in P or rotating the substring between X and Y of one symbol.

At any moment a string of the kind $X w Y$ can be spliced in compartment 1 by the evolution rules 5 and 6.

If $(|Z'_\lambda, X | w Y) \vdash_7 (w Y, X Z'_\lambda)$ takes place, then the first output string, sent out of the system, does not contribute to the language generated by Π as $Y \notin T$. The second output string remains in compartment 1 and cannot be involved in any splicing.

If $w = x B, x \in \{N \cup T\}^*$, then $(X x | B Y, Z_\lambda |) \vdash_5 (X x, Z_\lambda B Y)$ can take place. The first output string, remaining in the same compartment, can be involved in $(|Z'_\lambda, X | x) \vdash_6 (x, X Z'_\lambda)$. The resulting strings $x, Z_\lambda B Y$ and $X Z'_\lambda$ pass into the environment but only x can contribute to the language generated by Π.

If $x \in T^*$, then the system Π simulates a derivation of G.

Let us consider the three compartments in their initial configurations.

In compartment 1 it is possible to have $(X_{i-1} | Z_{X_{i-1}}, X_i | Z_{X_i}) \vdash_4 (X_{i-1} Z_{X_i}, X_i Z_{X_{i-1}})$ and $(X' | Z_{X'}, X_1 | Z_{X_1}) \vdash_5 (X' Z_{X_1}, X_1 Z_{X'})$. In both cases the first output strings pass into compartment 2 where no evolution rule can be applied on them. The second output string passes into the environment but it does not contribute to the language generated by Π.

In compartment 2 it is possible to have $(Z_{Y_i} | Y_i, Z_{Y_{i-1}} | Y_{i-1}) \vdash_8 (Z_{Y_i} Y_{i-1}, Z_{Y_{i-1}} Y_i)$. Both output strings pass into compartment 3 and no evolution rule can be applied on them.

In compartment 3 it is possible to have $(X_i | Z_{X_i}, X_i | Z_{X_i}) \vdash_{11} (X_i Z_{X_i}, X_i Z_{X_i})$. The first output string passes into compartment 1 while the second, remaining in compartment 3, belongs to L_3. In compartment 1 it is possible to have $(X_{i-1}|Z_{i-1}, X_i|Z_{X_i}) \vdash_4 (X_{i-1}Z_{X_i}, X_iZ_{i-1})$. The first output string passes into compartment 2 where it cannot be subject to any evolution rule. The second output string passes into the environment but it does not contribute to the language generated by Π. Other computations are possible but none of them results in strings over T passing to the environment.

All derivations in G can be simulated in Π. As only strings over T^* passing to the environment contribute to the language generated by Π, then we can say that $\mathsf{L}(G) = \mathsf{L}(\Pi)$. $\qquad\square$

A splicing P system has *global rules* when all its compartments have the same set of rules. We denote by $\mathsf{SPLG}_m(d, p_1, p_2, p_3, p_4)$ and $\mathsf{tSPLG}_m(p_1, p_2, p_3, p_4)$ the family of languages generated by splicing P systems having a cell-tree and a cell-graph, respectively, as underlying structure of degree at most m, depth at most d and diameter at most (p_1, p_2, p_3, p_4). If in the previous systems the set of rules are united so as to obtain one set, then the two theorems still hold.

So, we have:

Theorem 9.4 $\mathsf{SPLG}_2(2, 1, 2, 2, 1) = \mathsf{SPLG}_2(2, 2, 1, 1, 2) = \mathsf{RE}$.

Theorem 9.5 $\mathsf{tSPLG}_3(0, 2, 1, 0) = \mathsf{tSPLG}_3(1, 0, 0, 2) = \mathsf{RE}$.

It is possible to have a splicing P system with global rules having a cell-tree as underlying structure of degree 3, depth 3 and diameter $(1, 2, 1, 0)$.

Theorem 9.6 $\mathsf{SPLG}_3(3, 1, 2, 1, 0) = \mathsf{SPLG}_3(3, 1, 0, 1, 2) = \mathsf{RE}$.

Proof Here we define the splicing P system Π belonging to $\mathsf{SPLG}_3(3, 1, 2, 1, 0)$. The system Π simulates a type-0 grammar $G = (N, T, S, P)$ in Kuroda normal form in a way similar to the splicing P systems described in the proof of Theorem 9.3. For this reason we only give its formal description. The splicing P system $\Pi = (V, T, \mu, L_1, L_2, L_3, R)$ is such that:

$$
\begin{aligned}
V &= N \cup T \cup \{B, X, X', Y, Z_X, Z_{X'}, Z_Y, Z_\lambda, Z'_\lambda\} \cup \\
&\quad \{Y_i, Z_{Y_i} \mid 1 \le i \le n + m\} \cup \{X_i, Z_{X_i}, Z_i \mid 1 \le i \le n + m\} \cup \\
&\quad \{Y'_i, Z_{Y'_i} \mid n + m_1 + 1 \le i \le n + m\}; \\
\mu &= (\{0, 1, 2, 3\}, \{(0, 1), (1, 2), (2, 3)\}); \\
L_1 &= \{XSBY, Z_\lambda, Z'_\lambda\} \cup \{Z_{Y_i}Y_i \mid 1 \le i \le n + m\} \cup \\
&\quad \{Z_{Y'_i}Y'_i \mid n + m_1 + 1 \le i \le n + m\}; \\
L_2 &= \{X Z_X, X' Z_{X'}\} \cup \{X_i v_i Z_i \mid 1 \le i \le n + m\} \cup \\
&\quad \{X_i Z_{X_i} \mid 1 \le i \le n + m - 1\};
\end{aligned}
$$

$L_3 = \{Z_Y Y\};$
$R = \{(X'\#Z_{X'}\$X_1\#; in, in), (X\#Z_X\$X'\#; out, in),$
$\quad (\#Z'_\lambda\$X\#; out, out)\}\cup$
$\quad \{(\#u_i Y\$Z_{Y_i}\#; in, out) \mid u_i \to v_i \in P\}\cup$
$\quad \{(\alpha\#Y_i\$Z_{Y_{i-1}}\#; in, out) \mid 1 \le i \le n+m, \alpha \in N \cup T \cup \{B\}\}\cup$
$\quad \{(\alpha\#BY\$Z_\lambda\#; here, out) \mid \alpha \in T\}\cup$
$\quad \{(\alpha\#Z_i\$X\#; out, in) \mid 1 \le i \le n+m, \alpha \in N \cup T\}\cup$
$\quad \{(X_{i-1}\#Z_{X_{i-1}}\$X_i\#; out, in) \mid 2 \le i \le n+m\}\cup$
$\quad \{(\alpha\#Y_0\$Z_Y\#; out, out) \mid \alpha \in N \cup T \cup \{B\}\}\cup$
$\quad \{(u_1\#u_2 Y\$Z_{Y'}\#; here, out), (\#u_1 Y'_i\$Z_{Y_i}\#; in, out) \mid u_i = u_1 u_2,$
$\qquad n + m_1 + 1 \le i \le n+m, \ u_i \to v_1 \in P, \ u_1, u_2 \in N\}.$ □

9.7   Immediate communication

The fact that with splicing P systems computational completeness is reached with a few compartments and a small diameter allowed researchers in the field to introduce limitations in the definition of these systems. One of these restricted models sees evolution rules without a target indicator (that is, just splicing rules). The strings resulting from the application of a splicing rule present in compartment i have to pass into a compartment different from i but such that (i, j) is an arc in the underlying cell-tree.

The resulting restricted model of a splicing P system was then named *splicing P systems with immediate communication*. The family of languages generated by such systems with a cell-tree as underlying structure, of degree at most m, depth at most d and diameter at most (p_1, p_2, p_3, p_4) is denoted by $\mathsf{SPL}_{im,m}(d, p_1, p_2, p_3, p_4)$. As expected, these restrictions have effects on the computational power, degree, depth and diameter.

Theorem 9.7 $\mathsf{SPL}_{im,1}(1, 0, 1, 1, 0) = \mathsf{SPL}_{im,1}(1, 1, 0, 0, 1) = \mathsf{FIN}.$

Proof *Part I: (What is generated by such splicing P systems are finite languages)* This is obvious if we consider that the initial set of axioms in the splicing P system is finite. The output strings resulting from one application of splicing rules pass to the environment. These strings define a finite set.

Part II: (Such splicing P systems can generate any finite language) Here we prove that $\mathsf{SPL}_{im,1}(1, 0, 1, 1, 0) = \mathsf{FIN}$. Let V be an alphabet and $L = \{x_1, \ldots, x_n\}$ be a finite language over V and $y \notin V$. We construct the following splicing P system with immediate communication:

$$\Pi = (V \cup \{y\}, V, (\{0, 1\}, \{(0, 1)\}), L_1, R_1)$$

with:

$L_1 = \{yx_i \mid 1 \leq i \leq n\}$;
$R_1 = \{\#y\$y\#\}$.

The application of the splicing rule generates x_i and yyx_j, $1 \leq i, j \leq n$. Both strings pass to the environment but only x_i contributes to the language generated by Π. □

Theorem 9.8 $\mathrm{SPL}_{im,2}(2,1,1,1,1) = \mathrm{SPL}_{im,2}(2,1,1,1,1) \supseteq \mathrm{REG}$.

Proof It follows from Part I of the proof of Theorem 9.1 if the alphabet $V' = V \cup \{W\}$, $L_2 = \{U\}$ and $R_2 = \{U\#\$K\# \mid K \in N\}$. The function of compartment 2 is to pass into compartment 1 the strings ending with a non-terminal symbol. □

Theorem 9.9 $\mathrm{SPL}_{im,2}(2,1,1,1,1) \setminus \mathrm{REG} \neq \emptyset$.

Proof Let us consider the splicing P system $\Pi = (V, T, \mu, L_1, L_2, R_1, R_2)$ with:

$V = \{a, b, c, d, e, f\}$;
$T = \{a, b, c, d\}$;
$\mu = (\{0, 1, 2\}, \{(0, 1), (1, 2)\})$;
$L_1 = \{caabd, ebd\}$;
$L_2 = \{caf\}$;
$R_1 = \{b\#d\$e\#b\}$;
$R_2 = \{a\#f\$c\#a\}$.

In the initial configuration the splicing rule in compartment 1 can be applied to two $caabd$ and ebd. This generates $caabbd$ and ed which can pass to the environment or into compartment 2. Here $caabbd$ can be spliced with caf so that $caaabbd$ and cf are generated and pass into compartment 1. Only $caaabbd$ can be subject to a splicing.

In the initial configuration in compartment 1 the splicing rule can be applied to two instances of ebd obtaining $ebbd$ and ed, while in compartment 2 the splicing rule can be applied to two instances of caf so that the string $caaf$ and cf are generated. These stings do not contribute to the language generated by Π.

It should be clear that only strings of the form $ca^n b^n d$, $n \geq 2$, passing to the environment contribute to the language generated by Π and they do not belong to any family of languages in REG. □

Theorem 9.10 $\mathrm{SPL}_{im,*}(3,2,3,1,1) = \mathrm{SPL}_{im,*}(3,1,1,2,3) = \mathrm{RE}$.

Proof Here we prove that $\mathrm{RE} \subseteq \mathrm{SPL}_{im,*}(3,2,3,1,1)$.

Let $G = (N, T, S, P)$ be a type-0 grammar in Kuroda normal form with $|P| = m$ and $B \notin N \cup T$. The symbols in $N \cup T \cup \{B\}$ are numbered in a one-to-one manner so that $N \cup T \cup \{B\} = \{\alpha_1, \ldots, \alpha_n\}$. We consider $P' = \{u \to u \mid u \in \{\alpha_1, \cdots, \alpha_n\}\}$. Let $k = n + m$ and $\{X, Y, Y', Y'', Y_f, Z\} \cup \{Y_i \mid 1 \leq i \leq k\}$ be symbols not in $N \cup T \cup \{B\}$.

The splicing P system $\Pi = (V, T, \mu, L_1, L_2, L_{2'}, \ldots, L_k, L_{k'}, L_f, L_{f'}, R_1, R_2, R_{2'}, \ldots, R_k, R_{k'}, R_f, R_{f'})$ of degree $2k + 3$, depth 3 and diameter $(2, 3, 1, 1)$ simulating this grammar is defined in the following.

$V = N \cup T \cup \{B, X, Y, Y', Y'', Y_f, Z\} \cup \{Y_i \mid 1 \leq i \leq k\}$;

$\mu = (Q, E)$ with $Q = \{0, 0', 1, 1', \ldots, k, k', f, f'\}$;

The compartment representation is

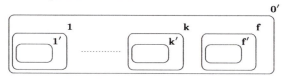

$L_1 = \{XBSY, Z, ZY_f\} \cup \{ZY_i \mid 1 \leq i \leq k\}$;

$L_j = \{ZY\} \cup \{XvZ \mid u \to v \in P \cup P'\}$ for $1 \leq j \leq k$;

$L_{j'} = \{ZY'\}$ for $1 \leq j \leq k$;

$L_f = \{Z, ZY''\}$;

$L_{f'} = \{ZY'\}$;

$R_{0'} = \{1 : \#uY\$Z\#Y_i \mid r_i : u \to v \in P \cup P', 1 \leq i \leq k\} \cup \{2 : \#BY\$Z\#Y_f, 3 : \#Y''\$Z\#\}$;

$R_j = \{4_j : Xv\#Z\$X\# \mid u \to v \in P \cup P'\} \cup \{5_j : \#Y\$Z\#Y\}$ for $1 \leq j \leq k$;

$R_{j'} = \{6_j : \#Y_j\$Z\#Y'\}$ for $1 \leq j \leq k$;

$R_f = \{7 : \#Z\$X\#, 8 : \#Y'\$Z\#Y''\}$;

$R_{f'} = \{9 : \#Y_f\$Z\#Y'\}$.

The environment is vertex 0, while the skin compartment is vertex $0'$. The proof is based on rotate-and-simulate.

Assume that in compartment $0'$ there is a string of the form $XwuY$ with $w, u \in \{N \cup T \cup \{B\}\}^*$ with either $|u| = 1$ or $|u| = 2$ (initially there is $XBSY$). Such a string can be spliced together with ZY_i, $1 \leq i \leq k$, by a rule of the kind 1. The output strings are XwY_i and ZuY. If any of these strings pass to the environment, then they do not contribute to the language generated by Π as they are not over T^*. If they pass into compartment $p \in \{1, \ldots, k, f\}$, then only XwY_i is spliced.

If $1 \leq p \leq k$, then $(Xv|Z, X|wY_i) \vdash_{4_i} (XvwY_i, XZ)$ takes place. Only the string $XvwY_i$ can be further spliced. This happens if it passes into compartment p' and if $p = i$. In this case $(Xvw|Y_i, Z|Y') \vdash_{6_i} (XvwY', ZY_i)$. Both output strings pass into compartment i but only the string $XvwY'$ is spliced: $(Xvw|Y', Z|Y) \vdash_{5_i} (XvwY, ZY')$. The simulation of $u \to v$ has been performed using rotate-and-simulate.

If $p = f$, then $(|Z, X|wY_i) \vdash_7 (wY_i, XZ)$. The output strings can pass into compartment $0'$ or f' but here they are not subject to any splicing.

If a string of the kind $XwBY$, $w \in \{N \cup T \cup \{B\}\}^*$, is present in compartment $0'$, then also $(Xw|BY, Z|Y_f) \vdash_2 (XwY_f, ZBY)$ can take place. The output strings do not belong to T^* and only XwY_f can be further spliced if it is passed to any compartment $p \in \{1, \ldots k, f\}$.

The splicing occurring to XwY_f in compartments $1 \leq p \leq k$ are similar to those described before. If instead $p = f$, then $(|Z, X|wY_f) \vdash_7 (wY_f, XZ)$ takes place. Only the string wY_f can be further spliced if it passes into compartment f'. Here $(w|Y_f, Z|Y') \vdash_9 (wY', ZY_f)$ takes place. Both output strings pass into compartment f but only wY' is spliced: $(w|Y', Z|Y'') \vdash_8 (wY'', ZY')$. Only the string wY'' can be spliced if it passes into compartment 1. This happens by $(w|Y'', Z| \vdash_3 (w, ZY'')$.

If $w \in T^*$, then w contributes to the language generated by Π only if it passes to the environment. □

The following result follows the proof techniques employed in Section 9.6; for this reason we only give the formal description of the considered P system with immediate communication.

Theorem 9.11 $\mathsf{SPL}_{im,2}(2,1,3,2,2) = \mathsf{SPL}_{im,2}(2,2,2,1,3) = \mathsf{RE}$.

Proof We define a splicing P system with immediate communication $\Pi = (V, T, \mu, L_1, L_2, R_1, R_2)$ of degree 2, depth 2 and diameter $(1, 3, 2, 2)$ simulating a type-0 grammar $G = (N, T, S, P)$. Let $N \cup T \cup \{B\} = \{a_1, \ldots, a_n\}$ with $a_n = B$ and $B \notin N \cup T$, then:

$$
\begin{aligned}
V &= N \cup T \cup \{B, X, Y, X_0, Y_0, X', Y', Z, Z_X, Z_Y\} \cup \{X_i, Y_i \mid 1 \leq i \leq n\} \cup \\
&\quad \{X'_j, Y'_j \mid 0 \leq j \leq n-1\}; \\
\mu &= \{(0,1,2), \{(0,1),(1,2)\}\}; \\
L_1 &= \{XSBY, ZY_0, X'Z, XZ, Z_X\} \cup \{ZY_i \mid 1 \leq i \leq n\} \cup \\
&\quad \{ZvY_j \mid u \to va_j \in P, \ 0 \leq j \leq n-1\} \cup \{ZY'_j \mid 0 \leq j \leq n-1\}; \\
L_2 &= \{ZY', ZY, Z_Y\} \cup \{X_i a_i Z \mid 1 \leq i \leq n\} \cup \{X'_j Z, X_j Z \mid 0 \leq j \leq n-1\}; \\
R_1 &= \{\#uY\$Z\#vY_j, \ \#a_i u'Y\$Z\#Y_i, \ \#a_i Y\$Z\#Y_i, \ x\#Y_i\$Z\#Y'_{i-1}, \\
&\quad x\#Y'_j\$Z\#Y_j, \ X_0\#x\$X'\#Z, \ X'\#x\$X\#Z, \ X'\#x\$\#Z_X \mid \\
&\quad u \to va_j, \ u' \to \epsilon \in P, \ 1 \leq i \leq n, \ 0 \leq j \leq n-1, \ x \in N \cup T \cup \{B\}\}; \\
R_2 &= \{X\#x\$X_i a_1\#Z, \ X_i\#x\$X'_{i-1}\#Z, \ X'_j\#x\$X_j\#Z, \ x\#Y_0\$Z\#Y', \\
&\quad x\#Y'\$Z\#Y, \ x\#BY_0\$Z_Y\# \mid \\
&\qquad 1 \leq i \leq n, \ 0 \leq j \leq n-1, \ x \in N \cup T \cup \{B\}\}.
\end{aligned}
$$

□

9.8 Direct universal splicing P system

To prove that splicing P systems can simulate type-0 grammars means proving that these models of membrane systems are computationally complete. It also means that splicing P systems are universal as a splicing P system can simulate a universal type-0 grammar. As a universal type-0 grammar is able to simulate any splicing P system, then a splicing P system Π_u able to simulate a universal type-0 grammar can also simulate any splicing P system, so Π_u is universal, too. The system Π_u would then get as input the encoding of a grammar (similarly to what was done in the previous section). For this reason we say that a splicing P system can simulate other splicing P systems in an *indirect* way (that is, through the encoding of a universal grammar). If instead a splicing P system Π_u could get as input the encoding of any other splicing P system Π and simulate it, then Π_u would perform a *direct* simulation.

Such a universal splicing P system Π_u would have a fixed alphabet, a fixed set of evolution rules and a fixed topological structure. One part of the axioms present in Π_u would be fixed, another part would encode the alphabet, axioms, evolution rules and topological structure of any other splicing P system. The language generated by Π_u would then be an encoding of the language generated by Π.

In this section we sketch such a universal splicing P system $\Pi_u = (V_u, T_u, (\{0, 1, 2, 3\}, \{(0, 1), (1, 2), (1, 3)\}, L_{1_u}, L_{2_u}, L_{3_u}, R_{1_u}, R_{2_u}, R_{3_u})$ simulating splicing P systems with a cell-tree or a cell-graph as underlying structure. We do not describe Π_u in its entirety as such a description would be long and tedious.

Let $\Pi = (V, T, \mu, L_1, \ldots, L_m, R_1, \ldots, R_m)$ be a splicing P system and let

$$C_L : V^* \to V_u^*;$$
$$C_R : V^* \# V^* \$ V^* \# V^* \times TAR \times TAR \to V_u^*;$$
$$D : T_u^* \to T^*;$$

(coding) injective functions with $TAR = \{0, 1, \ldots, m\}$. The function C_L encodes strings in $L_1, \ldots, L_m$ into strings in L_{1_u}; C_R encodes evolution rules in $R_1, \ldots, R_m$ into strings in L_{1_u} and D encodes strings over T_u into strings over T.

To indicate that Π_u gets as input an encoding of Π we write $\Pi_u(\Pi) = (V_u, T_u, \mu_u, L_{1_u} \cup \bigcup_{x \in L_1 \cup \ldots \cup L_m} C_L(x) \cup \bigcup_{r \in R_1 \cup \ldots \cup R_m} C_R(r), L_{2_u}, L_{3_u}, R_{1_u}, R_{2_u}, R_{3_u})$. The language generated by $\Pi_u(\Pi)$ is denoted by $L(\Pi_u(\Pi))$. When the strings in this language are given as input to the D function, then $L(\Pi)$, the language generated by Π, is obtained. So, we can write $\{D(y) \mid y \in L(\Pi_u(\Pi))\} = L(\Pi)$.

Let $N = V \setminus T$, $N = \{v_1, \ldots, v_p\}$, and $T = \{v_{p+1}, \ldots, v_q\}$. Moreover, let $f : V^* \to \{0, 1, 2\}^*$ and $f' : V \to \{0, 1, 2\}^*$ be functions such that $f(v) = f'(v_{j_1}) \ldots f'(v_{j_{|v|}})$ and

$$f'(v_j) = \begin{cases} 20^j & \text{for } j \leq p; \\ 10^j & \text{for } p + 1 \leq j \leq q; \end{cases}$$

with $v = v_{j_1} \ldots v_{j_{|v|}}$, $v_{j_1}, \ldots, v_{j_{|v|}} \in V$. So, the function f encodes strings over V into strings over $\{0, 1, 2\}$.

The axioms in Π_u are of the kind $h_k v t_k$ with $v \in V_u^*$, $h_k, t_k \in V_u$, $k \in \mathbb{N}$. If $x \in L_i$, $1 \le i \le m$, then $C_L(x) = h_1 z^i s_2 z^i B f(x) s_1 t_1$. We represent strings of this kind with:

$$h_1 z^i s_2 z^i \boxed{B\ldots\ldots} s_1 t_1$$

and we call them *input-axioms*. Notice that $C_L(x)$ contains the encoding of both x and i.

Let $\mu = (Q, E)$ be the cell-tree underlying Π. In order to define C_R we have to introduce:

$bar : \{0, 1, 2, \#\}^* \to \{\bar{0}, \bar{1}, \bar{2}, \bar{\#}\}^*$;
$mi : \{0, 1, 2, \#\}^* \to \{0, 1, 2, \#\}^*$;
$t : V^* \# V^* \$ V^* \# V^* \times \{here, out, in\} \times \{here, out, in\} \times \{1, 2\} \times$
$$\{1, \ldots, m\} \to \mathbb{N}_+.$$

If $x_1, \ldots, x_k \in \{0, 1, 2, \#\}$, then $bar(x_1 \ldots x_k) = \bar{x}_1 \ldots \bar{x}_k$ while $mi(x_1 \ldots x_k) = x_k \ldots x_1$. Let $r = (u_1 \# u_2 \$ u_3 \# u_4; tar_1, tar_2) \in R_j$, $1 \le j \le m$, then:

$$t(r, l, j) = \begin{cases} j & \text{if } tar_l = here; \\ k & \text{if } tar_l = out \text{ and } (k, j) \in E; \\ k & \text{if } tar_l = in \text{ and } (j, k) \in E; \end{cases}$$

and

$$C_R(r) = h_3 s_1 bar(mi(f(u_1) \# f(u_2))) z^{t(r,1,j)} s_2 z^{t(r,1,j)} \bar{\$} z^{t(r,2,j)} s_2 z^{t(r,2,j)} s_1$$
$$bar(mi(f(u_3) \# f(u_4))) z^j t_3.$$

We represent strings of this kind with:

$$h_3 s_1 \overset{site_1}{\boxed{\cdots}} \# \boxed{\cdots} z^p s_2 z^p \bar{\$} z^q s_2 z^q \overset{site_2}{\boxed{\cdots}} \# \boxed{\cdots} z^j t_3$$

and we call them *input-rules*. Notice that $C_R(r)$ contains the encoding of the evolution rule (splicing rule and target indicators) and j.

The decoding function D is such that

$$D(\gamma) = \begin{cases} f^{-1}(\gamma) & \text{if } \gamma \in 10^+; \\ \epsilon & \text{otherwise.} \end{cases}$$

In a splicing P system the strings present in a compartment can splice according to the evolution rules present in the same compartment. The generated strings are either sent to another compartment or they can remain in the same one. This is simulated by the universal splicing P system in the steps indicated in Fig. 9.10.

We give an overview of the simulation algorithm considering the splicing P system $\Pi = (V, T, \mu, L_1, R_1)$ with $V = \{a, b\}$, $T = \{c\}$, $\mu = (\{0, 1\}, \{(0, 1)\})$,

1. Modify input-axioms;
2. Create pairs of input-axioms;
3. Check pairs of input-axioms;
4. Rotate pairs of input-axioms;
5. Join an input-rule with a pair of input-axioms;
   (a *working string* is obtained)
6. Check working strings;
7. Rotate working strings;
8. Match first splicing site;
9. Wrong match input-rule pair and input-axioms;
10. Mark first splicing site;
11. End of matching. Ordering;
12. Match second splicing site;
13. Wrong match input-rule pair of input-axioms;
14. Mark second splicing site;
15. End of matching. Ordering;
16. Simulate splicing;
17. End simulation;
18. Two cases.

Fig. 9.10 Steps followed by the universal splicing P system

$L_1 = \{ac, ccb\}$, and $R_1 = \{r : (a\#c\$c\#b; here, out)\}$. We will see how Π is encoded in Π_u and how Π_u can simulate the splicing

$$(a|c, cc|b) \vdash_r (ab, ccc)$$

occurring in Π.

If we consider Π, then the functions f' and f are such that: $f'(a) = 20, f'(b) = 200, f'(c) = 1000, f(ac) = 201000$, and $f(ccb) = 10001000200$. The coding functions are such that the input-axioms are:

$C_L(ac) = h_1 z s_2 z B 201000 s_1 t_1$
$C_L(ccb) = h_1 z s_2 z B 10001000200 s_1 t_1$

while the input-rule is:

$C_R(r) = h_3 s_2 bar(mi(f(a)\#f(c)))z s_2 z\overline{\$} s_2 s_1 bar(mi(f(c)\#f(b)))z t_3 =$
$\qquad h_3 s_1 bar(mi(20\#1000))z s_2 z\overline{\$} s_2 s_1 bar(mi(1000\#200))z t_3 =$
$\qquad h_3 s_1 bar(0001\#02)z s_2 z\overline{\$} s_2 s_1 bar(002\#0001)z t_3 =$
$\qquad h_3 s_1 \overline{0001}\overline{\#}\overline{02}z s_2 z\overline{\$} s_2 s_1 \overline{002}\overline{\#}\overline{0001}z t_3$

In the following we say that a string is in *state* $i, i \in \mathbb{N}$, if it is flanked by h_i and t_i. We denote with an arrow ($\rightarrow$) the change of a string into another.

(*1*. Modify input-axioms) The leftmost z's and s_2 of some input-axioms can be removed, the remaining z's are moved to the right hand side, and the state is changed into 2. This is needed for the next step of the algorithm.

$$h_1 z s_2 z B 201000 s_1 t_1 \rightarrow h_2 B 201000 s_1 z t_2$$

(*2*. Create pairs of input-axioms) Two input-axioms, one in state 1 and another in state 2, are concatenated. The resulting string, a pair of input-axioms, is in state 4.

$h_1 z s_2 z B 10001000200 s_1 t_1$

$$\vdash \ldots \vdash h_4 z s_2 z B 10001000200 s_1 B 201000 s_1 z t_4$$

$h_2 B 201000 s_1 z t_2$

(*3*. Check pairs of input-axioms) It is checked if a pair of input-axioms is composed of axioms present in the same compartment in Π. If the number of z's present at the two ends of such strings is equal, then the string changes into state 5, otherwise the input-string is no longer spliced.

$$h_4 z s_2 z B 10001000200 s_1 B 201000 s_1 z t_4$$
$$\downarrow$$
$$h_5 s_2 z B 10001000200 s_1 B 201000 s_1 t_5$$

(*4*. Rotate pairs of input-axioms) Substrings of pairs of input-axioms can be rotated. What is in between a z and an s_1 or in between two s_1's can be rotated. The reason for this rotation will become clear in the explanation of step 5.

$$h_5 s_2 z B 10001000200 s_1 B 201000 s_1 t_5$$
$$\downarrow$$
$$h_5 s_2 z 1000200 B 1000 s_1 B 201000 s_1 t_5$$

(*5*. Join an input-rule with a pair of input-axioms) Pairs of input-axioms can be concatenated to input-rules, forming in this way a *working string* in state 7. It is on this kind of strings that the simulation of splicing is performed. Notice that in a working string the (encoded) axioms and the (encoded) splicing sites are one of the barred mirror images of the other (for instance, $2010000001 \# \overline{0}\overline{2}$). This can happen only if rotation of pairs of input-axioms (step *4*) is present.

$$h_5 s_2 z 1000200 B 1000 s_1 B 201000 s_1 t_5$$

$$h_3 s_1 \overline{0}\overline{0}\overline{0}\overline{1} \# \overline{0}\overline{2} z s_2 z \overline{\$} s_2 s_1 \overline{0}\overline{0}\overline{2} \# \overline{0}\overline{0}\overline{0}\overline{1} z t_3$$

$$\vdash$$
$$\vdots$$
$$\vdash$$

$$h_7 z 1000200 B 1000 s_1 B 2010000\overline{0}\overline{0}\overline{1} \# \overline{0}\overline{2} z s_2 z \overline{\$} s_2 s_1 \overline{0}\overline{0}\overline{2} \# \overline{0}\overline{0}\overline{0}\overline{1} z t_7$$

(*6*. Check working strings) It is checked if the working string is composed of a pair of input-axioms and an input-rule present in the same compartment in Π. If the number of z's present at the two ends of such strings is equal, then the working string changes into state 8, otherwise the working string is no longer spliced.

$$h_7z1000200B1000s_1B2010000\overline{001}\overline{\#}\overline{02}zs_2z\$s_2s_1\overline{002}\#\overline{0001}zt_7$$
$$\downarrow$$
$$h_81000200B1000s_1B2010000\overline{001}\overline{\#}\overline{02}zs_2z\$s_2s_1\overline{002}\#\overline{0001}t_8$$

(*7*. Rotate working strings) A working string in state 8 can be rotated. The rotation is needed to match (encoded) splicing sites with (encoded) axioms. Pairs of symbols $\alpha\bar{\alpha}, \alpha \in \{0,1,2\}$, present at the two ends of a working string are used to detect a match.

$$h_81000200B1000s_1B2010000\overline{001}\overline{\#}\overline{02}zs_2z\$s_2s_1\overline{002}\#\overline{0001}t_8$$
$$\downarrow$$
$$h_8\overline{00001}\overline{\#}\overline{02}zs_2z\$s_2s_1\overline{002}\overline{\#}\overline{00011}000200B1000s_1B20100t_8$$

(*8*. Match first splicing site) Rotating a working string it is possible to move $\alpha\bar{\alpha}$ to the left side of the working string. When this happens the state is changed, α is moved while $\bar{\alpha}$ is removed and the state is changed back to 8. When this operation is completed the working string is ready to match, after rotation, the next symbols.

$$h_8\overline{00001}\overline{\#}\overline{02}zs_2z\$s_2s_1\overline{002}\overline{\#}\overline{00011}000200B1000s_1B20100t_8$$
$$\downarrow$$
$$h_8B20100\overline{001}\overline{\#}\overline{02}zs_2z\$s_2s_1\overline{002}\overline{\#}\overline{00011}000200B1000s_10t_8$$

(*9*. Wrong match input-rule pair and input-axioms) Of course it is also possible to have a wrong match between a rule and a pair of axioms. This is detected by the presence of $\alpha\bar{\beta}$ where $\alpha \neq \beta, \alpha, \beta \in \{0,1,2\}$. In this case the working string changes state into 6 and it cannot take part in any splicing.

(*10*. Mark first splicing site) When $\alpha\#, \alpha \in \{0,1,2\}$, is present on the left side of the working string, then $\#$ is moved in the (encoded) axiom in order to keep track of the position where the splicing occurs. The matching goes on with the rest of the splicing site (if present).

$$h_8\overline{\#}\overline{02}zs_2z\$s_2s_1\overline{002}\overline{\#}\overline{00011}000200B1000s_11000B20t_8$$
$$\downarrow$$
$$h_81000B200\overline{2}zs_2z\$s_2s_1\overline{002}\overline{\#}\overline{00011}000200B1000s_1\overline{\#}t_8$$

(*11*. End of matching. Ordering) When the match between one (encoded) axiom and one (encoded) splicing site is completed the working string has either z or s_2 on its left side. When this happens the working string is ordered: the matched (encoded) axiom is rotated so as to have the B on its extreme right. Moreover, the (eventual) z's and the s_2 present on the left side of $\bar{\$}$ are moved to the left side of the rightmost s_1. When this is done the state of the working string is changed into 10.

$$h_8 z s_2 z \bar{\$} s_2 s_1 00\bar{2} \bar{\#} 000\bar{1} 1000200 B 1000 s_1 B 20 \bar{\#} 1000 t_8$$

$$\downarrow$$

$$h_9 z s_2 z \bar{\$} s_2 s_1 00\bar{2} \bar{\#} 000\bar{1} 1000200 B 1000 s_1 20 \bar{\#} 1000 B t_9$$

$$\downarrow$$

$$h_{10} \bar{\$} s_2 s_1 00\bar{2} \bar{\#} 000\bar{1} 1000200 B 1000 z s_2 z s_1 20 \bar{\#} 1000 B t_{10}$$

The steps *12*, ..., *15* are similar to the steps *8*, ..., *11*, respectively. At the end of these steps the working string is in state 20.

(*16.* Simulate splicing) If also the second (encoded) splicing site matches the other (encoded) axiom, then, after ordering, the working string is ready to simulate the splicing (of the encoded evolution rule). This is performed by removing and replacing symbols: B's are removed, s_1's are replaced by B's and $\bar{\#}$'s are replaced by s_1's. The obtained string, in state 21, encodes the strings produced by the simulated splicing together with an indication of the compartments in Π where they have to pass.

$$h_{20} B 10001000 \bar{\#} 200 z s_2 z s_1 20 \bar{\#} 1000 B s_2 s_1 t_{20}$$

$$\downarrow$$

$$h_{21} 10001000 s_1 200 z s_2 z B 20 s_1 1000 s_2 B t_{21}$$

(*17.* End simulation) A string in state 21 is split into two strings in state 22 encoding the strings produced by the simulated splicing together with an indication of the compartments in Π where they have to pass.

$$h_{21} 10001000 s_1 200 z s_2 z B 20 s_1 1000 s_2 B t_{21}$$

$$\downarrow$$

$$h_{22} z s_2 z B 20200 s_1 t_{22} \qquad h_{22} s_2 B 100010001000 s_1 t_{22}$$

(*18.* Two cases) Strings in state 22 can be of two kinds:

$h_{22} z^p s_2 z^p B \gamma s_1 t_{22}$ with $\gamma \in 10^+ \cup 20^+, p \in \mathbb{N}_+$, coding strings not sent to the environment of Π. The state of these strings is changed into 1 and they can be involved in other simulations as they were encoded axioms of Π.

$h_{22} s_2 B \gamma s_1 t_{22}$ with $\gamma \in 10^+ \cup 20^+$, coding strings sent to the environment of Π. The state of these strings is changed into 1, s_1 and s_2 are removed, and they are sent to the environment of Π_u.

$$h_{22} z s_2 z B 20200 s_1 t_{22} \qquad\qquad h_{22} s_2 B 100010001000 s_1 t_{22}$$

$$\downarrow \qquad\qquad\qquad\qquad \downarrow$$

$$h_1 z s_2 z B 20200 s_1 t_1 \qquad\qquad h_0 B 100010001000 t_0$$

The string $h_1 z s_2 z B 20200 s_1 t_1$ encodes ab present in compartment 1 while the string $h_0 B 100010001000 t_0$ encodes ccc sent to the environment of Π. So, the evolution rule r has been simulated.

The splicing P system $\Pi_u \in \mathsf{SPL}_3(2, 1, 2, 2, 1)$ and it has 594 evolution rules.

9.9 Final remarks and research topics

One may wonder why in this chapter we introduced target indicators first and then we considered a model of splicing P systems without them. Such an order seems unnatural but it follows what happened in this field: splicing P systems (with target indicators) were introduced before splicing P systems with immediate communication.

It is relevant to notice that we presented three ways to implement rotate-and-simulate:

subscripts of X and Y that are decreased (Theorem 9.3);
number of specific symbols (o's) on the two sides of the strings (Theorem 9.2);
use of different compartments (Theorem 9.10).

To the best of our knowledge, these are the only ways this technique has been implemented in systems based on splicing.

Differently from other models of membrane systems present in this monograph, there is no direct link between splicing P systems and Petri nets. There is no theorem linking global features present in splicing P systems to their computational power. Such theorems would be very general and useful as they could be applicable to other models of membrane systems based on splicing.

Suggestion for research 9.1 *Find laws linking some global features of splicing P systems to their computational power.*

Different modes of operation (maximal parallelism, maximal strategy, asynchronous, etc.) have never been studied in relation to splicing P systems. For this reason we formulate:

Suggestion for research 9.2 *Study how the computational power of splicing P systems changes according to different operational modes.*

Some of the theorems presented in this chapter could be improved. It could be possible to have splicing P systems equivalent in computing power to the presented ones but with smaller degree or diameter. Among these possible improvements we want to highlight one:

Suggestion for research 9.3 *Define a computationally complete splicing P system with immediate communication having a finite degree. If this is not possible, then indicate why.*

9.10 Bibliographical notes

Tom Head [106] was the first to formalise in mathematical terms biological splicing. In the same paper Head splicing systems were also defined for the first time. Păun splicing systems were introduced in [218], while Pixton splicing systems were introduced in [198]. See also [108].

The relationship between the three models of splicing described above was thoroughly investigated [32, 220] while the relationship between 2-splicing and 1-splicing was studied in [247]. Numerous computability models using the splicing operation were defined and their computational power studied (see, for instance, [54, 66, 107, 206, 219]).

Splicing P systems were introduced in [207]. Models of splicing P systems, different from those considered in this chapter have been studied in [246]. Rotate-and-simulate was introduced in [205] and since then it is commonly used to prove results concerning systems based on splicing. Splicing P systems such as those in Definition 9.2 were introduced in [223]. There it is proved that $SPL_3(3, 1, 2, 2, 1) = RE$ and that $SPL_5(2, 1, 2, 2, 1) = RE$. In [204] the authors show that $SPL_2(2, 2, 2, 2, 2) = RE$. Theorem 9.2 is from [77]. In [237] it is proved that this result holds also when $V = T$, that is, when the alphabet and the terminal alphabet of the splicing P system are equal.

Theorem 9.1 first appeared in [223], where the proof of Part II of that theorem is based on [220, Theorem 7.1]. This last proof originated in [127, 128] and it was then developed in [198]. Proofs in terms of closure properties of families of languages are given in [232, Lemma 3.14, Chapter 7, volume 2] and [199]. In [239] and [166] direct proofs are provided. Also interesting in this respect are Chapters 3 and 9 of [242] and [246].

Splicing P systems on asymmetric graphs were also introduced in [223]. There it is proved that $tSPL_*(1, *, 2, 1) = RE$. Theorem 9.3 is from [77]. Splicing P systems on symmetric graphs were studied in [230].

Splicing P systems with global rules were introduced in [202] where Theorem 9.6 was proved.

Splicing P systems with immediate communication were introduced in [173]. Theorem 9.11 is from [237] while the remaining results in Section 9.7 are from [173].

Splicing P systems were also studied in combination with other formalisms based on splicing [63, 169, 224].

One model of splicing P systems not included in this chapter, because the research is still in its initial phases, considers splicing rules associated with membranes [62].

The complete proof of the direct universal splicing P system can be found in [89].

9.10.1 About the notation

Splicing rules of the kind $r = u_1\#u_2\$u_3\#u_4$ have also been denoted by $r : \dfrac{u_1 \mid u_2}{u_3 \mid u_4}$ (see, for instance, [168, 244]).

Moreover, we denoted by $r = u_1\#u_2\$u_3\#$ splicing rules of the kind $r = u_1\#u_2\$u_3\#u_4$ where $u_4 = \epsilon$. Some authors (see, for instance, [168]) prefer to explicitly indicate the presence of empty strings: $r = u_1\#u_2\$u_3\#\epsilon$.

Different notations to refer to the families of languages generated by splicing P systems having a cell-tree as underlying structure are present in the literature of splicing P systems. For instance, in [207] SSC is adopted, while in [204] SPL(i/o) and SPL(tar) are used. The notation we used is that adopted in the latest papers on splicing P systems.

The use of t in tSPL is original to the present monograph and follows the style used to denote other families of sets generated by P systems having a cell-graph as underlying structure (see, for instance, Section 5.13.2).

10   Dynamic topological structures

In this chapter we consider models of membrane systems able to modify their underlying structure while computing: membranes, and related compartments, can be created or dissolved.

In Membrane Computing this direction of research has two sources: direct observation of the cell and Brane Calculi. The former source is mainly concerned with computational complexity issues while the latter source was driven by the idea to link Brane Calculi to Membrane Computing.

Computational complexity is the branch of computer science studying the amount of resources, mainly time and space, required for the execution of algorithms and devising algorithms using these resources in an efficient way. In this chapter we present results concerning the computational complexity of membrane systems.

We describe how some biological processes and phenomena have been of inspiration for the description of models of membrane systems that in turn allowed the definition of new computational complexity classes.

Brane calculi were also inspired by living cells but they abstract them differently from Membrane Computing. These calculi pay more attention to the fidelity of the biological reality and use the framework of process algebra to target problems arising in systems biology.

Differently from in Membrane Computing, where a membrane is mainly considered with respect to the compartment defined by it and the content of the compartment, in Brane Calculi membranes themselves have the main roles. The structure, properties, evolution of membranes and the proteins embedded in them are the hallmarks of Brane Calculi.

All these similarities and differences stimulated the bridging of Membrane Computing and Brane Calculi.

10.1   Biological motivations

The model of membrane systems considered in this chapters sees the creation and dissolution of compartments and it has been inspired by similar processes in living cells.

In Section 2.2, for instance, we already said that some vesicles are created by the endoplasmatic reticulum and Golgi apparatus in order to carry molecules.

Another process in which membrane compartments are created and destroyed is *cell division*. During cell division cells reproduce by duplicating their content and dividing into two. The original cell is called the *parent cell* while the two cells resulting from this process are called *daughter cells*.

Various morphological changes occur during cell division: the nuclear envelope breaks down, the endoplasmatic reticulum and Golgi apparatus reorganise, the chromosomes condense, etc. This process, guided by the cytoskeleton, is mediated by (membrane) proteins.

At the end of a cell division the two daughter cells contain one copy of the DNA present in the parent cell together with most of the organelles present in the parent cell. Several of these organelles are formed by growth and division. The mitochondria and the endoplasmatic reticulum fall in this category. Other organelles, such as the Golgi apparatus and lysosomes, are formed from the endoplasmatic reticulum.

In the daughter cells the only organelle that needs to be reconstructed is the nuclear envelope.

A biological process in which nested membrane compartments are created is the formation of membrane stacks (*thylakoids*) inside *chloroplasts*. Chloroplasts, organelle members of the *plastid* family, are present in living plant cells and they can form from undifferentiated precursor plastids, which contain an outer and an inner membrane. The inner membrane can bud off flat vesicles at its inner face, creating stacks of membrane structures. These membranes contain the photosystems that perform photosynthesis.

10.2 Basic definitions

In this section we define concepts that, in this monograph, are strictly related only to the present chapter. That is the reason why we did not include them in Chapter 3.

In the following we denote by *RDTM* a generic recognising deterministic Turing machine or any other recognising deterministic device having equivalent computing power. We recall that, in general, a *RDTM* receives an input ι (which can be a string, a multiset of symbols, etc.), denoted by $RDTM(\iota)$, and it goes through a sequence of configurations called *computation*. The passage from one configuration to another in a computation is called a *transition*. If the sequence of configurations is finite, then we say that $RDTM(\iota)$ *halts*. If the last configuration of a halting computation is of a certain kind (the computing device is in a certain state, a certain element is present in a specified compartment, etc.), then we say that *RDTM accepts* ι otherwise *RDTM rejects* ι.

We say that a function $f : A \rightarrow B$ is *computable* if there exist a *RDTM* such that for each a and b with $f(a) = b$, then $RDTM(a, b)$ halts. We define the function $t : \mathbb{N}_+ \rightarrow \mathbb{N}_+$ and we say that a *RDTM computes f in time t* if:

for each $a \in A$, $b \in B$ with $f(a) = b$, $RDTM(a, b)$ accepts after τ transitions with $\tau \leq t(|a| + |b|)$;

for each $a \in A$, $b \in B$ with $f(a) \neq b$, $RDTM(a, b)$ rejects after τ transitions with $\tau \leq t(|a| + |b|)$.

So, the number of transitions of a *RDTM* is bound in terms of a function of the length of the input.

Given a set of computing devices a *class* is defined by the elements in the set sharing some features. If, for instance, we consider P systems with symport/antiport, then the accepting and generating models define two classes, the models having degree 2 define another class, etc.

Definition 10.1 *A decision problem $\mathcal{D}$ is a pair $(I_\mathcal{D}, \theta_\mathcal{D})$ where $I_\mathcal{D}$ is a language whose elements are called* instances *of the problem $\mathcal{D}$ and $\theta_\mathcal{D} : I_\mathcal{D} \rightarrow \{\text{yes}, \text{no}\}$ is a total function.*

A decision problem $\mathcal{D} = (I_\mathcal{D}, \theta_\mathcal{D})$ is *solvable* by a *RDTM* if $I_\mathcal{D}$ is the set of inputs of the *RDTM*, $RDTM(\iota)$ halts for each $\iota \in I_\mathcal{D}$ and ι is accepted by the *RDTM* if and only if $\theta_\mathcal{D}(\iota) = \text{yes}$.

Definition 10.2 *Let $\mathcal{D} = (I_\mathcal{D}, \theta_\mathcal{D})$ be a decision problem, $t : \mathbb{N}_+ \rightarrow \mathbb{N}_+$ be a total computable function and $icod : I_\mathcal{D} \rightarrow (V \rightarrow \mathbb{N})$, $ncod : I_\mathcal{D} \rightarrow \mathbb{N}_+$ two functions computable in polynomial time.*

We say that $\mathcal{D}$ is solvable *by a class of recognising membrane systems with input compartment $\Pi = \{\Pi_{(i)} \mid i \in \mathbb{N}_+\}$ in time bounded by t if for each $\iota \in I_\mathcal{D}$ the class Π is:*

uniform: *there is a RDTM creating a description of $\Pi_{(ncod(\iota))}(icod(\iota))$ in polynomial time;*

bounded: *every computation of $\Pi_{(ncod(\iota))}(icod(\iota))$ halts after at most $t(|\iota|)$ transitions;*

sound: *if there is an accepting computation of $\Pi_{(ncod(\iota))}(icod(\iota))$, then $\theta_\mathcal{D}(\iota) = \text{yes}$;*

complete: *if $\theta_\mathcal{D}(\iota) = \text{yes}$, then every computation of $\Pi_{(ncod(\iota))}(icod(\iota))$ is an accepting computation.*

In this case we write $\mathcal{D} \in \mathbf{MC}^{\text{in}}{}_\Pi(t)$.

The previous definition crucially depends upon the functions t, *icod*, and *ncod*. Moreover:

ncod is the function returning the index j of $\Pi_{(j)} \in \mathbf{\Pi}$ associated with an instance ι of $I_\mathcal{D}$ (different instances can have the same j associated with them);

icod is the function returning the polynomial encoding of ι given as input to $\Pi_{ncod(\iota)}$;

the item *uniform* means that all the relevant information related to the simulation of $\Pi_{(ncod(\iota))}(icod(\iota))$ by the *RDTM* can be produced in polynomial time by *RDTM*.

confluence, that is, the fact that computations starting with the same input return the same output, is implicit;

we did not specify anything in relation to the size of the configurations in the computations. This size can grow exponentially.

We say that the family $\mathbf{\Pi}$ of Definition 10.2 solves the decision problem $\mathcal{D}$ in a *uniform way* (or that $\mathbf{\Pi}$ gives a *uniform* solution to $\mathcal{D}$).

Definition 10.3 *We denote by* $\mathbf{PMC}_{\mathbf{\Pi}}^{\text{in}}$ *the class of decision problems solvable in polynomial time by a class of recognising membrane systems* $\mathbf{\Pi}$ *with input compartment. That is,*

$$\mathbf{PMC}_{\mathbf{\Pi}}^{\text{in}} = \bigcup_t \mathbf{MC}^{\text{in}}{}_{\mathbf{\Pi}}(t)$$

for all functions t *computable in polynomial time.*

Recognising membrane systems with an input compartment allow us to solve all instances of the same size (according to a fixed polynomial time computable criterion) of a decision problem if an appropriate input is supplied. Of course it is possible to consider recognising membrane systems without an input compartment. Such systems are built with a special-purpose algorithmic solution for each instance of a decision problem.

Definition 10.4 *Let* $\mathcal{D} = (I_\mathcal{D}, \theta_\mathcal{D})$ *be a decision problem,* t $: \mathbb{N}_+ \to \mathbb{N}_+$ *be a total computable function.*

We say that $\mathcal{D}$ *is* solvable *by a class of recognising membrane systems without input compartment* $\mathbf{\Pi} = \{\Pi_{(\iota)} \mid \iota \in I_\mathcal{D}\}$ *in time bounded by* t *if the class* $\mathbf{\Pi}$ *is:*

uniform: *there is a* RDTM *creating a description of* $\Pi_{(\iota)}$ *in polynomial time;*

bounded: *every computation of* $\Pi_{(\iota)}$ *halts and it consists of at most* $t(|\iota|)$ *transitions;*

sound: *if there is an accepting computation of* $\Pi_{(\iota)}$, *then* $\theta_\mathcal{D}(\iota) = $ **yes**;

complete: *if* $\theta_\mathcal{D}(\iota) = $ **yes**, *then every computation of* $\Pi_{(\iota)}$ *is an accepting computation.*

In this case we write $\mathcal{D} \in \mathbf{MC}_{\mathbf{\Pi}}(t)$.

We say that the family $\mathbf{\Pi}$ of Definition 10.4 solves the decision problem $\mathcal{D}$ in a *semi-uniform way* (or that $\mathbf{\Pi}$ gives a *semi-uniform solution* to $\mathcal{D}$).

Definition 10.5 *We denote by* $\mathbf{PMC_\Pi}$ *the class of decision problems solvable in polynomial time by a class of recognising membrane systems without input compartment* $\mathbf{\Pi}$. *That is,*

$$\mathbf{PMC_\Pi} = \bigcup_t \mathbf{MC_\Pi}(t)$$

for all functions t *computable in polynomial time.*

In the literature of Membrane Computing also $\mathbf{LMC_\Pi^{in}} = \bigcup_t \mathbf{MC^{in}}_\Pi(t)$ and $\mathbf{LMC_\Pi} = \bigcup_t \mathbf{MC_\Pi}(t)$, for all functions t computable in linear time, are defined. These classes identify the decision problems solvable in linear time by a class of recognising membrane systems with or without an input compartment, respectively.

The name of such a class can be misleading as readers may regard the term *linear* in a strict sense. The classes $\mathbf{LMC_\Pi^{in}}$ and $\mathbf{LMC_\Pi}$ cannot be linear (in a strict sense) because of item *uniform* in Definitions 10.2 and 10.4: if a *RDTM* can create the description in polynomial time, then the class cannot be linear. The point here is that the term *linear* refers to the time needed by the membrane systems to solve some decision problems, and not to the time needed to create a description of the membrane system. To avoid such a misunderstanding, we decide not to refer to $\mathbf{LMC_\Pi^{in}}$ and $\mathbf{LMC_\Pi}$.

Conforming to what is present in the literature of Computational Complexity Theory, we denote by:

P the class of decision problems solvable in polynomial time by Turing machines;

NP the class of decision problems solvable in polynomial time by non-deterministic Turing machines;

co-NP the class of decision problems whose complements are in **NP**;

NP-complete the class subset of **NP** problems such that any problem in **NP** can be reduced in polynomial time to any problem in this subset;

PSPACE the class of decision problems solvable using a polynomial amount of memory by deterministic Turing machines;

PSPACE-complete the subset of **PSPACE** problems such that any problem in **PSPACE** can be reduced in polynomial time to any problem in this subset.

In the following definition we use a *bracket representation* to picture compartments and the cell-tree structure of a membrane system. In this representation a compartment in a membrane system is represented by a pair of matching square brackets each with a subscript indicating the vertex in the structure. For instance, a compartment i is represented by $[_i \ ]_i$. If in a cell-tree a vertex i has $j_1, \ldots, j_k$ as children, then this is represented by $[_i[_{j_1} \ ]_{j_1} \ldots [_{j_k} \ ]_{j_k}]_i$. The structure depicted in Fig. 3.2 is then represented by

$$[_{env} \ [_{m_1} \ [_{m_2} \ [_{m_5}]_{m_5} \ [_{m_6} \ [_{m_8}]_{m_8} \ ]_{m_6} \ ]_{m_2} \ [_{m_3} \ [_{m_7}]_{m_7} \ ]_{m_3} \ [_{m_4}]_{m_4} \ ]_{m_1} \ ]_{env}$$

where env denotes the environment. Eventually, symbols present in the compartments are placed between the appropriate brackets. The next definition introduces, among others, compartments with labels (polarities). In the bracket representation polarities are denoted by a superscript in the closing bracket.

In the present chapter when multisets are represented by strings, the empty multiset is denoted by the empty string ϵ.

Definition 10.6 *A* (generating) P system with active membranes *is a construct*

$$\Pi = (V, \mu, L_1, \ldots, L_m, R)$$

where

V is an alphabet.

$\mu = (Q, E, pol)$ *is a vertex-labelled cell-tree underlying* Π *where:*

$Q \subseteq W \times \mathbb{N}_+$ *contains vertices of the form* (q, n) *with* $q \in W$, $W = \{0, 1, \ldots, m\}$, $n \in \mathbb{N}_+$. *The root of the cell-tree is vertex* 0 *and each vertex in* Q *defines a compartment of* Π.

$E \subseteq Q \times Q$ *defines edges between vertices denoted by* (i, j), $i, j \in Q$, $i \neq j$.

$pol : Q \to \{+, -, 0\}$ *is a labelling function, called a* polarity function, *associating elements of* $Q_P = \{+, -, 0\}$ *to the vertices.*

$L_i : V \to \mathbb{N}, i \in Q \setminus \{0\}$, *are multisets of symbols associated with the vertices of* Q.

R is a finite set of rules of the form:

(a) $[_i \ v \to w \ ]_i^p$
with $i \in W$, $p \in Q_P$, $v \in V$ and $w : V \to \mathbb{N}$.
Informally: productions associated with compartments.

(b) $v \ [_i \ ]_i^p \to [_i \ w \ ]_i^{p'}$
with $i \in W$, $p, p' \in Q_P$, $v \in V$ and $w \in V \cup \{\epsilon\}$.
Informally: a symbol can pass into a compartment.

(c) $[_i \ v \ ]_i^p \to [_i \ ]_i^{p'} \ w$
with $i \in W$, $p, p' \in Q_P$, $v \in V$ and $w \in V \cup \{\epsilon\}$.
Informally: a symbol can pass outside a compartment.

(d) $[_i \ v \ ]_i^p \to w$
with $i \in W \setminus \{0, 1\}$, $p \in Q_P$, $v \in V$ and $w \in V \cup \{\epsilon\}$.
Short name: dissolution.

Informally: the presence of a symbol in a compartment allows the membrane defining the compartment to disappear.

(e) $[_i \, v \,]_i^p \to [_{i'} \, w \,]_{i'}^{p'} \, [_{i'} \, z \,]_{i'}^{p''}$

with $i, i' \in W \setminus \{0, 1\}$, $p, p', p'' \in Q_P$, $v \in V$ and $w, z \in V \cup \{\epsilon\}$.

Short name: division of a (simple or composite) compartment.

Informally: the presence of a symbol in a compartment allows the compartment to duplicate.

(f) $[_{i_1} [_{i_2}]_{i_2}^{p^{(1)}} \cdots [_{i_r}]_{i_r}^{p^{(1)}} [_{i_{r+1}}]_{i_{r+1}}^{p^{(2)}} \cdots [_{i_q}]_{i_q}^{p^{(2)}}]_{i_1}^{p^{(3)}} \to$

$$[_{i_1'} [_{i_2}]_{i_2}^{p^{(4)}} \cdots [_{i_r}]_{i_r}^{p^{(4)}}]_{i_1'}^{p^{(5)}} [_{i_1'} [_{i_{r+1}}]_{i_{r+1}}^{p^{(6)}} \cdots [_{i_q}]_{i_q}^{p^{(6)}}]_{i_1'}^{p^{(7)}}$$

with $i_j, i_1', \in W \setminus \{0, 1\}$, $1 \leq j \leq q$, $p^{(1)}, \ldots, p^{(7)} \in Q_P$ and $p^{(1)} = +$, $p^{(2)} = -$.

Short name: division of a composite compartment.

Informally: if a compartment contains membranes with different polarities, these these membranes can be parted in two compartments depending on their polarities;

(g) $[_i \, v \,]_i^p \to [_i [_{i'} \, w \,]_{i'}^{p'}]_i^p$

with $i, i' \in W \setminus \{0, 1\}$, $p, p' \in Q_P, v \in V$ and $w \in V \cup \{\epsilon\}$.

Short name: creation of a compartment.

Informally: the presence of a symbol in a compartment allows a new compartment to be created.

Vertex 0 identifies the environment, while vertex 1 identifies the skin compartment of μ. The set $Q_P = \{+, -, 0\}$ is called the *set of polarities* and its elements are called *polarities*. Specifically, $+$ is called *positive*, $-$ is called *negative* and 0 is called *neutral* polarity. A compartment having a positive polarity is said to be a *positive compartment*, similarly for the remaining polarities.

A *configuration* of a P system Π with active membranes $\Pi = (V, \mu, L_1, \ldots, L_m, R)$ is given by a tuple of the kind $(\mu', M'_{z_0}, \ldots, M'_{z_{|Q'|}})$, $\mu' = (Q', E', pol')$, $z_i \in Q'$, $0 \leq i \leq |Q'|$, where μ' is a vertex-labelled cell-tree and $M'_{z_i} : V \to \mathbb{N}$, $0 \leq i \leq |Q'|$, are multisets of symbols.

Given a configuration $c = (\mu', M'_{z_0}, \ldots, M'_{z_{|Q'|}})$, $\mu' = (Q', E', pol')$, $Q \subseteq W' \times \mathbb{N}_+$, $W' = \{0, 1, \ldots, m\}$, the application of the rules in R can change μ' and M'_{z_i}, $0 \leq i \leq |Q'|$, in the following way:

(a) if $[_i \, v \to w \,]_i^p \in R$, $i \in W'$, $p \in Q_P$, $v \in V$, $w : V \to \mathbb{N}$ and there is $k \in \mathbb{N}_+$ such that $v \in M'_{(i,k)}$, $pol'(i, k) = p$, then

$M'_{(i,k)}$ changes into $M''_{(i,k)} = M'_{(i,k)} \setminus \{v\} \cup \{w\}$

and the remaining components of c are unchanged.

(b) if $v \, [_i \,]_i^p \to [_i \, w \,]_i^{p'} \in R$, $i \in W'$, $p, p' \in Q_P$, $v \in V$, $w \in V \cup \{\epsilon\}$ and there is $k \in \mathbb{N}_+$ such that $v \in M'_{\text{parent}(i,k)}$, $pol'(i, k) = p$, then

$M'_{\text{parent}(i,k)}$ changes into $M''_{\text{parent}(i,k)} = M'_{\text{parent}(i,k)} \setminus \{v\}$,
$M'_{(i,k)}$ changes into $M''_{(i,k)} = M'_{(i,k)} \cup \{w\}$,
$pol(i,k)$ changes into p'

and the remaining components of c are unchanged.

(c) if $[_i \, v \,]_i^p \to [_i \, v \,]_i^{p'} \, w \in R$, $i \in W'$, $p, p' \in Q_P$, $v \in V$, $w \in V \cup \{\epsilon\}$ and there
is $k \in \mathbb{N}_+$ such that $v \in M'_{(i,k)}$, $pol'(i,k) = p$, then

$M'_{\text{parent}(i,k)}$ changes into $M''_{\text{parent}(i,k)} = M'_{\text{parent}(i,k)} \cup \{w\}$,
$M'_{(i,k)}$ changes into $M''_{(i,k)} = M'_{(i,k)} \setminus \{v\}$,
$pol'(i,k)$ changes into p'

and the remaining components of c are unchanged.

(d) if $[_i \, v \,]_i^p \to w \in R, i \in W' \setminus \{0,1\}$, $p \in Q_P$, $v \in V$, $w \in V \cup \{\epsilon\}$ and there is
$k \in \mathbb{N}_+$ such that $v \in M'_{(i,k)}$, $pol'(i,k) = p$, then

Q' changes into $Q'' = Q' \setminus \{(i,k)\}$,

E' changes into $E'' = E' \setminus \{(\text{parent}(i,k), (i,k))\} \cup \bigcup_{((i,k),x)\in child}\{(\text{parent}(i,k), x)\}$,

$M'_{\text{parent}(i,k)}$ changes into $M''_{\text{parent}(i,k)} = M'_{\text{parent}(i,k)} \cup M'_{(i,k)} \setminus \{v\} \cup \{w\}$,

and the remaining components of c are unchanged. When such a rule is
applied all the symbols present in the compartment defined by the dissolved
membrane pass into the parent compartment. Notice that the environment
and the skin compartment cannot be dissolved.

(e) there are two semantics:

(e') if $[_i \, v \,]_i^p \to [_{i'} \, w \,]_{i'}^{p'} [_{i'} \, z \,]_{i'}^{p''} \in R$, $i, i' \in W' \setminus \{0,1\}$, $p, p', p'' \in Q_P$, $v \in$
V, $w, z \in V \cup \{\epsilon\}$, there is $k \in \mathbb{N}_+$ such that $v \in M'_{(i,k)}$, $pol'(i,k) = p$,
there are not $k' \in \mathbb{N}_+$, $j \in W'$ such that $(j,k') \in Q'$, $((i,k),(j,k')) \in$
E', then

Q' changes into $Q'' = Q' \setminus \{(i,k)\} \cup \{(i',k'''), (i',k''' + 1)\}$ with $k''' =$
$max(\{k'''' \mid (i',k'''') \in Q'\}) + 1$,

E' changes into $E'' = E' \setminus \{(\text{parent}(i,k),(i',k''')), (\text{parent}(i,k),(i',$
$k''' + 1))\} \cup \{(\text{parent}(i,k),(i,k))\}$,

$pol'(i',k''') = p'$,

$pol'(i',k''' + 1) = p''$,

$M'_{(i,k)}$ changes into $M''_{(i,k)} = \phi$,

$M'_{(i',k''')} = M'_{(i,k)} \setminus \{v\} \cup \{w\}$,

$M'_{(i',k'''+1)} = M'_{(i,k)} \setminus \{v\} \cup \{z\}$

and the remaining components of c are unchanged.

(e'') if $[_i \, v \,]_i^p \to [_{i'} \, w \,]_{i'}^{p'} [_{i'} \, z \,]_{i'}^{p''} \in R$, $i, i' \in W' \setminus \{0,1\}$, $p, p', p'' \in Q_P$, $v \in$
V, $w, z \in V \cup \{\epsilon\}$, there is $k \in \mathbb{N}_+$ such that $v \in M'_{(i,k)}$, $pol'(i,k) = p$,
there are $k' \in \mathbb{N}_+$, $j \in W'$ such that $(j,k') \in Q'$, $((i,k),(j,k')) \in E'$,
then

Q' changes into $Q'' = Q' \setminus \{(i,k)\} \cup \{(i',k'''), (i',k'''+1)\}$ with $k''' = max(\{k'''' \mid (i',k'''') \in Q'\}) + 1$,

E' changes into $E'' = E' \setminus \{(parent(i,k),(i',k''')), (parent(i,k),(i', k'''+1))\} \cup \{(parent(i,k),x) \mid ((i,k),x) \in child\} \cup \{(parent(i,k), (i,k))\}$,

$pol'(i',k''') = p'$,

$pol'(i',k'''+1) = p''$,

$M'_{(i,k)}$ changes into $M''_{(i,k)} = \phi$,

$M'_{(i',k''')} = M'_{(i,k)} \setminus \{v\} \cup \{w\}$,

$M'_{(i',k'''+1)} = M'_{(i,k)} \setminus \{v\} \cup \{z\}$

and the remaining components of c are unchanged.

When such a rule is applied (both semantics), then all the compartments and symbols present in the original compartment (with the exception of v) are also duplicated in the two new compartments. Notice that the environment and the skin compartment cannot divide.

(f) if $[_{i_1} [_{i_2}]_{i_2}^{p^{(1)}} \cdots [_{i_k}]_{i_k}^{p^{(1)}} [_{i_{k+1}}]_{i_{k+1}}^{p^{(2)}} \cdots [_{i_q}]_{i_q}^{p^{(2)}}]_{i_1}^{p^{(3)}} \to$

$\qquad [_{i_1'} [_{i_2}]_{i_2}^{p^{(4)}} \cdots [_{i_k}]_{i_k}^{p^{(4)}}]_{i_1'}^{p^{(5)}} [_{i_1''} [_{i_{k+1}}]_{i_{k+1}}^{p^{(6)}} \cdots [_{i_q}]_{i_q}^{p^{(6)}}]_{i_1''}^{p^{(7)}} \in R$,

$i_j, i_1' \in W' \setminus \{0,1\}$, $1 \le j \le q$, $p^{(1)}, \ldots, p^{(7)} \in Q_P$, $p^{(1)} = +$, $p^{(2)} = -$, there are $k, k_{j_1} \in \mathbb{N}_+$, $2 \le j_1 \le r$ and $k_{j_2} \in \mathbb{N}_+$, $r+1 \le j_2 \le q$ such that $pol'(i_1,k_1) = p^{(3)}$, $pol'(i_{j_1},k_{j_1}) = p^{(1)}$, $pol'(i_{j_2},k_{j_2}) = p^{(2)}$, then

Q' changes into $Q'' = Q' \setminus \{(i_1,k_1)\} \cup \{(i_1',k'), (i_1'',k''')\}$ with $k' = max(\{k'' \mid (i_1',k'') \in Q'\}) + 1$ and $k''' = max(\{k'''' \mid (i_1'',k'''') \in Q'\}) + 1$,

E' changes into $E'' = E' \setminus \{(parent(i_1,k_1),(i_1,k_1))\} \setminus \{((i_1,k_1),(i_j,k_j)) \mid 2 \le j \le r\} \cup \{((i_1',k'),(i_{j_1},k_{j_1})) \mid 2 \le j_1 \le r\} \cup \{((i_1'',k'''),(i_{j_2},k_{j_2})) \mid r+1 \le j_2 \le q\} \cup \{(parent(i_1,k_1),(i_1',k')),(parent(i_1,k_1),(i_1'',k'''))\} \cup \{((i_1',k'),x),((i_1'',k'''),x) \mid ((i_1,k_1),x) \in child, \ x \ne (i_j,k_j) \ 2 \le j \le q\}$,

$pol'(i_{j_1},k_{j_1}) = p^{(4)}$,

$pol'(i_1',k') = p^{(5)}$,

$pol'(i_{j_2},k_{j_2}) = p^{(6)}$,

$pol'(i_1'',k''') = p^{(7)}$,

$M'_{(i_1,k_1)}$ changes into $M''_{(i_1,k_1)} = \phi$,

$M'_{(i_1',k')} = M'_{(i_1,k_1)}$,

$M'_{(i_1'',k''')} = M'_{(i_1,k_1)}$,

and the remaining components of c are unchanged. Notice that the environment and the skin compartment cannot divide.

(g) if $[_i \ v \ ]_i^p \to [_i [_i \ w \ ]_{i'}^{p'}]_i^p \in R$, $i, i' \in W' \setminus \{0,1\}$, $p, p' \in Q_P$, $v \in V$, $w \in V \cup \{\epsilon\}$, there is $k \in \mathbb{N}_+$ such that $v \in M'_{(i,k)}$, $pol(i,k) = p$, then

Q' changes into $Q'' = Q' \cup \{(i',k')\}$ with $k' = max(\{k'' \mid (i',k'') \in Q'\}) + 1$,

E' changes into $E'' = E' \cup \{((i,k),(i',k'))\}$,

$pol'(i', k') = p'$,

$M'_{(i,k)}$ changes into $M''_{(i,k)} = M'_{(i,k)} \setminus \{v\}$,

$M'_{(i',k')} = \{w\}$

and the remaining components of c are unchanged. Notice that compartments cannot be created either in the environment nor in the skin compartment.

If the tree $\mu = (Q, E, pol)$ underlying a P system with active membranes Π is such that pol always returns the same polarity and no rule changing polarities is present in Π, then $\mu = (Q, E)$ is a cell-tree.

Because of rules of the kind *(d)*, *(e)*, *(f)* and *(g)* the cell-tree underlying Π can change. For this reason the vertex-labelled cell-tree is called *dynamic*.

The fact that the vertex-labelled cell-tree is dynamic can create some confusion on the application of the rules. What happens, for instance, if in one configuration both rules $[_i \; v \;]_i^p \to [_{i'} \; w \;]_{i'}^{p'} [_{i''} \; z \;]_{i''}^{p''}$ and $[_i \; v' \to x \;]_i^p$ can be applied? In this case they are both applied so that x is present in both compartments i' and i''.

If in one configuration both rules $[_i \; v' \;]_i^p \to w'$ and $[_i \; v \;]_i^p \to [_{i'} \; w \;]_{i'}^{p'} [_{i''} \; z \;]_{i''}^{p''}$ can be applied, then only one of the two rules is applied. Similarly for other groups of rules.

For two configurations $(\mu', M'_0, \ldots, M'_z)$ and $(\mu'', M''_0, \ldots, M''_{z'})$ of Π we write $(\mu', M'_0, \ldots, M'_z) \Rightarrow (\mu'', M''_0, \ldots, M''_{z'})$ to denote a transition from $(\mu', M'_0, \ldots, M'_z)$ to $(\mu'', M''_0, \ldots, M''_{z'})$, that is, the application of a multiset of rules associated with each compartment following what was indicated in the above and under the requirement of maximal parallelism. The reflexive and transitive closure of $\Rightarrow$ is denoted by $\Rightarrow^*$.

A *computation* is a sequence of transitions between configurations of a system Π starting from the *initial configuration* $(\mu, \phi, L_1, \ldots, L_m)$. If a computation is finite, then the last configuration is called *final* and we say that the system *halts*. The result of a finite computation is given by the number of symbols passing to the environment as a consequence of the application of a rule of type *(c)*. The set $N(\Pi)$ denotes the set of numbers generated by a P system Π with active membranes.

A *recognising P system with active membranes and input compartment* is a construct

$$\Pi = (V, \mu, L_1, \ldots, L_m, R, comp)$$

with $\mu = (Q, E, pol)$ a vertex-labelled cell-tree, $comp \in Q$ and yes, no $\in V$. In the initial configuration $comp$ contains the input of Π. The system Π is said to *accept* the input if there is at least one sequence of configurations such that the symbol yes passes to the environment in the last transition of a finite computation. Such sequences of configurations are called *accepting computations*. The system Π is said to *reject* the input if for all sequences of configurations the symbol no passes to the environment in the last transition of a finite computation. Such sequences are called *rejecting computations*. Moreover, either yes or no can pass to the environment in the last transition of a finite computation.

A *recognising P system with active membrane without input compartment* is similar to the above but without *comp*.

10.3 Examples

The examples presented in this section are rather simple; they do not exploit the full potential of P systems with active membranes. Here our aim is to introduce readers to the way these systems operate.

Example 10.1 A P systems with active membranes generating non-negative even numbers.

The formal definition of such a P system is:

$$\Pi_1 = (V, \mu, L_1, R)$$

with

$$
\begin{aligned}
V &= \{a, b\}; \\
\mu &= (\{0, 1\}, \{(0, 1)\}); \\
L_1 &= \{a\}; \\
R &= \{1 : [_1\, a \to abb\,]_1, 2 : [_1\, a \to \epsilon\,]_1, \\
&\qquad 3 : [_1\, b\,]_1 \to [_1]_1\, b\}.
\end{aligned}
$$

In order to facilitate the explanation, rules have been numbered. This system lacks many features present in Definition 10.6. The vertices are not labelled and not all possible kinds of rules are used. Specifically, rules 1 and 2 are of type (*a*) while rule 3 is of type (*c*). The fact that no rule of types (*d*), (*e*) and (*f*) is present in R implies that μ is not a dynamic cell-tree.

In the initial configuration either rule 1 or 2 can be applied. If rule 2 is applied, then the system halts, so that the number zero is generated; if rule 1 is applied, then two occurrences of b are present in compartment 1. In the next configuration the symbol a can still be subject to rules 1 or 2, while the b symbols pass to the environment because of the application of rule 3.

This process halts only when rule 2 is applied. It should be clear that Π_1 generates non-negative even numbers. $\diamond$

The next example aims to show the use of different kinds of rules.

Example 10.2 A(nother) P systems with active membranes generating non-negative even number.

The formal definition of this P system is:

$$\Pi_2 = (V, \mu, L_1, L_2, R)$$

with

$V = \{a, b, c, d, e\};$
$\mu = (\{0, 1, 2\}, \{(0, 1), (1, 2)\});$
$L_1 = \phi;$
$L_2 = \{a\};$
$R = \{1 : [_2 \, a \,]_2 \rightarrow [_2 \, a \,]_2[_2 \, b \,]_2, 2 : [_2 \, a \rightarrow \epsilon \,]_2,$
$\qquad 3 : [_2 \, b \rightarrow cdd \,]_2, 4 : [_2 \, c \,]_2 \rightarrow e,$
$\qquad 5 : [_1 \, e \rightarrow \epsilon \,]_1, 6 : [_1 \, d \,]_1 \rightarrow [_1]_1 \, d\}.$

Rules 2, 3 and 5 are of type (a), rule 1 is of type (e), rule 4 is of type (d) and rule 6 is of type (c).

The bracket representation $[_0[_1[_2 \, a \,]_2]_1]_0$ denotes the initial cell-tree structure of Π_2. In the initial configuration either rule 1 or 2 can be applied. If rule 2 is applied, then the system halts, so that the number zero is generated. Every time rule 1 is applied, then, in the subsequent configurations, two occurrences of e are generated and pass into the environment. If in the initial configuration rule 1 is applied, then the underlying graph changes into $[_0[_1[_2 \, a \,]_2[_2 \, b \,]_2]_1]_0$. Either rule 1 or 2 can be applied together while rule 3. If we assume that rules 2 and 3 are applied in parallel, then the underlying graph changes into $[_0[_1[_2 \, cdd \,]_2]_1]_0$. In such a configuration the presence of the symbol c in a compartment 2 allows the compartment to be dissolved by the application of rule 4 so that the underlying graph changes into $[_0[_1 \, edd \,]_1]_0$. Notice that the dissolution of compartment 2 allows the two occurrences of d present in it to pass into compartment 1. Rules 5 and 6 are now applied. Because of maximal parallelism rule 6 is applied twice. When this takes place, then two applications of rule 6 let the underlying graph change into $[_0[_1]_1 \, dd \,]_0$.

If when the underlying graph is $[_0[_1[_2 \, a \,]_2[_2 \, b \,]_2]_1]_0$ the rules 2 and 3 are applied, then the underlying graph changes into $[_0[_1[_2 \, a \,]_2[_2 \, a \,]_2[_2 \, cdd \,]_2]_1]_0$.

It should be clear that Π_2 generates non-negative even numbers and that rule 1, of type (e), is applied according to the (e') semantics. $\qquad \diamond$

Example 10.3 A P systems with active membranes generating $\{2^n \mid n \geq 2\}$.

The formal definition of this P system is:

$$\Pi_3 = (V, \mu, L_1, L_2, L_3, R)$$

with

$V = \{a, b\}$;

$\mu = (\{0, 1, 2, 3\}, \{(0, 1), (1, 2), (1, 3)\})$;

$L_1 = L_3 = \{b\}$;

$L_2 = \{a\}$;

$R = \{1 : [_2\, a \rightarrow aa\,]_2, 2 : [_3\, b\,]_3 \rightarrow [_3]_3\, b\} \cup$
$\quad\quad \{3 : b\, [_\alpha]_\alpha \rightarrow [_\alpha\, b\,]_\alpha, 4 : [_\alpha\, b\,]_\alpha \rightarrow \epsilon \mid \alpha \in \{2, 3\}\} \cup$
$\quad\quad \{5 : [_1\, a\,]_1 \rightarrow [_1]_1\, a\}$.

Similarly to Example 10.1, this system lacks many features present in Definition 10.6. The underlying structure is a cell-tree (not dynamic and without labelling).

The bracket representation $[_0[_1b[_2\, a\,]_2[_3\, b\,]_3]_1]_0$ denotes the initial cell-tree of Π_3. The systems works in the following way: the number of occurrences of a can double in any configuration in compartment 2. In the meantime the b symbols can pass into and out of compartment 3 or one of the b's can pass into compartment 2. When an occurrence of b passes into compartment 2, then this compartment is dissolved (by rule 4 with $\alpha = 2$) so that all the occurrences of a can pass from the skin compartment to the environment. The passage of b into and out of compartment 3 ends when this compartment is dissolved (by rule 4 with $\alpha = 3$). Notice that the application of rule 4 allows one occurrence of b to disappear (that is, to be replaced by the empty string ϵ).

The minimum number that can be generated is 4. This happens in the following way. In the initial configuration rules 1, 2 and 3 can be applied in parallel. Rule 1 allows the number of occurrences of a to double, rule 2 allows the occurrence of b present in compartment 3 to pass into compartment 1, and rule 3 allows the occurrence of b present in compartment 1 to pass into compartment 2. The bracket representation of the resulting system is $[_0[_1\, b\, [_2\, aab\,]_2[_3]_3]_1]_0$.

In this configuration rules 1, 3 and 4 can be applied in parallel. Rule 1 allows the number of occurrences of a to double, rule 3 allows the occurrence of b present in compartment 1 to pass into compartment 3, and rule 4 allows compartment 2 to dissolve and the b present in this compartment to disappear. Because of the application of rule 4 the symbols present in compartment 2 pass into compartment 1. The bracket representation of the resulting system is $[_0[_1\, aaaa[_3b]_3]_1]_0$.

In this configuration rules 4 and 5 are applied in parallel. Rule 5 (applied four times because of maximal parallelism) allows the occurrences of a to pass from the skin compartment to the environment, while rule 4 allows compartment 3 to dissolve and the instance of b to disappear. The bracket representation of the resulting system is $[_0\, aaaa\, [_1]_1]_0$.

It should be clear that Π_3 generates $\{2^n \mid n \geq 2\}$. $\qquad\qquad \diamond$

10.4   Trading space for time

In this section we see that membrane systems with active membranes can solve **NP**-complete problems in a time efficient way. As the title of the present section suggests, this time efficiency comes at the cost of an inefficient use of space. In this chapter the space used by the considered P systems refers to the number of compartments used by these systems during the computations.

Definition 10.7 *Let us consider a set of Boolean variables $\{v_1, \ldots, v_n\}$ and Boolean formula ψ written in* conjunctive normal form, *that is, $\psi = C_1 \wedge \cdots \wedge C_m$. Each C_j, $1 \leq j \leq m$, is a* clause *written as a disjunction of literals $C_j = l_{j_1} \vee \cdots \vee l_{j_{k_j}}$, each literal being a variable v_i or its negation $\bar{v}_i$, $1 \leq i \leq n$.*

   The SAT *problem is: given such a Boolean formula ψ is there an assignment of the variables such that ψ is satisfied?*

   It is known that if $k_j \geq 3$, $1 \leq j \leq m$, then the problem raised by the previous question is an **NP-complete** problem identified by k-CNF or k-SAT with $k \geq 3$.

Theorem 10.1 *k-SAT, $k \geq 3$, can be solved by a recognising P system with active membranes without input compartment using only rules of the kind (a), (b), (c), (d) and (e), according to the (e'') semantics, having as underlying structure a dynamic cell-tree and no polarities. The P system requires linear time and exponential space.*

Proof   Let us consider an instance of k-SAT, $k \geq 3$, based on a Boolean formula ψ with m clauses and n variables $\{v_1, \ldots, v_n\}$. Moreover, we define two functions *true* and *false*, both from $\{v_1, \ldots, v_n\}$ to $\mathcal{P}\{C_1, \ldots, C_m\}$ such that:

   $true(v)$ $= \{C_j \mid \exists r, \ 1 \leq r \leq k_i \text{ such that } l_{j_r} = v\};$
   $false(v) = \{C_j \mid \exists r, \ 1 \leq r \leq k_i \text{ such that } l_{j_r} = \bar{v}\}$

   for $v \in \{v_1, \ldots, v_n\}$. So, these functions return the set of clauses verified and falsified by v, respectively.

   We define the recognising P system with active membranes $\Pi = (V, \mu, L_1, \ldots, L_{m+4}, R)$ with:

   $V = \{v_i, F_i, T_i \mid 1 \leq i \leq n\} \cup \{c_j, d_j \mid 1 \leq j \leq m\} \cup$
      $\{p_i \mid 1 \leq i \leq 2n + 1\} \cup \{r_1, r_2, r_3, r_4, b_1, b_2, \mathbf{yes}, \mathbf{no}\};$
   $\mu = (\{0, \ldots, m + 4\}, \{(0, 1), (1, 2), (2, 3)\} \cup \{(3, t) \mid 4 \leq t \leq 4 + m\});$
   $L_3 = \{v_1\};$
   $L_{4+m} = \{p_1\};$
   $L_r = \phi, \ r \in \{1, 2, 4, \ldots, 3 + m\}.$

   The bracket representation of the initial underlying cell-tree of Π is

$$[_0[_1[_2[_3 \ v_1 \ [_4]_4 \cdots [_{m+4} \ p_1 \ ]_{m+4}]_3]_2]_1]_0.$$

The set R consists of:

1: $[_{m+4}\, p_i \to p_{i+1}\,]_{m+4}$ for $1 \le i \le 2n,$

$[_3\, r_i \to r_{i+1}\,]_3$ for $1 \le i \le 3,$

$[_2\, r_i \to r_{i+1}\,]_2$ for $4 \le i \le 6;$

2: $[_3\, v_i\,]_3 \to [_3\, F_i\,]_3[_3\, T_i\,]_3$ for $1 \le i \le n,$

$[_3\, F_i \to false(v_i)v_{i+1}\,]_3$ for $1 \le i \le n-1,$

$[_3\, T_i \to true(v_i)v_{i+1}\,]_3$ for $1 \le i \le n-1,$

$[_3\, F_n \to false(v_n)\,]_3$ for $1 \le i \le n-1,$

$[_3\, T_n \to true(v_n)\,]_3$ for $1 \le i \le n-1;$

3: $c_j\, [_j]_j \to [_j\, c_j\,]_j,$

$[_j\, c_j\,]_j \to d_j$ for $4 \le j \le m+3;$

4: $[_{m+4}\, p_{2n+1}\,]_{m+4} \to [_{m+4}]_{m+4}\, p_{2n+1}$

$[_3\, p_{2n+1} \to b_1 r_1]_3;$

5: $b_1\, [_j]_j \to [_j\, b_1\,]_j$ for $4 \le j \le m+3,$

$[_j\, b_1\,]_j \to b_2$ for $4 \le j \le m+3,$

$[_3\, b_2\,]_3 \to b_2;$

6: $[_3\, r_4\,]_3 \to r_4,$

$[_2\, r_4 \to \mathbf{yes}\,]_2,$

$[_\beta\, \mathbf{yes}\,]_\beta \to [_\beta]_\beta\, \mathbf{yes}$ for $\beta \in \{1, 2\},$

$[_2\, r_7\,]_2 \to r_7,$

$[_1\, r_7 \to \mathbf{no}\,]_1,$

$[_1\, \mathbf{no}\,]_1 \to [_1]_1\, \mathbf{no}.$

In order to facilitate the explanation, rules have been grouped and groups of rules have been numbered.

The symbols p_i and r_i are used as registers; their subscripts increments of 1 in each transition. This is performed by the rules in group 1.

In compartment 3 the truth assignments for the n variables are generated; this is performed by the rules in group 2. In the first $2n$ transitions compartment 3 (which is composite) alternates between its division and the introduction of symbols. During division a value (T for *true* and F for *false*) is associated with variable v_i; these symbols are those associated (by the functions *false* and *true*) with the clause satisfied by v_i and $\bar{v}_i$. It is important to notice that during the division of compartment 3 all the compartments present in it are duplicated. After $2n$ transitions all the 2^n truth assignments for ψ are generated and present in compartments 3 (called *3-named compartments*, in the following).

In the first $2n + 2$ transitions if a clause C_j is satisfied by a variable v_i, then the symbol c_j passes into compartment j and dissolves it allowing a symbol d_j in the relative 3-named compartment. This is performed by the rules in group 3.

In the transitions $2n + 1$ and $2n + 2$ the two rules in group 4 are applied. This allows an instance of b_1 to be present in each 3-named compartment.

So, after $2n + 2$ transitions the underlying cell-tree of Π contains 2^n 3-named compartments, each of which contains:

symbols c_j and d_j, $1 \leq j \leq m$;

compartments $4, \ldots, 3 + m$ corresponding to the clauses which are not satisfied
   by the truth assignment generated in the relative 3-named compartment;

the empty compartment $m + 4$.

Moreover, compartments 1 and 2 are empty.

So, the Boolean formula ψ is satisfied if and only if there is a 3-named compartment without any compartment with name $4, \ldots, 3 + m$ in it. In order to check this the system proceeds as follows.

In transitions $2n + 3$ the symbol b_1 present in a 3-named compartment can pass into one of its compartments with names $4, \ldots, 3 + m$. If this happens, then in transition $2n + 4$ such a compartment is dissolved and an instance of b_2 appears in the 3-named compartment. This is performed by the rules in group 5.

In transition $2n + 4$ the symbol r_4 is generated and each 3-named compartment with b_2 is dissolved. So, if after this transition there is at least one 3-named compartment, then the formula ψ is satisfiable. If such a compartment exist, then in the next transition the symbol **yes** is generated and passes to the environment. This is performed by the rules in group 6.

If the formula ψ is not satisfiable, then in transition $2n + 7$ the compartment 2 has not been dissolved and the symbol r_6 in that compartment evolves into r_7 in the next transition. Then the symbol r_7 dissolves membrane 2 producing **no** in the skin compartment. In the next transition **no** passes to the environment. Otherwise, after $2n + 8$ transitions the symbol **yes** passes to the environment indicating that ψ is satisfiable.   □

P systems with active membranes have been used to solve **NP** problems in ways different from that just described (see Section 10.12). All these results differ in two respects: the used features of P systems with active membranes (presence or absence of polarities and their number, etc.) and the time complexity (linear or polynomial). All these results use an exponential number of compartment in relation to the size of the instance they solve.

10.5 About the classes **P** and **NP**

In this section we characterise the classes **P** and **NP** in terms of P systems with active membranes. P systems with active membranes operating under maximal parallelism, using polarities and only rules of the kind *(a)*, *(b)* and *(c)*, can generate RE. This proof is based on the simulation of matrix grammars. In the following we prove this result using models of P/T systems and, for this purpose, we introduce a few more concepts.

Definition 10.8 *A recognising dynamic P/T system is a tuple* $N = (P, P_{\text{dyn}}, T,$ $R_{\text{dyn}}^+, R_{\text{dyn}}^-, F, W, C_{\text{in}}, P_{\text{in}}, p_{\text{acc}}, p_{\text{rej}})$ *where:*

$P, T, F,$ *and* C_{in} *are as in Definition 4.3 with the only difference that the underlying net* (P, T, F) *is not supposed to be connected;*

$W : (P \times T) \cup (T \times P) \to \mathbb{N}_+;$

$P_{\text{dyn}}, P_{\text{in}} \subset P;$

$R_{\text{dyn}}^+ \subseteq P_{\text{dyn}} \times ((P \times T) \cup (T \times P));$

$R_{\text{dyn}}^- \subseteq P_{\text{dyn}} \times ((P \times T) \cup (T \times P));$

$p_{\text{acc}}, p_{\text{rej}} \in P,\ p_{\text{acc}} \neq p_{\text{rej}};$

$C_{\text{in}}(p_{\text{acc}}) = C_{\text{in}}(p_{\text{rej}}) = 0.$

If possible such systems alternate between two modes of operation: firing transitions and changing its underlying net. The firing of transitions is similar to what was present in Section 4.1. The changing of the underlying net occurs as follows: for $C, D \subseteq P$, $p \in P_{\text{dyn}}$, $p \notin C$, $p \in D$, if there is $U \subseteq T$ such that $C[U\rangle D$, then in configuration D the net underlying N changes such that $F = F \cup r_{\text{dyn}}^+(p) \setminus r_{\text{dyn}}^-(p)$ with $r_{\text{dyn}}^+(p) = \{(f_1, f_2) \mid (p, (f_1, f_2)) \in R_{\text{dyn}}^+\}$ and $r_{\text{dyn}}^-(p) = \{(f_1, f_2) \mid (p, (f_1, f_2)) \in R_{\text{dyn}}^-\}$.

If there are conflicting changes in the underlying net, then only one change non-deterministically is applied. Formally, for $C, D \subseteq P$, $p, p' \in P_{\text{dyn}}$, $p, p' \notin C$, $p, p' \in D$ if there is $U \subseteq T$ such that $C[U\rangle D$, then $r_{\text{dyn}}^+(p) \cap r_{\text{dyn}}^-(p') \neq \emptyset$ or $r_{\text{dyn}}^+(p') \cap r_{\text{dyn}}^-(p) \neq \emptyset$, then the net underlying N changes such that either $F = F \cup r_{\text{dyn}}^+(p) \setminus r_{\text{dyn}}^-(p)$ or $F = F \cup r_{\text{dyn}}^+(p') \setminus r_{\text{dyn}}^-(p')$.

If no changing of the underlying net is possible, then N fires again (if possible).

The definition of sequential configuration graph, configuration graph, maximal strategy configuration graph and maximally parallel configuration graph are similar to those given for P/T systems in Section 4.1.

A configuration $C \in \mathbb{C}_N$ is said to be *final* if no firing and no change in the underlying net is possible from C. In the following we use DG to refer to sequential configuration graphs, configuration graphs and maximal strategy configuration graphs. We say that a recognising dynamic P/T system with $P_{\text{in}} = (p_{\text{in},1}, \ldots, p_{\text{in,k}})$ *accepts* the vector $(C_{\text{in}}(p_{\text{in},1}), \ldots, C_{\text{in}}(p_{\text{in,k}}))$ if in DG there is a final configuration C_{fin} such that:

$C_{\text{fin}}(p_{\text{rej}}) = 0;$

there is at least one path from C_{in} to C_{fin} and $C_{\text{fin}}(p_{\text{acc}}) > 0$;

no other configuration D in the paths from C_{in} to C_{fin} is such that $D(p_{\text{acc}}) > 0$
   or $D(p_{\text{rej}}) > 0$.

We say that a recognising dynamic P/T system with $P_{\text{in}} = (p_{\text{in},1}, \ldots, p_{\text{in,k}})$ *rejects* the vector $(C_{\text{in}}(p_{\text{in},1}), \ldots, C_{\text{in}}(p_{\text{in,k}}))$ if in DG there is a final configuration C_{fin} such that:

$C_{\text{fin}}(p_{\text{acc}}) = 0;$

for each path from C_{in} to C_{fin} and $C_{\text{fin}}(p_{\text{rej}}) > 0$;

no other configuration D in the paths from C_{in} to C_{fin} is such that $D(p_{\text{rej}}) > 0$
   or $D(p_{\text{acc}}) > 0$.

In all the other cases we say that the recognising dynamic P/T system has an *undefined behaviour* on the vector $(C_{\text{in}}(p_{\text{in},1}), \ldots, C_{\text{in}}(p_{\text{in,k}}))$.

It is important to notice a few things:

the weight of the elements introduced by R_{dyn}^+ is defined by the function W;

in a dynamic P/T system the flow relation changes, not the number of places
   and transitions.

Example 10.4 A dynamic P/T system recognising even numbers $m \geq 2$.

Let us consider the recognising dynamic P/T system $N = (P, P_{\text{dyn}}, T, R_{\text{dyn}}^+, R_{\text{dyn}}^-, F, W, C_{\text{in}}, P_{\text{in}}, p_{\text{acc}}, p_{\text{rej}})$ with:

$P = \{p_1, \ldots, p_4, p_{\text{in}}, p_{\text{acc}}, p_{\text{rej}}\};$
$P_{\text{dyn}} = \{p_3, p_{\text{rej}}\};$
$T = \{t_1, \ldots, t_4\};$
$R_{\text{dyn}}^+ = \{(p_3, (p_{\text{in}}, t_2)), (p_3, (p_1, t_2)), (p_3, (p_1, t_3)), (p_3, (p_1, t_3))\};$
$R_{\text{dyn}}^- = \{(p_{\text{rej}}, (p_4, t_4))\};$
$W(p_{\text{in}}, t_1) = 2$ and W returns 1 in all the other cases;
$P_{\text{in}} = \{p_{\text{in}}\}.$

The initial net underlying N is depicted in Fig. 10.1(a).

The initial configuration is such that $C_{\text{in}}(p_1) = 2$, $C_{\text{in}}(p_2) = 1$, p_{in} contains an arbitrary number $m \geq 2$ of tokens and the remaining places are initially

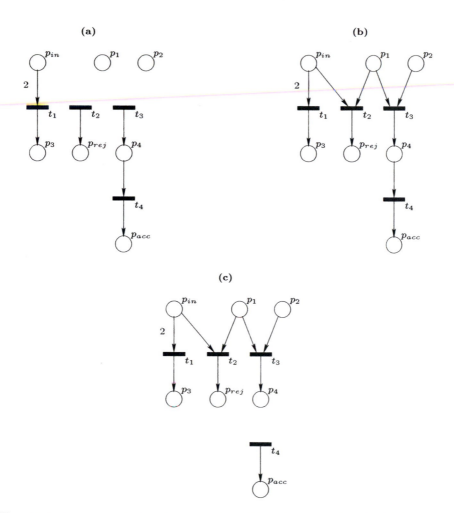

Fig. 10.1 Nets underlying the dynamic P/T system associated with Example 10.4

empty. The P/T system accepts m only if it is an even number. The $MPCG(N)$ for m being an even number is depicted in Fig. 10.2(a) while the $MPCG(N)$ for $m = m'$ being an odd number is depicted in Fig. 10.2(b) (where, for the sake of clarity, some abuse of notation in the indication of the configurations is present).

In the initial configuration only t_1 can fire. The result of this firing is that the number of tokens initially present in p_{in} is divided by 2 (as $W(p_{in}, t_1) = 2$). It is important to notice that now p_3 has at least one token. When this happens the underlying structure of N changes into that depicted in Fig. 10.1(b).

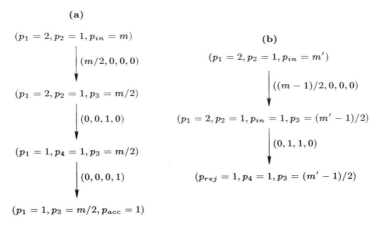

(a)

$(p_1 = 2, p_2 = 1, p_{in} = m)$

$\downarrow (m/2, 0, 0, 0)$

$(p_1 = 2, p_2 = 1, p_3 = m/2)$

$\downarrow (0, 0, 1, 0)$

$(p_1 = 1, p_4 = 1, p_3 = m/2)$

$\downarrow (0, 0, 0, 1)$

$(p_1 = 1, p_3 = m/2, p_{acc} = 1)$

(b)

$(p_1 = 2, p_2 = 1, p_{in} = m')$

$\downarrow ((m-1)/2, 0, 0, 0)$

$(p_1 = 2, p_2 = 1, p_{in} = 1, p_3 = (m'-1)/2)$

$\downarrow (0, 1, 1, 0)$

$(p_{rej} = 1, p_4 = 1, p_3 = (m'-1)/2)$

Fig. 10.2 $MPCGs$ associated with Example 10.4, $m \geq 2$ is an even number, $m' \geq 3$ is an odd number

The following two things can happen. If p_{in} contains one token, then t_2 and t_3 fire so that p_{rej} and p_4 get a token each. If this happen, then, as $p_{rej} \in P_{dyn}$, the underlying structure of N changes into that depicted in Fig. 10.1(c), and the P/T system halts rejecting m.

If instead no token is present in p_{in}, then t_3 and t_4 fire one after the other, a token is present in p_{acc} and the P/T system halts accepting m.

It should be clear that N accepts only even numbers bigger than 1. $\diamond$

Theorem 10.2 *For each deterministic recognising dynamic P/T system N there is a RDTM weakly simulating $MPCG(N)$ with a polynomial slowdown.*

Proof Let $N = (P, P_{dyn}, T, R^+_{dyn}, R^-_{dyn}, F, W, C_{in}, P_{in}, p_{acc}, p_{rej})$ be a deterministic recognising dynamic P/T system, with $P = \{p_1, \ldots, p_{|P|}\}$ places and $T = \{t_1, \ldots, t_{|T|}\}$ transitions.

The recognising deterministic Turing machine simulating $MPCG(N)$ has $|P| + 2|T| + 1$ tapes infinite in one direction. We assume that the tapes are uniquely numbered.

The tapes from 1 to $|P|$ are associated with the places. These tapes, called *place tapes*, initially contain $\downarrow$ followed by as many x's as the number of tokens present in the associated place.

If, for instance, a place p_1 contains three tokens, then the place tape associated with it contains $\downarrow xxx$.

The tapes from $|P| + 1$ to $2|T|$ are associated with the transitions. The tapes with an odd number (called *input set tapes*) contain an encoding of the places defining the input set of the associated transition and the value returned by the weight function, the tapes with an even number (called *output set tapes*) contain the encoding of the places defining the output set of the associated transition and the value returned by the weight function.

These tapes use the symbol $\odot$ as right end marker.

If, for instance, a transition t is such that $^\bullet t = \{p_1, p_2\}$, $t^\bullet = \{p_3\}$, $W(p_1, t) = 2$, $W(p_2, t) = 1$ and $W(t, p_3) = 3$, then the input set tape associated with t contains $p1xxp2x\odot$ while the output set tape associated with t contains $p3xxx\odot$.

The remaining tape, called the *dyn tape*, is associated with the relations R^+_{dyn} and R^-_{dyn}. Also this tape uses the symbol $\odot$ as right end marker.

If $(p_a, (p_b, t_c)) \in R^\alpha_{\text{dyn}}$, $\alpha \in \{+, -\}$, then $\alpha p_a \alpha p_b t c x^{W(p_b, t_c)}$ is present on the *dyn tape*. If, for instance, $(p_4, (p_1, t_1)), (p_5, (p_6, t_1)) \in R^+_{\text{dyn}}$, $(p_4, (t_1, p_3)) \in R^-_{\text{dyn}}$, $W(p_1, t_1) = 2$ and $W(p_6, t_1) = 3$, then this tape contains $+p4 + p1t1xx + p_6 + p6t1xxx - p_4 - t2p3$.

The alphabet of the Turing machine is $\{p, t, x, +, -, \flat, \downarrow, \uparrow, \infty, \odot, 0, 1, 2, 3, 4, 5, 6, 7, 8, 9\}$, where $\flat$ denotes the space (that is, what is present in an empty cell of a tape). In the initial configuration the place tapes encode the initial configuration of N.

In the following we indicate with $maxW$ the maximum values returned by the weight function W.

The Turing machine tries to simulate the firing of the transitions of N in a specific order. For each transition $t \in T$ the simulation of its firing is preceded by the check of the input set.

Check input set. For each transition $t \in T$ the Turing machine checks if the content of the place tapes specified in the input set tape associated with t is at least equal to what indicated in this input set tape (that is, it checks that $^\bullet t \subseteq C$ where C is a configuration of N). Notice that the Turing machine has to read in the input set tape associated with t also the encoding of the places. If this is not the case, then the Turing machine performs the same check for the next transition in the order.

In the worst case the input set of a transition has at most $|P|$ elements. The encoding of the places uses at most $log_{10}|P|$ symbols. For each transition the number of steps required to perform this check is then of the order of $O(maxW \cdot |P|log_{10}|P|)$. So, for all transitions this check requires a number of steps of the order of $O(|T| \cdot maxW \cdot |P|log_{10}|P|)$.

Simulate firing. Otherwise the Turing machine effectively simulates the firing of transition t. Similarly to the checks, the number of steps required to perform the simulation is of the order of $O(|T| \cdot maxW \cdot |P|log_{10}|P|)$.

For each place tape whose content passes from 0 to another number, the flag ↓ is changed into ↑. When the simulation of the firing of a transition is completed the Turing machine checks the input set of the same transition. In this way the multiple firing of the same transition, feature of maximal parallelism, is simulated.

If no transition has been simulated (this can be recorded in the state of the Turing machine), then if the place tape associated with p_{acc} contains at least 1 the Turing machine halts in a specific final (accepting) state; if the place tape associated with p_{rej} contains at least 1 the Turing machine halts in a specific (rejecting) state. Similarly the Turing machine could write either **yes** or **no** in one (output) tape.

If no transition has been simulated and the place tapes associated with p_{acc} and p_{rej} are empty, then the Turing machine keeps applying two rules (for instance: moving left and right of one cell in a tape) so as to never halt. If instead at least one transition has been simulated, then the Turing machine simulates the changes in the flow relation of N.

Simulate change in the underlying net. The Turing machine reads the *dyn tape*. For each place between two + or −, it checks if the content of the associated place tape starts with ↑ (indicating that that place had no token in the previous configuration). If not, then the Turing machine goes on reading the *dyn tape* until the next place between two + or − or until ⊙ is reached. Otherwise, what is present in the *dyn tape* is used to change the input set tapes and the output set tapes. If, for instance, $+p4 + p1t1xx + p_6 + p6t1xxx - p_4 - t2p3$ is present on the *dyn tape* and the place tape associated with place p_4 starts with ↑, then $p1xx$ is added to the input set tape associated with transition t_1 and $p3$ is removed from the output set tape associated with transition t_2. The length of the string present in the *dyn tape* depends on:

The number of places in the P_{dyn} set. At most $|P|$ places can be in P_{dyn} so that $O(|P|log_{10}|P|)$ symbols are needed to encode their names.

The number of elements in the flow relation F added or removed. For each place at most $|P||T|$ (that is, all) elements in the flow relation can be added or removed. The encoding of the names of the places and transitions needs at most $|P|log_{10}|P|$ and $|T|log_{10}|T|$ symbols, respectively. The number of symbols is then of the order of $O(|P|^2 \cdot |T|^2 \cdot log_{10}|P|log_{10}|T|)$.

The number of steps required for the simulation of the change in the underlying net is of the order of $O(|P|^2 \cdot |T|^2 \cdot log_{10}|P|log_{10}|T|)$.

When the end of the *dyn tape* is reached, all the ↑ flags are changed into ↓ and the Turing machines tries to simulate the firing of transitions again.

The number of steps required by the Turing machine to simulate N is then of the order of $O(|P|^3 \cdot |T|^3 \cdot maxW \cdot log_{10}|P|log_{10}|T|)$.

The theorem then follows from the above and from the fact that multiple tape Turing machines can be simulated by single tape Turing machines with a polynomial slowdown. $\qquad\qquad\qquad\qquad\qquad\qquad\qquad\qquad\qquad\qquad\qquad\qquad\square$

Readers may justifiably object that the previous proof does not follow the standard way to describe Turing machines when the purpose is to study computational complexity. For instance, we did not count a few steps in the overall computational complexity and we did not use a binary alphabet to encode places and transitions. Such imprecisions do not effect in a relevant way the main outcome of the previous theorem, that is, the overall time complexity needed for the simulation.

Theorem 10.3 *For each recognising P systems Π with active membranes and with input compartment having only rules of the kinds (a), (b), (c) and (d) there is a dynamic recognising P/T system N such that $MPCG(N)$ weakly simulates Π.*

Proof Let $\Pi = (V, \mu, L_1, \ldots, L_m, R, comp)$ be a recognising P system with active membranes having rules only of the kinds (a), (b), (c) and (d). Polarities are not present. We assume that $V = \{v_1, \ldots, v_{|V|}\}$, yes, no $\in V$ and $\mu = (Q, E)$ with $Q = \{0, 1, \ldots, m\}$.

The recognising dynamic P/T system $N = (P, P_{\text{dyn}}, T, R_{\text{dyn}}^+, R_{\text{dyn}}^-, F, W, C_{\text{in}}, P_{\text{in}}, p_{\text{acc}}, p_{\text{rej}})$ is such that $|P| = m|V| + 2$. There is one place for each symbol in each compartment. These places, called *symbol-compartment places*, are $p_{i,j}$, $1 \leq i \leq |V|$ and $1 \leq j \leq m$ and $P_{\text{in}} = \{p_{i,comp} \mid 1 \leq i \leq |V|\}$. The two remaining places are ph_1 and ph_2 called *phase places*.

The number of tokens present in a symbol-compartment place $p_{v_i,j}$ represents the occurrences of symbol v_i present in compartment j. The phase places contain at most one token and their function is described in the following.

We assume that the rules in R are numbered in a unique way. The transitions in T are of the kind $t_{num}^{(\alpha)}$, $\alpha \in \{a, b, c, d\}$, with num being the unique number associated with a rule in R. The transitions in T and the elements of the flow relation F are:

For each rule of the kind $num : [_j v \rightarrow w]_j$ with $j \in Q$, $w \in V \cup \{\epsilon\}$, there is the transition $t_{num}^{(a)}$. The flow relation contains $(p_{v,j}, t_{num}^{(a)})$ and $(t_{num}^{(a)}, p_{k,j})$ for each $k \in supp(x)$.

For each rule of the kind $num : v [_j]_j \rightarrow [_j w]_j$ with $j \in Q, v \in V$ and $w \in V \cup \{\epsilon\}$, there is the transition $t_{num}^{(b)}$. The flow relation contains $(p_{v,l}, t_{num}^{(b)})$ and $(t_{num}^{(b)}, p_{w,j})$ if $w \in V$.

For each rule of the kind $num : [_j \; v \;]_j \to [_j]_j \; w$ with $j \in Q$, $v \in V$ and $w \in V \cup \{\epsilon\}$, there is the transition $t^{(c)}_{num}$. The flow relation contains $(p_{v,j}, t^{(c)}_{num})$ and $(t^{(c)}_{num}, p_{w,l})$ if $w \in V$.

For each rule of the kind $num : [_j \; v \;]_j \to w$ with $l, j \in Q$, $v \in V$ and $w \in V \cup \{\epsilon\}$, there are transitions $t^{(d)}_{num}$ and $t_{v_i,j}$ for each $1 \le i \le |V|$. The flow relation contains $(p_{v,j}, t^{(d)}_{num})$, $(t_{v_i,j}, p_{v_i,l})$ for each $1 \le i \le |V|$ and $(t^{(d)}, p_{w,l})$ if $w \in V$.

The set T contains also t_1 and t_2, while the flow relation contains also (t_1, ph_2), (ph_2, t_2) and (t_2, ph_1).

The weight function W always returns 1.

The transitions of the kind $t^{(\alpha)}_{num}$, $\alpha \in \{a, b, c, d\}$, $num \in \mathbb{N}_+$, are called *simulation transitions* and they fire when rules in Π are simulated. The transitions of the kind $t_{v_i,j}$, $i \in \{1, \dots, |V|\}$, $j \in \{0, \dots, m\}$, are called *traversal transitions* and they fire to simulate the passage of symbols from one dissolved compartment to its parent compartment. The transitions t_1 and t_2 are called *phase transitions*.

The set P_{dyn} is such that $P_{\mathrm{dyn}} = \{p_{v,j} \mid [_j \; v \;]_j \to w \in R\} \cup \{p_{\mathrm{yes},0}, p_{\mathrm{no},0}, ph_2\}$. The relations R^+_{dyn} and R^-_{dyn} are informally introduced in the following.

While alternating between firing transitions and changing the underlying net the simulation of N cycles between two phases:

abc phase: rules of the kind *(a)*, *(b)* and *(c)* are simulated;

d phase: rules of the kind *(d)* are simulated.

In the initial configuration the symbol-compartment places reflect the initial configuration of Π, that is, $p_{v_i,j} = L_j(v_i)$ for $v_i \in V$ and ph_1 has one token. The remaining places have no tokens.

The presence of one token in ph_1 indicates that the P/T system is in the *abc phase*. The system remains in this phase until a place in P_{dyn} gets a token. When this happens, then the P/T system changes its underlying net.

The changes in the underlying net caused by a token in a place in P_{dyn} are:

blocking the abc phase: all the present edges incoming simulation transitions related to rules of the kinds *(a)*, *(b)* and *(c)* are removed;

allowing the d phase: all the edges incoming traversal transitions are introduced (these edges are not present in the initial configuration of N);

let t_1 to fire: the edge (ph_1, t_1) is introduced.

Notice that these changes are not conflicting even if more than one place in P_{dyn} gets a token in the same configuration. These changes in the flow relation F allow the subsequent firing:

not to fire simulation transitions related to rules of the kind *(a)*, *(b)* and *(c)*;

to fire simulations transitions related to rules of the kind *(d)*;

to fire traversal transitions; in this way the passage of all the occurrences of symbols from a dissolved compartment to the parent one is simulated;

to fire transition t_1 so that a token is present in ph_2.

The simulation of rules of the kind $[_j \ v \ ]_j \rightarrow w$ is already done when a token is present in ph_2. This simulation is composed of the simulation that all the symbols in compartment j pass into the compartment parent of j, that v disappears from compartment j and that one occurrence of $w \in V$ appears in the compartment parent of l.

When a token is present in ph_2 then the P/T system again changes its underlying structure so that simulation transitions can fire again.

The changes in F are then:

restoring the abc phase: the edges incoming simulation transitions related to rules of the kinds *(a)*, *(b)* and *(c)* are reintroduced;

blocking the d phase: all the edges incoming traversal transitions related to rules of the kind *(d)* are removed;

do not let t_1 to fire: the edge (ph_1, t_1) is removed.

The net underlying the P/T system is equal to what it was at the beginning. The simulation transitions can then fire again. Transition t_1 cannot fire as its only incoming edge is not present.

It should be clear that the P/T system simulates Π faithfully. The P/T system halts when a token is present in either $p_{\text{yes},0}$ or $p_{\text{no},0}$. When this happens all the edges are removed so that the P/T system halts. □

Example 10.5 An example related to Theorem 10.3.

Let $\Pi = (V, \mu, L_1, L_2, R)$ be a P system with active membranes where:

$V = \{v_1, v_2, v_3, v_4, v_5, \text{yes}, \text{no}\}$;

$\mu = (\{0, 1, 2\}, \{(0, 1), (1, 2)\})$;

$L_1 = \phi$;

$L_2 = \{v_1\}$;

$R = \{1 : [_2 \ v_1 \rightarrow v_2 v_3 v_4 \ ]_2, 2 : [_1 [_2 \ v_2 \ ]_2]_1 \rightarrow [_1 \ v_5 \ ]_1,$
$\qquad 3 : [_0 [_1 \ v_3 \ ]_1]_0 \rightarrow [_0 [_1]_1 \ \text{yes}]_0\}$.

Rule 1 is of type *(a)*, rule 2 is of type *(d)* and rule 3 is of type *(c)*.

We do not define the simulating recognising dynamic P/T system N but part of its initial underlying net is depicted in Fig. 10.3 while its $MPCF$ is depicted in Fig. 10.4.

It is clear that the net underlying N is not connected. The places ph_1 and ph_2 are phase places, the remaining places are symbol-compartment places. The transitions $t_1^{(a)}$, $t_2^{(d)}$ and $t_3^{(c)}$ are simulation transitions, the transitions $t_{v_3,2}$ and $t_{v_4,2}$ are traversal transitions, while the transitions t_1 and t_2 are phase transitions. The set $P_{\text{dyn}} = \{p_{v_2,2}, p_{\text{yes},0}, p_{\text{rej},0}, ph_2\}$.

$p_{v_1,2}$
$t_1^{(a)}$
$p_{v_2,2}$
$p_{v_3,2}$
$p_{v_4,2}$
$t_2^{(d)}$
$t_{v_3,2}$
$t_{v_4,2}$
$p_{v_5,1}$
$p_{v_3,1}$
$p_{v_4,1}$
$t_3^{(c)}$
$p_{\mathrm{yes},0}$
t_1
ph_1
ph_2
t_2

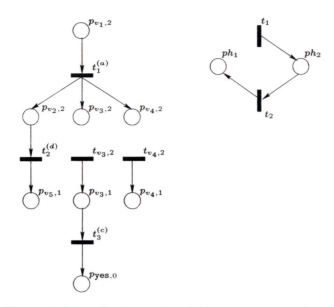

Fig. 10.3 Net underlying the dynamic P/T system associated with Example 10.5

In the initial configuration there is one token in $p_{v_1,2}$, indicating the presence of one occurrence of v_1 in compartment 1, and one token in ph_1, indicating that the P/T system is in the *abc phase*. In this configuration only $t_1^{(a)}$ can fire so that the places $p_{v_2,2}$, $p_{v_3,2}$ and $p_{v_4,2}$ get a token. This simulates the application of rule 1.

As there is a token in $p_{v_2,2}$ and this place belongs to P_{dyn}, then the P/T system changes its underlying structure. This sees the removal of $(p_{v_1,2}, t_1^{(a)})$ and $(p_{v_3,1}, t_3^{(c)})$, and the introduction of $(p_{v_3,2}, t_{v_3,2})$, $(p_{v_4,2}, t_{v_4,2})$ and (ph_1, t_1).

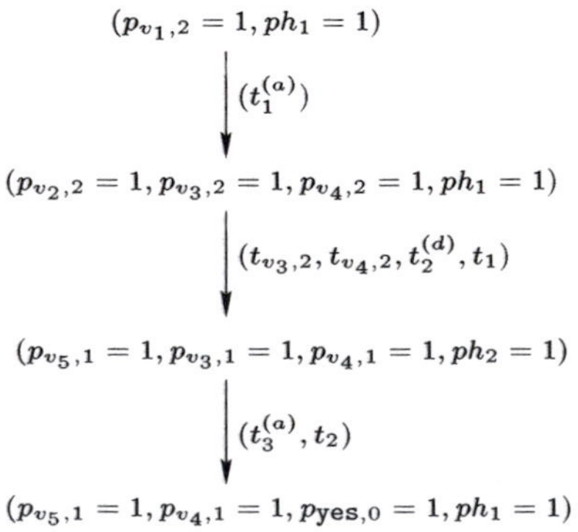

$(p_{v_1,2} = 1, ph_1 = 1)$
$(t_1^{(a)})$
$(p_{v_2,2} = 1, p_{v_3,2} = 1, p_{v_4,2} = 1, ph_1 = 1)$
$(t_{v_3,2}, t_{v_4,2}, t_2^{(d)}, t_1)$
$(p_{v_5,1} = 1, p_{v_3,1} = 1, p_{v_4,1} = 1, ph_2 = 1)$
$(t_3^{(a)}, t_2)$
$(p_{v_5,1} = 1, p_{v_4,1} = 1, p_{\mathrm{yes},0} = 1, ph_1 = 1)$

Fig. 10.4 *MPCG* associated with with Example 10.5

After this the transitions $t_{v_3,2}$ and $t_{v_4,2}$ fire (simulating the passage of the symbols present in the dissolved compartment 2 into its parent compartment 1), the transition $t_2^{(d)}$ fires (simulating the removal of v_2 from compartment 2 and the introduction of v_5 in compartment 1) and transition t_1 fires. When this happens the places with tokens are $p_{v_5,1}$, $p_{v_3,1}$, $p_{v_4,1}$ and ph_2.

As ph_2 has a token and this place belongs to P_{dyn}, then the P/T system changes its underlying structure. This time this sees the removal of the edges introduced before (that is, $(p_{v_3,2}, t_{v_3,2})$, $(p_{v_4,2}, t_{v_4,2})$ and (ph_1, t_1)) and the introduction of the edges removed before (that is, $(p_{v_1,2}, t_1^{(a)})$ and $(p_{v_3,1}, t_3^{(c)})$).

Now $t_3^{(c)}$ and t_2 can fire so that the final configuration is $p_{v_5,1}$, $p_{\text{yes},0}$, $p_{v_4,1}$ and ph_1. As $p_{\text{yes},0}$ has a token and this place belongs to P_{dyn}, then the P/T system changes its underlying structure removing all the edges so that no firing is possible. The output of the system is the token in $p_{\text{yes},0}$ indicating the passage of one occurrence of **yes** to the environment of Π. $\diamond$

From the previous two theorems we have:

Corollary 10.1 *Recognising P systems with active membranes and with input compartment operating under maximal parallelism having only rules of the kinds (a), (b), (c) and (d) can be simulated by RDTM with a polynomial slowdown.*

It should be clear that the dynamical changes possible in the net underlying the kind of P systems with active membranes considered in Theorem 10.3 are only one specific case of changes that can be simulated by dynamic P/T systems (that is, removing vertices in the underlying cell-tree). One could imagine other models of membrane systems having a dynamical underlying structure. These other models could see the removal but also the creation (up to a certain limit) of the number of compartments (having either a cell-tree or a cell-graph as underlying structure).

Theorem 10.1 and Corollary 10.1 indicate that it is the creation of new compartments that allows P systems with active membranes to solve **NP-complete** problems in linear time and exponential space. We discuss this further in Section 10.11. In the next section we see how rules of type *(d)* play a role in the class of problems that can be solved by P systems with active membranes.

From the above we have:

Theorem 10.4 *For every decision problem $\mathcal{D}$ solvable in polynomial time by a family of recognising P systems with active membranes and with input compartment operating under maximal parallelism having only rules of the kinds (a), (b), (c) and (d) there exists a RDTM solving $\mathcal{D}$ in polynomial time.*

Now we characterise $\mathbf{P} \neq \mathbf{NP}$ showing that **NP-complete** problems cannot be solved by recognising P systems with active membranes with input compartment.

Definition 10.9 *We denote by* $\Pi_{(x_1,\ldots,x_n)}$, $1 \leq n \leq 8$, *the class of recognising P systems with active membranes with input compartment operating under maximal parallelism, having only rules of the kind* $x_i \in \{a, b, c, d, e, f, g\}$, $x_i \neq x_j$ *for* $i \neq j$, $1 \leq i, j \leq n$ *and no polarities.*

We denote by $\mathbf{PMC}^{\mathrm{in}}_{\Pi_{(\mathbf{x_1},\ldots,\mathbf{x_n})}}$ *the class of decision problems solvable in polynomial time by* $\Pi_{(x_1,\ldots,x_n)}$.

We denote by $\mathbf{PMC}_{\Pi_{(\mathbf{x_1},\ldots,\mathbf{x_n})}}$ *the class of decision problems solvable in polynomial time by recognising P systems with active membranes without input compartment operating under maximal parallelism, having only rules of the kind* $x_i \in \{a, b, c, d, e, f, g\}$, $x_i \neq x_j$ *for* $i \neq j$, $1 \leq i, j \leq n$ *and no polarities.*

We denote by $\mathbf{PMC}_{\Pi^+_{(\mathbf{x_1},\ldots,\mathbf{x_n})}}$ *the class of decision problems solvable in polynomial time by recognising P systems with active membranes without input compartment operating under maximal parallelism, having only rules of the kind* $x_i \in \{a, b, c, d, e, f, g\}$, $x_i \neq x_j$ *for* $i \neq j$, $1 \leq i, j \leq n$, *and polarities.*

Theorem 10.5 *The following statements are equivalent:*

(i) $\mathbf{P} \neq \mathbf{NP}$;

(ii) there exists $\mathcal{D}$, $\mathcal{D} \in \mathbf{NP\text{-}complete}$, *and* $\mathcal{D} \notin \mathbf{PMC}^{\mathrm{in}}_{\Pi_{(\mathbf{a},\mathbf{b},\mathbf{c},\mathbf{d})}}$;

(iii) for each $\mathcal{D}$, $\mathcal{D} \in \mathbf{NP\text{-}complete}$, *then* $\mathcal{D} \notin \mathbf{PMC}^{\mathrm{in}}_{\Pi_{(\mathbf{a},\mathbf{b},\mathbf{c},\mathbf{d})}}$.

Proof *((i) implies (iii))* Let us assume that there is an **NP-complete** problem $\mathcal{D}$, $\mathcal{D} \in \mathbf{PMC}^{\mathrm{in}}_{\Pi_{(\mathbf{a},\mathbf{b},\mathbf{c},\mathbf{d})}}$. Then from Theorem 10.4 there is a $RDTM$ solving $\mathcal{D}$ in polynomial time. So $\mathcal{D} \in \mathbf{P}$, which means that $\mathbf{P} = \mathbf{NP}$, a contradiction.

((iii) implies (ii)) This is trivial as the class **NP-complete** is not empty.

((ii) implies (i)) Let $\mathcal{D} = (I_{\mathcal{D}}, \theta_{\mathcal{D}})$ be an **NP-complete** problem such that $\mathcal{D} \notin \mathbf{PMC}^{\mathrm{in}}_{\Pi_{(\mathbf{a},\mathbf{b},\mathbf{c},\mathbf{d})}}$ and let us assume that $\mathbf{P} = \mathbf{NP}$. Then $\mathcal{D} \in \mathbf{P}$ meaning that there is a $RDTM$ solving $\mathcal{D}$ in polynomial time.

We denote, then, by $\mathcal{D}_{RDTM} = (I_{\mathcal{D}_{RDTM}}, \theta_{\mathcal{D}_{RDTM}})$ the decision problem associated with $RDTM$ where $I_{\mathcal{D}_{RDTM}}$ is the set of inputs of $RDTM$. For each $\iota \in I_{\mathcal{D}_{RDTM}}$, $\theta(\iota) = 1$ if and only if $RDTM$ accepts ι.

Because of Theorem 10.4 $X_{RDTM} \in \mathbf{PMC}^{\mathrm{in}}_{\Pi_{(\mathbf{a},\mathbf{b},\mathbf{c},\mathbf{d})}}$. Then, there is a class of recognising membrane systems Π_{RDTM} simulating $RDTM$ in polynomial time. Because of Definition 10.2 there are functions *icod* and *ncod* associated with Π_{RDTM}. Here we denote these functions by c_{RDTM} and s_{RDTM}, respectively.

We consider also the function $c_{\mathcal{D}} : I_{\mathcal{D}} \to \bigcup_{k \in \mathbb{N}_+} I_{\Pi_{RDTM}(k)}$, given by $c_{\mathcal{D}}(\iota) = c_{RDTM}(\iota)$, and the function $s_{\mathcal{D}} : I_{\mathcal{D}} \to \mathbb{N}_+$, given by $s_{\mathcal{D}}(\iota) = |\iota|$. Then, with regard to $\mathcal{D}$, the class $\mathbf{\Pi}_{RDTM}$ is:

uniform;

sound. Let $\iota \in I_{\mathcal{D}}$ be such that there is a computation of $\Pi_{RDTM}(s_{\mathcal{D}}(\iota)) = \Pi_{RDTM}(s_{RDTM}(\iota))$ with input $c_{\mathcal{D}}(\iota) = c_{RDTM}(\iota)$ being an accepting computation. Then $\theta_{RDTM}(\iota) = 1$, so $\theta_{\mathcal{D}}(\iota) = 1$.

complete. Let $\iota \in I_{\mathcal{D}}$ be such that $\theta(\iota) = 1$. Then $RDTM$ accepts ι. Therefore $\theta_{RDTM}(\iota) = 1$ and every computation of $\Pi_{RDTM}(s_{RDTM}(\iota)) = \Pi_{RDTM}(s_{\mathcal{D}}(\iota))$ with input $c_{RDTM}(\iota) = c_{\mathcal{D}}(\iota)$ is an accepting computation.

So, $\mathcal{D} \in \mathbf{PMC}_{\mathbf{\Pi}_{(a,b,c,d)}}^{\text{in}}$, a contradiction. $\square$

10.6 Using polarities

In this section we define a class of deterministic recognising P systems with active membranes solving an **NP-complete** problem, called KNAPSACK, using the vertex labelling function (*polarity function*).

Definition 10.10 *Let us consider a finite set A, the functions* weight: $A \to \mathbb{N}$ *and* value: $A \to \mathbb{N}_+$, *and the constants $c, k \in \mathbb{N}_+$.*

The KNAPSACK problem is the following one: given such a set, functions and constants, is there $B \subseteq A$ such that $\sum_{b \in B}$ weight$(b) \leq k$ and $\sum_{b \in B}$ value$(b) \geq c$?

Theorem 10.6 KNAPSACK $\in \mathbf{PMC}_{\mathbf{\Pi}_{(a,c,e)}^+}$.

Proof Given an instance of KNAPSACK as in Definition 10.10 we define a P system in $\mathbf{PMC}_{\mathbf{\Pi}_{(a,c,e)}}$ solving it. The algorithm followed by the P system runs in linear time and it uses exponential space as a function of the length of the instance.

This algorithm is a brute force one that can be logically divided into five phases:

creation phase: using division of compartments all the subsets B of A are created;

weight and value phase: the *weight* and *value* functions are computed for the subsets B present in each of the newly created compartments;

checking weight phase: the condition $\sum_{b \in B} weight(b) \leq k$ is verified in each of the newly created compartments;

checking value phase: the condition $\sum_{b \in B} value(b) \geq c$ is verified in each of the newly created compartments;

output phase: the P system lets either **yes**, indicating that there is at least one subset B verifying the instance, or **no**, indicating that there is no subset B verifying the instance, pass to the environment.

We define the P system $\Pi = (V, \mu, L_1, L_2, R, 2)$ with

$$V = \{a_0, a, \bar{a}_0, \bar{a}, b_0, b, \bar{b}_0, \bar{b}, \hat{b}_0, \hat{b}, d_+, d_-, q_0, \ldots, q_{2k+1}, q, \bar{q}, \bar{q}_0, \ldots, \bar{q}_{2c+1},$$
$$z_0, \ldots, z_{2|A|+2k+2c+6}, \#, \textbf{yes}, \textbf{no}\} \cup \{e_i, x_i, y_i \mid 0 \le i \le |A|\};$$
$$\mu = (\{0, 1, 2\}, \{(0, 1), (1, 2)\}, pol) \text{ with } pol(0) = pol(1) = pol(2) = 0;$$
$$L_1 = \{z_0\};$$
$$L_2 = \{e_0 \bar{a}^k \bar{b}^c x_1^{w_1} y_1^{v_1} \ldots x_{|A|}^{w_{|A|}} y_{|A|}^{v_{|A|}}\}.$$

The set R consists of:

1: $[_2 \ e_i \ ]_2^0 \to [_2 \ q \ ]_2^- \ [_2 \ e_i \ ]_2^+$ for $1 \le i \le |A|$,

   $[_2 \ e_i \ ]_2^+ \to [_2 \ e_{i+1} \ ]_2^0 \ [_2 e_{i+1} \ ]_2^+$ for $1 \le i \le |A| - 1$.

These rules, defining the *creation phase*, create as many compartments as the subsets of A. These compartments, called *2-named compartments*, are all positive. When the symbol q is present in a 2-named compartments, then that compartment passes to the *checking weight phase.*

2: $[_2 \ x_0 \to \bar{a}_0 \ ]_2^0$,

   $[_2 \ x_0 \to \epsilon \ ]_2^+$,

   $[_2 \ x_i \to x_{i-1} \ ]_2^+$ for $1 \le i \le |A|$,

   $[_2 \ y_0 \to \bar{b}_0 \ ]_2^0$,

   $[_2 \ y_0 \to \epsilon \ ]_2^+$,

   $[_2 \ y_i \to y_{i-1} \ ]_2^+$ for $1 \le i \le |A|$.

These rules define the *weight and value phase.* In the initial configuration each 2-named compartment has w_j copies of x_j and v_j copies of y_j, $1 \le j \le |A|$, that is, the multiplicity of the x_j and y_j symbols encodes the weight and value, respectively, of $a_j \in A$. During the *creation phase* these rules allow $\bar{a}_0$ and $\bar{b}_0$ to be created in the compartments 2. The multiplicity of $\bar{a}_0$ and $\bar{b}_0$ encodes the weights and values, respectively, of the symbols defining the subset of A encoded in the newly 2-named compartments.

3: $[_2 \ q \to \bar{q} q_0 \ ]_2^-$,

   $[_2 \ \bar{a}_0 \to a_0 \ ]_2^-$,

   $[_2 \ \bar{a} \to a \ ]_2^-$,

   $[_2 \ \bar{b}_0 \to \hat{b}_0 \ ]_2^-$,

   $[_2 \ \bar{b} \to \hat{b} \ ]_2^-$.

These rules are applied when a 2-named compartment gets a negative polarity, that is, when the symbol q is present in it. When this happens the symbols $\bar{a}_0$ are replaced by a_0 and the symbols $\bar{a}$ are replaced by a. The symbols $\bar{q}$, $\hat{b}_0$ and $\hat{b}$ are also introduced.

4: $[_2\, a_0\,]_2^- \rightarrow [_2\,]_2^0\, \#,$
$\quad [_2\, a\,]_2^0 \rightarrow [_2\,]_2^-\, \#.$

These rules define the *checking weight phase* for the subset of A encoded in each 2-named compartment. They do this by alternating the polarity of the compartment between negative and neutral, removing one instance of a_0 and a each time and creating instances of $\#$ in the skin compartment.

5: $[_2\, q_{2j} \rightarrow q_{2j+1}\,]_2^-$ for $0 \leq j \leq k,$
$\quad [_2\, q_{2j+1} \rightarrow q_{2j+2}\,]_2^0$ for $0 \leq j \leq k - 1.$

The symbol q_0, created by the rules in group 3, behaves as a counter controlling the *checking weight phase* defined by the rules in group 4. The subscript of the q symbol increases by one every time an a_0 or an a is removed from the compartment.

6: $[_2\, q_{2j+1}\,]_2^- \rightarrow [_2\,]_2^+\, \#$ for $0 \leq j \leq k.$

If $\sum_{b \in B} weight(b) \leq k$ is verified by the subset B of A encoded in one 2-named compartment, then in that compartment there will be fewer a_0 symbols than a symbols. If this is the case the rules in group 5 can no longer be applied. In particular the rule $[_2\, a_0\,]_2^- \rightarrow [_2\,]_2^0\, \#$ cannot be applied. Thus, one symbol q_{2j+1} will be present in a negative compartment. When this happens one rule in group 6 is applied and the *checking value phase* starts.

7: $[_2\, \bar{q} \rightarrow \bar{q}_0\,]_2^+,$
$\quad [_2\, \hat{b}_0 \rightarrow b_0\,]_2^+,$
$\quad [_2\, \hat{b} \rightarrow b\,]_2^+,$
$\quad [_2\, a \rightarrow \epsilon\,]_2^+.$

Similar to the rules of group 3 (but for the *checking value phase*).

8: $[_2\, b_0\,]_2^+ \rightarrow [_2\,]_2^0\, \#,$
$\quad [_2\, b\,]_2^0 \rightarrow [_2\,]_2^+\, \#.$

Similar to the rules of group 4 but now the 2-named compartments alternate between positive and neutral polarities.

9: $[_2 \; \bar{q}_{2j} \rightarrow \bar{q}_{2j+1} \;]_2^+$ for $0 \leq j \leq c$,

$[_2 \; \bar{q}_{2j+1} \rightarrow \bar{q}_{2j+2} \;]_2^0$ for $0 \leq j \leq c - 1$.

Similar to the rules of group 5.

10: $[_2 \; \bar{q}_{2c+1} \;]_2^+ \rightarrow [_2 \;]_2^0$ yes,

$[_2 \; \bar{q}_{2c+1} \;]_2^0 \rightarrow [_2 \;]_2^0$ yes.

If the subset B of A encoded in one 2-named compartment verifies $\sum_{b \in B} value(b) \geq c$, then in that compartment there will be fewer b symbols than b_0 symbols. If this is the case, then the rules in group 8 are applied c times each and $\bar{q}_{2c}$ will be present in that compartment. When this happens the rule $[_2 \; \bar{q}_{2c} \rightarrow \bar{q}_{2c+1} \;]_2^+$ is applied (possibly together with the rule $[_2 \; b_0 \;]_2^+ \rightarrow [_2 \;]_2^0 \; \#$). The subsequent application of one rule in group 10 ends the *checking value phase*.

11: $[_1 \; z \rightarrow z_{i+1} \;]_1^0$ for $0 \leq i \leq 2|A| + 2k + 2c + 5$,

$[_1 \; z_{2|A|+2k+2c+6} \rightarrow d_+ d_- \;]_1^0$.

The symbols z behave as counters. Before the end of the computation all the 2-named compartments should have either ended their checking phases successfully or no rule could be applied in them. In the worst case the only solution to the instance of the problem is A, that is, the whole set, and the number of transitions needed to check this solution is a maximum when the weight is exactly k.

12: $[_1 \; d_+ \;]_1^0 \rightarrow [_1 \;]_1^+ \; d_+$,

$[_1 \; d_- \rightarrow$ no $]_1^+$,

$[_1 \;$ yes $]_1^+ \rightarrow [_1 \;]_1^0$yes,

$[_1 \;$ no $]_1^+ \rightarrow [_1 \;]_1^0$no.

The output passes to the environment only when the skin compartment has positive polarity.

It should be clear that the P system creates an exponential number of compartments (all the subsets of A) and that the output is given after a liner number of transitions $(2|A| + 2k + 2c + 6)$. It is important to notice that the rules of type (e) are applied only according to the (e') semantics. $\square$

10.7 The power of dissolution rules

Also in this section we characterise the **P** and **NP** classes but, differently from the previous section, this is based on the absence or presence of dissolution rules (that is, rules of type *(d)*).

Some of the definitions and results presented in the following are valid for P systems with active membranes with or without input compartment. When the presence of such compartment is not relevant, then we do not specify it. Let $\Pi_{(a,b,c,e)} = (V, \mu, L_1, \ldots, L_m, R)$ be a P system with active membranes with $\mu = (Q, E)$, having only rules of the kind *(a)*, *(b)*, *(c)* and *(e)* and no polarities. These kinds of rules can be rewritten as $(v, h) \to (\alpha_1, h'_1) \ldots (\alpha_s, h'_s)$ with $v, \alpha_1, \ldots, \alpha_s \in V$ and $h, h'_i, \ldots, h'_s \in Q$ where:

rules of type *(a)* correspond to $h = h'_i$, $1 \le i \le s$;

rules of type *(b)* correspond to $s = 1$ and $h = parent(h'_1)$;

rules of type *(c)* correspond to $s = 1$ and $h'_1 = parent(h)$;

rules of type *(e)* correspond to $s = 2$, $h'_1 = h'_2 = h$.

This rewriting can be interpreted as: from symbol v in compartment h the symbols $\alpha_1, \ldots, \alpha_s$ in compartments $h'_1, \ldots, h'_s$, respectively, can be reached.

We formalise these concepts in the following definition.

Definition 10.11 *Let* $\Pi_{(a,b,c,e)} = (V, \mu, L_1, \ldots, L_m, R)$ *be a P system with active membranes having only rules of the kind* (a), (b), (c) *and* (e) *with* $\mu = (Q, E)$ *and no polarities. The* dependency graph *associated with* $\Pi_{(a,b,c,e)}$ *is the graph* $\mu_{\Pi_{(a,b,c,e)}} = (Q_{\Pi_{(a,b,c,e)}}, E_{\Pi_{(a,b,c,e)}})$ *with:*

$Q_{\Pi_{(a,b,c,e)}} = Q^l_{\Pi_{(a,b,c,e)}} \cup Q^r_{\Pi_{(a,b,c,e)}}$;

$$Q^l_{\Pi_{(a,b,c,e)}} = \{(v, h) \in V \times Q \mid \exists\, w : V \to \mathbb{N},\ [_h\, v \to w\,]_h \in R \text{ or}$$
$$\exists\, w \in V \cup \{\epsilon\},\ \exists\, h' \in Q,\ h' = parent(h),\ (h, h') \in E,\ v\,[_{h'}]_{h'} \to [_{h'}\, w\,]_{h'} \in R \text{ or}$$
$$\exists\, w \in V \cup \{\epsilon\},\ [_h\, v\,]_h \to [_h]_h\, w \in R \text{ or}$$
$$\exists\, w, z \in V \cup \{\epsilon\},\ [_h\, v\,]_h \to [_h\, w\,]_h[_h\, z\,]_h \in R\};$$

$$Q^r_{\Pi_{(a,b,c,e)}} = \{(w, h) \in V \times Q \mid$$
$$\exists\, v \in V,\ \exists\, w \in V \cup \{\epsilon\},\ w \in supp(x),\ [_h\, v \to x\,]_h \in R \text{ or}$$
$$\exists\, v \in V,\ \exists\, h' \in Q,\ h = parent(h'),\ (h, h') \in E,\ v\,[_{h'}]_{h'} \to [_{h'}\, w\,]_{h'} \in R \text{ or}$$
$$\exists\, v \in V,\ \exists\, h' \in Q,\ h = parent(h'),\ (h, h') \in E,\ [_{h'}\, v\,]_{h'} \to [_{h'}]_{h'}\, w \in R \text{ or}$$
$$\exists\, v \in V,\ z \in V \cup \{\epsilon\},\ [_h\, v\,]_h \to [_h\, w\,]_h[_h\, z\,]_h \in R\};$$

$$E_{\Pi_{(a,b,c,e)}} = \{((v, h), (w, h')) \in Q^l_{\Pi_{(a,b,c,e)}} \times Q^r_{\Pi_{(a,b,c,e)}} \mid$$
$$\exists\, v \in V \cup \{\epsilon\},\ w \in supp(x),\ [_h\, v \to x]_h \in R,\ h = h' \text{ or}$$
$$\exists\, (h, h') \in E,\ v\,[_{h'}]_{h'} \to [_{h'}\, w\,]_{h'} \in R \text{ or}$$
$$\exists\, (h, h') \in E,\ [_{h'}\, v\,]_{h'} \to [_{h'}]_{h'}\, w \in R \text{ or}$$
$$\exists\, z \in V \cup \{\epsilon\},\ [_h\, v\,]_h \to [_h\, w\,]_h[_h\, z\,]_h \in R\}.$$

The rules present in a P system with active membranes can be regarded as strings, so each rule has a length associated with it. This is the view considered in the next lemma.

Lemma 10.1 *Let* $\Pi_{(a,b,c,e)} = (V, \mu, L_1, \dots, L_m, R)$ *be a P system with active membranes having only rules of the kind* (a), (b), (c) *and* (e) *and no polarities. There is a deterministic Turing machine constructing the dependency graph* $\mu_{\Pi_{(a,b,c,e)}} = (Q_{\Pi_{(a,b,c,e)}}, E_{\Pi_{(a,b,c,e)}})$ *associated with* $\Pi_{(a,b,c,e)}$ *in polynomial time as a function of* $|R|$ *and the maximum length of the rules in* R.

Proof Given $\Pi_{(a,b,c,e)}$ we present the deterministic algorithm followed by the Turing machine.

Input: R
$Q_{\Pi_{(a,b,c,e)}} = \emptyset$, $E_{\Pi_{(a,b,c,e)}} = \emptyset$
for each rule $r \in R$ do {
    if $r = [_h \, v \to w \,]_h$ then {
        $Q_{\Pi_{(a,b,c,e)}} = Q_{\Pi_{(a,b,c,e)}} \cup \{(v,h)\} \cup \bigcup_{w \in supp(x)} \{(w,h)\}$
        $E_{\Pi_{(a,b,c,e)}} = E_{\Pi_{(a,b,c,e)}} \cup \bigcup_{w \in supp(x)} \{((v,h), (w,h))\}$
    }
    if $r = [_{h'} \, v \,]_{h'} \to [_{h'}]_{h'} \, w$ then {
        $Q_{\Pi_{(a,b,c,e)}} = Q_{\Pi_{(a,b,c,e)}} \cup \{(v,h'), (w,h)\}$
        $E_{\Pi_{(a,b,c,e)}} = E_{\Pi_{(a,b,c,e)}} \cup \{((v,h'), (w,h))\}$
    }
    if $r = [_h \, v \,]_h \to [_h]_h \, w$ then {
        $Q_{\Pi_{(a,b,c,e)}} = Q_{\Pi_{(a,b,c,e)}} \cup \{(v,h), (w,h')\}$
        $E_{\Pi_{(a,b,c,e)}} = E_{\Pi_{(a,b,c,e)}} \cup \{((v,h), (w,h'))\}$
    }
    if $([_h \, v \,]_h \to [_h \, z \,]_h[_h \, w \,]_h \in R) \wedge (h$ **not the skin compartment**) then {
        $Q_{\Pi_{(a,b,c,e)}} = Q_{\Pi_{(a,b,c,e)}} \cup \{(v,h), (z,h), (w,h)\}$
        $E_{\Pi_{(a,b,c,e)}} = E_{\Pi_{(a,b,c,e)}} \cup \{((v,h), (z,h)), ((v,h), (w,h))\}$
    }
}
The running time of this algorithm is of the order of $O(|R| \cdot maxR)$ where $maxR = max(\{|r| \mid r \in R\})$.   □

Definition 10.12 *The* REACHABILITY *problem is the following one: given a directed graph* $\mu = (Q, E)$ *and* $x, y \in Q$ *is there a path from* x *to* y?

Given $\mu = (Q, E)$ and $x, y \in Q$ as an instance of REACHABILITY, it is possible to design an algorithm for a Turing machine returning 1 or 0 if there is a path from x to y or not, respectively. Such an algorithm could perform a depth-first search having x as starting vertex and checking if y is in any of the found paths.

The total running time of this algorithm is of the order of $O(|Q| + |E|)$, that is, in the worst case it is quadratic in the number of vertices. Moreover, this algorithm needs to store a linear number of items.

Definition 10.13 *Given a P system* Π *and its dependency graph* $\mu_\Pi = (Q_\Pi, E_\Pi)$ *we define the set of successful pairs as*

$$\Delta_\Pi = \{(v,h) \in Q_\Pi \mid \text{ there is a path in } \mu_\Pi \text{ from } (v,h) \text{ to } (\textbf{yes}, 0)\}.$$

Lemma 10.2 *Let* $\Pi_{(a,b,c,e)} = (V, \mu, L_1, \ldots, L_m, R)$ *be a recognising P system with active membranes having* $\mu_{\Pi_{(a,b,c,e)}} = (Q_{\Pi_{(a,b,c,e)}}, E_{\Pi_{(a,b,c,e)}})$ *as dependency graph.*

There is a deterministic Turing machine that constructs $\Delta_{\Pi_{(a,b,c,e)}}$ *in polynomial time as a function of* $|R|$ *and the maximum length of the rules in* R.

Proof Given $\Pi_{(a,b,c,e)}$ we present the deterministic algorithm followed by the Turing machine.

```
Construct the dependency graph μΠ(a,b,c,e)
Input:  μΠ(a,b,c,e) = (QΠ(a,b,c,e), EΠ(a,b,c,e))
ΔΠ(a,b,c,e) = ∅
for each (v, h) ∈ QΠ(a,b,c,e) do {
    if REACHABILITY(μΠ(a,b,c,e), (v, h), (yes, 0)) = yes then {
        ΔΠ(a,b,c,e) = ΔΠ(a,b,c,e) ∪ (v, h)
    }
}
```

The running time of this algorithm is of the order of $O(|V|^3 \cdot m^3 \cdot |R| \cdot maxR)$ where $maxR = max(\{|r| \mid r \in R\})$. $\qquad\square$

Given a family of recognising P systems with active membranes solving a decision problem $\mathcal{D} = (I_\mathcal{D}, \theta_\mathcal{D})$ we are going to characterise the acceptance of an instance $\iota \in I_\mathcal{D}$ using the dependency graph $\mu_{\Pi(\iota)}$ associated with the system $\Pi(\iota)$. The instance ι is accepted by Π if and only if there is a configuration of Π such that **yes** passes in the last transition of the computation to the environment.

Lemma 10.3 *Let* $\mathcal{D} = (I_\mathcal{D}, \theta_\mathcal{D})$ *be a decision problem, let* $\Pi = \{\Pi_{(i)} \mid i \in \mathbb{N}_+\}$ *be a family of recognising P systems with active membranes with input compartment solving* $\mathcal{D}$ *(as indicated by Definition 10.2) and let icod and ncod be the polynomial encodings associated with that solution.*

For each instance $\iota \in I_\mathcal{D}$ *the following statements are equivalent:*

(i) $\theta_{\mathcal{D}}(\iota) = $ yes;

(ii) $X \cap supp(icod(\iota) \cup \bigcup_{j=1}^{m} L_j) \neq \emptyset$, where

$X = \{x \mid (x, y) \in \Delta_{\Pi_{(ncod(\iota))}}\}$;

$L_1, \ldots, L_m$ are the multisets of symbols present in the initial configuration of $\Pi(\iota)$.

Proof Let $\iota \in I_{\mathcal{D}}$. Then $\theta_{\mathcal{D}}(\iota) = $ yes if and only if there is an accepting computation of $\Pi_{(ncod(\iota))}(icod(\iota))$. But this is equivalent to the presence in the dependency graph $\mu_{\Pi_{(ncod(\iota))}(icod(\iota))}$ of a path from (v, h) to (yes, 0) for $\Pi_{(ncod(\iota))}(icod(\iota))$ having $v \in V$ in compartment h, $1 \leq h \leq m$ in its initial configuration.

Thus, item (i) holds if and only if item (ii) holds.   □

Theorem 10.7 $\mathbf{PMC}^{in}_{\Pi_{(a,b,c,e)}} = \mathbf{P}$.

Proof The inclusion $\mathbf{PMC}^{in}_{\Pi_{(a,b,c,e)}} \supseteq \mathbf{P}$ follows from $\mathbf{PMC}^{in}_{\Pi_{(a,b,c,e)}}$ being closed under polynomial time reduction.

To show that $\mathbf{PMC}^{in}_{\Pi_{(a,b,c,e)}} \subseteq \mathbf{P}$ we consider $\mathcal{D} \in \mathbf{PMC}^{in}_{\Pi_{(a,b,c,e)}}$, $\Pi = \{\Pi_{(i)} \mid i \in \mathbb{N}_+\}$ a family of recognising P systems of the kind considered in this lemma solving $\mathcal{D} = (I_{\mathcal{D}}, \theta_{\mathcal{D}})$ (as indicated by Definition 10.2) and let *icod* and *ncod* be the polynomial encodings associated with that solution.

We consider the following deterministic algorithm:

```
Input: ι ∈ I_D
construct the system Π_ncod(ι)(icod(ι))
construct the dependency graph μ_Π_ncod(ι)
construct the set Δ_Π_ncod(ι)
construct the set X = {x | ∃y, (x,y) ∈ Δ_Π_(ncod(ι))}
output = no, j = 1
while (j≤m and output==no) do {
    if X ∩ supp(L_j) ≠ ∅ then {
        output = yes
    }
    j = j + 1
}
if X ∩ supp(icod(ι)) ≠ ∅ then {
    output = yes
}
return(output)
```

It should be clear that this algorithm solves $\mathcal{D}$.

From what was said above we know that the running time of this algorithm is polynomial in the size of ι.   □

If $\Pi \in \mathbf{\Pi}_{(a,b,c,e,f)}$, then the dependency graph associated with Π is as in Definition 10.11, that is, the rules of type *(f)* do not add any vertex nor any edge to this graph.

From Theorem 10.7 we get:

Theorem 10.8 $\mathbf{PMC}^{\text{in}}_{\mathbf{\Pi}_{(a,b,c,e,f)}} = \mathbf{P}$.

Now we consider P systems with active membranes without the input compartment.

Lemma 10.4 *Let $\mathcal{D} = (I_{\mathcal{D}}, \theta_{\mathcal{D}})$ be a decision problem and let $\mathbf{\Pi} = \{\Pi_{(\iota)} \mid \iota \in I_{\mathcal{D}}\}$ be a class of recognising P systems with active membranes without the input compartment solving $\mathcal{D}$ (as indicated by Definition 10.4).*

For each instance $i \in I_{\mathcal{D}}$ the following statements are equivalent:

(i) $\theta_{\mathcal{D}}(\iota) = 1$;

(ii) $X \cap supp(\bigcup_{j=1}^{m} L_j) \neq \emptyset$, *where*
$\qquad X = \{x \mid \exists y, \ (x,y) \in \Delta_{\Pi_{(\iota)}}\}$;
$\qquad L_1, \ldots, L_m$ *are the multisets of symbols present in the initial configuration of $\Pi(\iota)$.*

Proof Similar to that of Lemma 10.3. $\qquad\qquad\qquad\qquad\qquad\qquad\qquad\qquad$ □

Theorem 10.9 $\mathbf{PMC}_{\mathbf{\Pi}_{(a,b,c,e)}} = \mathbf{P}$.

Proof This follows from that of Theorem 10.7. $\qquad\qquad\qquad\qquad\qquad\qquad$ □

Considering that the presence of rules of type *(f)* does not influence the dependency graph we have:

Theorem 10.10 $\mathbf{PMC}_{\mathbf{\Pi}_{(a,b,c,e,f)}} = \mathbf{P}$.

Now we are ready to show how the presence of dissolution rules (that is, rules of type *(d)*) allows some classes of membrane systems to solve **NP-complete** problems.

Definition 10.14 *The* SUBSET SUM *problem is the following one: given a finite set A, a (weight) function $w : A \to \mathbb{N}_+$ and a constant $k \in \mathbb{N}_+$ is there a set $B \subseteq A$ such that $w(B) = k$?*

SUBSET SUM is an **NP-complete** problem.

Theorem 10.11 SUBSET SUM $\in \mathbf{PMC}_{\mathbf{\Pi}_{(a,d,e,f)}}$.

Proof Let us consider an instance of SUBSET SUM as in Definition 10.14.

Given u we are going to describe a P system Π_u in $\mathbf{PMC}_{\mathbf{\Pi}_{(a,d,e,f)}}$ following a brute force algorithm to decide u. This algorithm can be logically divided into four phases:

creation phase: for every subset of A a compartment is created;

weight phase: in each compartment created in the previous phase the weight of the subset of A encoded in it is computed;

checking phase: for each created compartment it is checked whether or not the weight is k;

output phase: the P system lets either **yes**, indicating a positive answer, or **no**, indicating a negative answer, pass to the environment.

We define the P systems $\Pi_u = (V, \mu, L_1, \dots, L_{k+4}, R)$ with

$$V = \{a_i, e_i \mid 1 \le i \le |A|\} \cup$$
$$\{d_0, \dots, d_{2|A|+1}, z_0, \dots, z_{2|A|+k+5}, b, s, c, c', \text{yes}, \text{no}\};$$
$$\mu = (\{0, 1, \dots, k+4\}, \{(i, i+1) \mid 0 \le i \le k+3\});$$
$$L_{k+4} = \{d_0\};$$
$$L_2 = \{z_0\};$$
$$L_i = \phi, \; i \in \{1, 3, \dots, k+3\}.$$

The set R consists of:

1: $[_{k+4} \, d_{2i} \rightarrow a_{i+1}d_{2i+1} \,]_{k+4}$ for $1 \le i \le |A|$,

   $[_{k+4} \, d_{2i+1} \rightarrow a_{i+1}d_{2i+2} \,]_{k+4}$ for $1 \le i \le |A| - 1$.

The symbols d_i behave as registers: only when the index i is an odd number is a symbol a_j generated.

2: $[_{k+4} \, a_i \,]_{k+4} \rightarrow [_{k+4} \, e_i \,]_{k+4}[_{k+4} \, b \,]_{k+4}$ for $1 \le i \le |A|$,

   $[_{k+4} \, e_i \rightarrow s^{w_i} \,]_{k+4}$ for $1 \le i \le |A|$.

The symbols a_i allow compartments to divide. The two resulting compartments contain the symbols e_i and b, respectively. The symbol b is not subject to any rule; the symbol e_i is used to generated as many occurrences of the symbol s as the weight w_i. It is important to notice that the rules of type (e) are applied according to the (e') semantics.

3: $[_{i-1}[_i[_i[_i]_i]_i]_{i-1} \rightarrow [_{i-1}[_i[_i[_i]_i]_i]_{i-1}[_{i-1}[_i[_i[_i]_i]_i]_{i-1}$ for $1 \le i \le k$.

This is a set of rules of type (f). After the application of the rules in groups 1, 2 and 3 the cell-tree has $2^{|A|}$ branches. On each of the leaves of the cell-tree there is a compartment with as many symbols s as the weight of a subset of A.

4: $[_{k+4} \, d_{2|A|+1} \,]_{k+4} \rightarrow b$,

   $[_\beta \, s \,]_\beta \rightarrow c$ for $3 \le \beta \le k+3$.

The symbol $d_{2|A|+1}$ dissolves the compartments $k + 4$. At this point the symbols s start to dissolve compartments. If there are enough symbols s the compartments of a branch $3, \ldots, k+3$ are dissolved. Otherwise, no rule is applied in these compartments.

5: $[_3 \, c \,]_3 \rightarrow c'$.

This is a key rule in Π_u. If in a branch the encoded weight $w(S)$ of $S \subseteq A$ is less than k, then no rule is applied in that branch. Otherwise, all the compartments in the branch are dissolved until the compartment 3 is reached. If $w(S) = k$ then in this compartment there are no symbols s dissolving it and the symbol c' remains in this compartment. Otherwise, if $w(S) > k$ then the compartment is dissolved in the same transition in which c' is generated. The symbol c' then passes into compartment 2.

6: $[_2 \, z_i \rightarrow z_{i+1} \,]_2$ for $0 \leq i \leq 2|A| + k + 4$,

$[_3 \, c' \,]_3 \rightarrow$ **yes**,

$[_2 \, \mathbf{yes} \,]_2 \rightarrow$ **yes**,

$[_2 \, z_{2|A|+k+5} \,]_2 \rightarrow$ **no**.

If one of the subsets of A has weight k, then the symbol c' is generated in a compartment 3. This symbol dissolves the compartment and allows **yes** to be generated in compartment 2. In this last compartment symbols z, behaving as registers, are present during the entirety of the computation. If the symbol **yes** is present in compartment 2, then it dissolves the compartment avoiding $z_{2|A|+k+5}$ to be generated in the same compartment. Otherwise, if symbol c' (that generated **yes**) is never generated, then the symbol $z_{2|A|+k+5}$ will be present in compartment 2. In the following transition this compartment is dissolved and the symbol **no** is generated in compartment 1.

7: $[_1 \, \mathbf{yes} \,]_1 \rightarrow$ **yes**,

$[_1 \, \mathbf{no} \,]_1 \rightarrow$ **no**.

In each computation compartment 1 (that is, the skin compartment) contains either **yes** or **no**. These rules let that symbol pass to the environment in the last transition of the computation. □

Theorem 10.12 $\mathbf{PMC}_{\Pi_{(a,d,e,f)}} \supseteq \mathbf{NP} \cup \mathbf{co\text{-}NP}$.

Proof SUBSET SUM is an **NP-complete** problem belonging to $\mathbf{PMC}_{\Pi_{(a,d,e,f)}}$ and this class is stable under polynomial reduction and closed under complement. □

10.8   About PSPACE-complete problems

In this section we define a class of recognising P systems with active membranes solving **PSPACE-complete** problems.

Definition 10.15 *Let us consider a set of Boolean variables $\{v_1, \ldots, v_n\}$ and a Boolean formula $\psi(v_1, \ldots, v_n)$ written in conjunctive normal form, that is,*
$$\psi = C_1 \wedge \cdots \wedge C_m.$$
  The sentence $\psi^* = \exists v_1 \forall v_2 \ldots Q_n v_n \psi(v_1, \ldots, v_n)$, where Q_n is $\exists$ for n odd and Q_n is $\forall$ for n even, is said to be a fully quantified *formula associated with $\psi(v_1, \ldots, v_n)$.*
  The QSAT *problem is the following one: given a ψ^* is there a truth assignment verifying the extended fully quantified formula ψ^*?*

It is known that QSAT is a **PSPACE-complete** problem.

Theorem 10.13 $\text{QSAT} \in \mathbf{PMC}^{\text{in}}_{\Pi_{(\mathbf{a},\mathbf{b},\mathbf{c},\mathbf{d},\mathbf{g})}}$.

Proof   Given an instance of QSAT we define a P system in $\mathbf{PMC}^{\text{in}}_{\Pi_{(\mathbf{a},\mathbf{b},\mathbf{c},\mathbf{d},\mathbf{g})}}$ solving it. The solution provided is uniform. The algorithm, similar to that described in the proof of Theorem 10.11, followed by the P system is a brute force one. It can be logically divided into three phases:

creation phase: Using compartment creation all the truth assignments on the instance of QSAT are created and evaluated. The underlying structure of the P system becomes a cell-tree in which the leaves encode all the possible truth assignments.

checking phase: It is checked if the formula evaluates to *true* for all possible truth assignments.

output phase: the P system lets either **yes**, indicating that there is at least one truth assignment verifying the instance, or **no**, indicating that there is no such truth assignment, pass to the environment.

  Let us consider the polynomial time computable function $p : \mathbb{N}_+^2 \to \mathbb{N}_+$ such that $p(a, b) = (a+b)(a+b+1)/2+a$. For any given Boolean formula in conjunctive normal form $\psi(v_1, \ldots, v_a) = C_1 \wedge \cdots \wedge C_b$ with a variables and b clauses we define the P system $\Pi_{(p(a,b))} = (V_{p(a,b)}, \mu, L_1, L_{<t,\vee>}, R, comp)$ processing the fully quantified formula ψ^* associated with ψ when an appropriate input is supplied.
  The system $\Pi_{(p(a,b))}$ is such that:
$$V_{p(a,b)} = \{v_{i,j}, \bar{v}_{i,j}, v_{i,j,l}, \bar{v}_{i,j,l} \mid 1 \leq i \leq b,\ 1 \leq j \leq a,\ l \in \{t,f\}\}\cup$$
$$\{z_{j,c}, z_{j,c,l} \mid 0 \leq j \leq a,\ c \in \{\wedge, \vee\},\ l \in \{t,f\}\}\cup$$
$$\{r_i, r_{i,t}, r_{i,f}, d_i \mid 1 \leq i \leq b\}\cup$$
$$\{q, t_0, \ldots, t_4, s_0, \ldots, s_5, y, n, y_\vee, n_\vee, y_\wedge, n_\wedge, \tilde{y}, \tilde{n}, \underline{y}_\wedge, \underline{n}_\wedge, \underline{n}_\vee, \text{yes},$$
$$\text{no}\};$$

$\mu = (Q, E)$ with

$Q = \{< l, c > | \, l \in \{t, f\}, \, c \in \{\wedge, \vee\}\} \cup \{0, 1, u, m_1, \dots, m_b\};$

$E = \{(0, 1), (1, < t, \vee >)\};$

$L_1 = \phi;$

$L_{<t,\vee>} = \{z_{0,\wedge,t}, z_{0,\wedge,f}\};$

$comp = < t, \vee > .$

In the following if $c = \vee$, then $\bar{c} = \wedge$ and if $c = \wedge$, then $\bar{c} = \vee$. The set R consists of:

1: $\left[_{<l,\bar{c}>} z_{j,c} \rightarrow z_{j,c,t} z_{j,c,f} \right]_{<l,\bar{c}>};$

  $\left[_{<l',\bar{c}>} z_{j,c,l} \right]_{<l',\bar{c}>} \rightarrow \left[_{<l',\bar{c}>} \left[_{<l,c>} z_{j+1,\bar{c}} \right]_{<l,c>} \right]_{<l',\bar{c}>}$

$$\text{for } l, l' \in \{t, f\}, \, c \in \{\vee, \wedge\}, \, 0 \le j \le a - 1.$$

These rules create one compartment for each truth assignment of the formula. Firstly, the symbol $z_{j,c}$ generates two symbols: $z_{j,c,t}$, for the assignment *true*, and $z_{j,c,f}$ for the assignment *false*. Secondly, these symbols create new compartments. A newly created compartment $< t, c >$ represents the assignment $v_{j+1} = true$, while a compartment $< f, c >$ represents the assignment $v_{j+1} = false$.

2: $\left[_{<l,c>} v_{i,j} \rightarrow v_{i,j,t} v_{i,j,f} \right]_{<l,c>},$

  $\left[_{<l,c>} \bar{v}_{i,j} \rightarrow \bar{v}_{i,j,t} \bar{v}_{i,j,f} \right]_{<l,c>},$

  $\left[_{<l,c>} r_i \rightarrow r_{i,t} r_{i,f} \right]_{<l,c>}$

$$\text{for } l \in \{t, f\}, \, c \in \{\vee, \wedge\}, \, 1 \le j \le a, \, 1 \le i \le b.$$

These rules duplicate the symbols representing the formula so that it can be evaluated on the two possible assignments: $v_i = true$ (represented by $v_{i,j,t}$ and $\bar{v}_{i,j,t}$) and $v_j = false$ (represented by $v_{i,j,f}$ and $\bar{v}_{i,j,f}$). The symbols r_i are also duplicated into $r_{i,t}$ and $r_{i,f}$ in order to keep track of the clauses that evaluate to true on the previous assignment of the variables.

3: $v_{i,1,t} \left[_{<t,c>}\right]_{<t,c>} \rightarrow \left[_{<t,c>} r_i \right]_{<t,c>},$

  $\left[_{<t,c>} \bar{v}_{i,1,t} \rightarrow \epsilon \right]_{<t,c>},$

  $\bar{v}_{i,1,f} \left[_{<f,c>}\right]_{<f,c>} \rightarrow \left[_{<f,c>} r_i \right]_{<f,c>},$

  $\left[_{<f,c>} \bar{v}_{i,1,f} \rightarrow \epsilon \right]_{<f,c>} \text{ for } c \in \{\vee, \wedge\}, \, 1 \le i \le b.$

These rules allow the formula to be evaluated in the two possible truth assignments of the variable that is being analysed. The symbols $v_{i,1,t}$ ($\bar{v}_{i,1,f}$) allow the symbols r_i to be generated in compartment $< t, c >$ ($< f, c >$) indicating that clause C_i evaluates *true* to the assignment $v_{j+1} = true$ ($v_{j+1} = false$). Similarly, the symbols $\bar{v}_{i,1,t}$ ($v_{i,1,t}$) disappear indicating that the clause C_i does not evaluate to *true* to the assignment $v_{j+1} = true$ ($v_{j+1} = false$).

$$4\text{: } v_{i,j,l} \, [_{<l,c>}]_{<l,c>} \to [_{<l,c>} v_{i,j-1}]_{<l,c>},$$
$$\bar{v}_{i,j,l} \, [_{<l,c>}]_{<l,c>} \to [_{<l,c>} \bar{v}_{i,j-1}]_{<l,c>}.$$
$$r_{i,t} \, [_{<l,c>}]_{<l,c>} \to [_{<l,c>} r_i]_{<l,c>}$$
$$\text{for } l \in \{t, f\}, \ c \in \{\vee, \wedge\}, \ 1 \le i \le b, \ 2 \le j \le a.$$

In order to analyse the next variable the second subscript of the symbols $v_{i,j,l}$ and $\bar{v}_{i,j,l}$ is decreased by 1 when they pass into the compartment $< l, c >$, $c \in \{\vee, \wedge\}$. Moreover, the symbol $r_{i,j}$ passes into the compartment $< l, c >$ to keep track of the clauses that evaluate to *true* in the previous truth assignment.

$$5\text{: } [_{<l,\bar{c}>} z_{a,c} \to d_i \ \ldots \ d_b q]_{<l,\bar{c}>} \text{ for } l \in \{t, f\}, \ c \in \{\vee, \wedge\}.$$

At the end of the *creation phase* the symbols generated by the application of this rule takes part to the *checking phase*.

$$6\text{: } d_i \, [_i]_i \to [_i t_0]_i,$$
$$r_{i,t} \, [_i]_i \to [_i r_i]_i,$$
$$[_i r_i \to \epsilon]_i,$$
$$[_i t_k \to t_{k+1}]_i,$$
$$[_i t_2]_i \to t_3 \text{ for } k \in \{0, 1\}, \ c \in \{\vee, \wedge\}, \ i \in \{m_1, \ldots, m_b\}.$$

These rules allow symbol d_i to create a compartment i with t_0, a symbol behaving as a register, in it. The symbol r_i passes into compartment i dissolving it and preventing in this way the register t from becoming t_2. If the symbol t_2 is present in compartment i, meaning that there is no symbol r_i, than the clause C_i does not evaluate *true* on the truth assignment associated with that compartment. Therefore the formula does not evaluate to *true* on that truth assignment.

$$7\text{: } q \, [_u]_u \to [_u s_0]_u,$$
$$t_3 \, [_u]_u \to [_u t_4]_u,$$
$$[_u t_4]_u \to \epsilon,$$
$$[_u s_h \to s_{h+1}]_u,$$
$$[_u s_5]_u \to y,$$
$$[_{<l,c>} s_5 \to n]_{<l,c>} \text{ for } l \in \{t, f\}, \ c \in \{\vee, \wedge\}, \ 0 \le h \le 4.$$

When the symbol q is present in compartment $< l, c >$ the compartment u is created with the symbol s_0 in it. This symbol evolves into s_{h+1} while the symbol

t_3 can pass into compartment u dissolving it and avoiding in this way for **yes** to pass to compartment $< l, c >$.

8: $[_{<l,c>} y]_{<l,c>} \to y_{\bar{c}},$

$\quad [_{<l,c>} n]_{<l,c>} \to n_{\bar{c}},$

$\quad [_{<l,v>} y_v]_{<l,v>} \to \tilde{y},$

$\quad [_{<l,v>} n_v \to \underline{n}_v]_{<l,v>},$

$\quad [_{<l,\wedge>} \tilde{y} \to y_\wedge]_{<l,\wedge>},$

$\quad [_{<l,v>} \underline{n}_v]_{<l,v>} \to n_\wedge,$

$\quad [_{<l,\wedge>} \underline{n}_v \to \epsilon]_{<l,\wedge>},$

$\quad [_{<l,\wedge>} y_v \to \epsilon]_{<l,\wedge>},$

$\quad [_{<l,\wedge>} n_\wedge]_{<l,\wedge>} \to \tilde{n},$

$\quad [_{<l,\wedge>} y_\wedge \to \underline{y}_\wedge]_{<l,\wedge>},$

$\quad [_{<l,v>} \tilde{n} \to n_v]_{<l,v>},$

$\quad [_{<l,\wedge>} \underline{y}_\wedge]_{<l,\wedge>} \to y_v,$

$\quad [_{<l,v>} \underline{n}_\wedge \to \epsilon]_{<l,v>},$

$\quad [_{<l,v>} \underline{y}_\wedge \to \epsilon]_{<l,v>},$

$\quad [_1 \tilde{y}]_1 \to [_1]_1 \textbf{yes},$

$\quad [_1 n_\wedge]_1 \to [_1]_1 \textbf{no}$ for $l \in \{t, f\}$.

After the rules in group 7 are applied either the symbol y or n is present in each compartment previously created. These symbols indicate whether the assignment associated with that compartment satisfies or not, respectively, the formula. The symbol **yes** passes to the environment if and only if at least one appropriate combination of truth assignments according to the quantifiers $\exists$ and $\forall$ satisfying the formula is found.

Let $\{v_1, \ldots, v_a\}$ be a set of Boolean variables and $\psi(v_1, \ldots, v_a)$ a Boolean formula written in conjunctive normal form, that is, $\psi = C_1 \wedge \cdots \wedge C_b$. Moreover, let ψ^* be the fully quantified Boolean formula associated with ψ. We define $ncod(\psi^*) = p(a, b)$ and $icod(\psi^*) = \{v_{i,j} \mid v_j \in C_i\} \cup \{\bar{v}_{i,j} \mid \bar{v}_j \in C_i\}$.

The recognising P system $\Pi_{ncod(\psi^*)}$ with input $icod(\psi^*)$ works as follows. In the initial configuration the symbols $icod(\psi^*)$ are in the input compartment $< t, \vee >$ together with the symbols $z_{0,\wedge,t}$ and $z_{0,\wedge,f}$.

In the first transition the symbol $z_{0,\wedge,t}$ is used to create the compartment $< t, \vee >$ representing the assignment $x_1 = true$ and the symbol $z_{0,\wedge,f}$ is used to create the compartment $< f, \vee >$ representing the assignment $x_1 = false$. In these two compartments the symbol $z_{1,\vee}$ is present.

In the following transitions the input multiset encoding the formula ψ is duplicated following the first two rules in group 2. Then, following the rules in group 3, the formula is evaluated on the two truth assignments of x_1. In the same transition the rules in group 4 decrease the second subscript of the symbols representing the formula in order to perform a similar evaluation for the subsequent variable. Moreover, the symbol $z_{1,c}$ produces the symbols $z_{1,c,t}$ and $z_{1,c,f}$, $c \in \{\vee, \wedge\}$.

In this way the *creation phase* goes on until all the possible assignments to the variables are generated and the formula is evaluated for each of them. As it takes two transitions to generate the possible assignments for a variable and to evaluate the formula on these assignments, then the *creation phase* consists of $2a$ transitions.

The *checking phase* starts when the symbol $z_{a,c}$ produces the symbols $d_1, \dots,$ d_b, q. In the first transition in this phase each symbol d_i, $1 \leq i \leq b$ creates a new compartment u with the symbol y_0 in it.

The symbols $r_{i,t}$, indicating that the clause C_i evaluates *true* for the truth assignment associated with that compartment, pass in these compartments by the application of the last rule in (4). In this way Π keeps track of the clauses evaluating to *true*. The symbols $r_{i,t}$ pass into the compartment i dissolving it in the following two transitions. This prevents symbol t_2 from dissolving the same compartment producing t_3 (last rule in group 6).

If for some i there is no r_i, meaning that clause C_i does not evaluate *true* on the associated assignment, then symbol t_2 dissolves compartment i producing t_3. This last symbol passes into compartment u where the symbol s_h evolves following the rules in group 7. The symbol t_4 dissolves the compartment u preventing s_5 from being generated. So, the *checking phase* consists of six transitions.

Finally, the *output phase* occurs because of the rules in group 8. If some symbol s_5 is present in any compartment $< l, c >$, $l \in \{t, f\}$, $c \in \{\vee, \wedge\}$, then there is at least one clause not satisfied by the truth assignment associated with that compartment. As a consequence of the application of the last rule in group 7 the symbol n occurs in that compartment. Otherwise the symbol s_5 occurs in compartment u dissolving it and letting y to pass into the parent compartment.

In such a configuration in each of the created 2^a compartments there is either y or n depending if the associated truth assignment satisfies ψ.

In the last transitions the flow of the symbols y and n to the environment is controlled. If there is one symbol y in a compartment $< l, \vee >$, $l \in \{t, f\}$, the compartment is dissolved and another y symbol passes into the parent compartment; otherwise n symbol passes into the parent compartment. Similarly, if there is one symbol n in a compartment $< l, \wedge >$, $l \in \{t, f\}$, the compartment is dissolved and another n symbol passes into the parent compartment; otherwise a y symbol passes into the parent compartment.

It should be clear that Π solves ψ^* in a linear number of transitions and using an exponential number of compartments as a function of a and b. □

Theorem 10.14 $\textbf{PMC}^{\text{in}}_{\Pi_{(a,b,c,d,g)}} \supseteq \textbf{PSPACE}$.

Proof This follows from the fact that QSAT is **PSPACE-complete**, QSAT $\in \textbf{PMC}^{\text{in}}_{\Pi_{(a,b,c,d,g)}}$ and the class $\textbf{PMC}^{\text{in}}_{\Pi_{(a,b,c,d,g)}}$ is closed under polynomial time reduction. □

10.9 Brane Calculi

We consider the idea of bridging Brane Calculi with Membrane Computing a very interesting one. We also think that this line of research misses the uniformity and volume of results present in the other aspects of Membrane Computing we considered. It is for this reason that we limit our presentation on the links between Brane Calculi and membrane systems to a minimum.

In the following we give an informal introduction to the basic operations present in Brane Calculi, we translate them in terms of Membrane Computing and then we prove a result.

In Brane Calculi a *membrane structure* is a collection of nested membranes. Membranes are formed of *patches* and a patch can be a *composition* of *subpatches*. If u_1 and u_2 are patches, then their composition is denoted by $u_1 \mid u_2$. A patch which is not a sub-patch consists of an *action* followed by another patch.

If a is an action and s is a patch, then $a.s$ is the action followed by the patch. Actions can come in *complementary pairs* causing *interactions* between sub-structures. Actions whose name is preceded by *co* are called *co-actions*, that is, *complementary actions*. A structure consisting of a membrane with patch u_1, another membrane with patch u_2 both nested in a membrane with patch u_3, is denoted by

$$[[\,]u_1[\,]u_2]u_3$$

Each specific brane calculus has a fixed set of actions with specific operational meaning in terms of reductions. In the following we introduce some of these actions using the symbol $\rightarrow$ to denote a *reduction*, α and β to denote arbitrary sub-structures and u_1, u_2, u_3, u_4 and u_5 to denote patches.

pino: $[\alpha]u_1 \mid (pino.(u_2).u_3) \rightarrow [\alpha[\,]u_2]u_1 \mid u_3$

This action creates a membrane in that in which the *pino* action resides. The patch to the newly created membrane is a parameter to *pino*.

exo: $[[\alpha]u_1 \mid (exo.u_3)\beta]u_3 \mid (co\text{--}exo.u_4) \rightarrow \alpha[\beta]u_1 \mid u_3 \mid u_2 \mid u_4$

This action merges two nested membranes. As a consequence of this the sub-structure α passes to the outside of the resulting membrane and the patches of the two membranes become contiguous.

phago: $[\alpha]u_1 \mid (phago.u_2)[\beta]u_3 \mid (co\text{--}phago(u_4).u_5) \rightarrow [[[\alpha]u_1 \mid u_2]u_4\beta]u_3 \mid u_5$

This action models a membrane, β, engulfing another membrane, α. An additional membrane is created around the engulfed membrane, the patch of the additional membrane is specified by the patch u_4 parameter of the *co--phago* action.

drip: $[\alpha]u_2 \mid (drip(u_2).u_3) \rightarrow [\alpha]u_1 \mid u_3[\ ]u_2$

This action produces a membrane outside another membrane.

mate: $[\alpha]u_1 \mid (mate.u_2)[\beta]u_3 \mid (co\text{--}mate.u_4) \rightarrow [\alpha\beta]u_1 \mid u_2 \mid u_3 \mid u_4$

This action merges two membranes that are not nested.

bud: $[[\alpha]u_1 \mid (bud.u_2)\beta]u_3 \mid (co\text{--}bud(u_4).u_5) \rightarrow [[\alpha]u_1 \mid t_2]u_4[\beta]u_3 \mid u_5$

This action expels a membrane from another one adding to this last an additional membrane.

The *drip*, *mate* and *bud* operations are expressible in the *pino*, *exo*, *phago* calculus by simple operations. Moreover, the *pino*, *exo*, *phago* calculus is a complete set, that is, it is Turing complete under opportune conditions, while the *drip*, *mate* and *bud* calculus is not Turing complete under the same conditions.

10.10 From Brane Calculi to membrane computing

The Brane Calculi actions just indicated can be translated into rules for membrane systems. For instance, the *mate* and *drip* actions can become:

MC-mate: $[_j\alpha]_j^{u_1 v}[_k\beta]_k^{u_2} \rightarrow [_j\alpha\beta]_j^{u_1 u_3 u_2}$ and

MC-drip: $[_j\alpha]_j^{u_1 v u_2} \rightarrow [_j]_j^{u_1 u_3}[_k\alpha]_k^{u_2}$

respectively, where, if V is an alphabet, then $v \in V$, $u_1, u_3 : V \rightarrow \mathbb{N}$, $u_2 : V \rightarrow \mathbb{N}_+$ and j and k are vertices.

The symbol v and the multisets u_1, u_2 and u_3 represent the actions while α and β represent the arbitrary sub-structures present in the definitions of *mate* and *drip*. The actions can be regarded as multisets of proteins associated with the relative membranes, while the sub-structures can be regarded as other membranes (and associated multisets) in a bracket representation of a membrane system.

It is possible to define models of membrane systems based on these rules.

Definition 10.16 *A* P system with mate and drip *is a construct*

$$\Pi = (V, \mu, L_1, \ldots, L_m, R)$$

where

V *is an alphabet.*

$\mu = (Q, E)$ *is a cell-tree underlying* Π *with:*

$Q \subset \mathbb{N}$ *contains* vertices. *For simplicity we define* $Q = \{0, 1, \ldots, m\}$. *Each vertex in* Q *defines a* compartment *of* Π.

$E \subseteq Q \times Q$ *defines directed labelled* edges *between vertices, denoted by* (i, j), $i, j \in Q, i \neq j$.

L_i, $1 \leq i \leq m$, *are multisets from* V *to* $\mathbb{N}$.

R *is a set of* MC-mate *and* MC-drip *rules.*

A *configuration* of a P system with mate and drip is a tuple $(\mu', M'_1, \ldots, M'_{q'})$ where μ' is a cell-tree and M'_i, $1 \leq i \leq q'$, are multisets from V to $\mathbb{N}$. The $(m+1)$-tuple $(\mu, L_1, \ldots, L_m)$ is called the *initial configuration*.

Given a configuration $(\mu', M'_1, \ldots, M'_{q'})$ with $\mu' = (Q', E')$, the application of the rules in R can change the configuration in the following way:

if $[_j\alpha]_j^{u_1 v}[_k\beta]_k^{u_2} \rightarrow [_j\alpha\beta]_j^{u_1 u_3 u_2} \in R$, $i, j, k \in Q'$, $u_1, v \in M'_j$, $u_2 \in M'_k$, then

Q' changes into $Q'' = Q' \setminus \{k\}$;

E' changes into $E'' = E' \setminus \{(i, k)\} \setminus \{(j, p) \mid (k, p) \in E'\}$;

M'_j changes into $M''_j = M'_j \setminus \{v\} \cup M'_k \cup \{u_3\}$;

$u_3 : V \rightarrow \mathbb{N}$;

if $[_j\alpha]_j^{u_1 v u_2} \rightarrow [_j]_j^{u_1 u_3}[_k\alpha]_k^{u_2} \in R$, $i, j \in Q'$, $j \neq 1$, $u_1, v, u_2 \in M'_j$, then

Q' changes into $Q'' = Q' \cup \{|Q'| + 1\}$;

E' changes into $E'' = E' \cup \{(i, |Q'| + 1)\}$;

M'_j changes into $M''_j = M'_j \setminus \{v, u_2\} \cup \{u_3\}$;

$M''_{|Q'|+1} = \{u_2\}$;

$u_3 : V \rightarrow \mathbb{N}$.

Notice that the skin membrane cannot be subject to an *MC-drip* rule. Given two configurations $(\mu', M'_1, \ldots, M'_{q'})$ and $(\mu'', M''_1, \ldots, M''_{q''})$ we write $(\mu', M'_1, \ldots, M'_{q'}) \Rightarrow (\mu'', M''_1, \ldots, M''_{q''})$ to denote a *transition* from $(\mu', M'_1, \ldots, M'_{q'})$ to $(\mu'', M''_1, \ldots, M''_{q''})$, that is, the application of a multiset of rules associated with each membrane under the requirement of maximal parallelism. The reflexive and transitive closure of $\Rightarrow$ is denoted by $\Rightarrow^*$.

A sequence of transitions defines a *computation* of a P system with mate and drip. A computation starting from the initial configuration is *successful* if it is finite and if the last configuration, called *final*, is (μ^f, M_1^f, M_2^f) with $u_i^f = \phi$.

So, in a successful computation the last configuration sees only two membranes: the skin, with no multiset of proteins associated with it, and and another, *inner*, membrane. The *result* of a successful computation is the cardinality of the multiset associated with the inner membrane present in the final configuration. Computations which are not successful render no result.

The *weight* of a *MC-mate* rule $[\alpha]^{u_1 v}[\beta]^{u_2} \to [\alpha\beta]^{u_1 u_3 u_2}$ is $|u_1 v u_2|$, while the *weight* of a *MC-drip* rule is $[\alpha]^{u_1 v u_2} \to [\,]^{u_1 u_3}[\alpha]^{u_2}$ is $|u_1 v u_2|$.

The set $N(\Pi)$ denotes the set of numbers generated by a P system with mate and drip Π. Formally, if $\Pi = (V, \mu, u_1, \ldots, u_m, R)$, then $N(\Pi) = \{|u_2^f| \mid (\mu, L_1, \ldots, L_m) \Rightarrow^* (\mu^f, \phi, M_2^f)$ and there is no configuration $(\mu', M_1', \ldots, M_{q'}')$ such that $(\mu^f, \phi, M_2^f) \Rightarrow (\mu', M_1', \ldots, M_{q'}')\}$. The family of all sets $N(\Pi)$ generated by a P system with mate and drip using at most m compartments, *MC-mate* rules of at most weight p and using *MC-drip* rules of at most weight q is denoted by $N\,OP_m(\text{mate}_p, \text{drip}_q)$.

What performed by the *MC-mate* and *MC-drip* rules seem very similar to what performed by the building blocks *join* and *fork* (see Fig. 4.15), respectively. So, considering Theorem 4.2, a P system operating under maximal parallelism equipped with *MC-mate* and *MC-drip* rules could be computationally complete. The next theorem confirms this intuition.

Theorem 10.15 $N_1\,OP_9(\textit{mate}_3, \textit{drip}_3) = N_1\,RE$.

Proof Let us consider a generating register machine $M = (S, I, s_1, s_f)$ with n registers $\gamma_1, \ldots, \gamma_n$ such that when it halts only one specific (output) register is not empty. Moreover, let us assume that register γ_n is the output register.

We define the P system with mate and drip $\Pi = (V, \mu, u_{skin}, L_1, \ldots, L_n, L_{state}, L_\star, R)$ simulating M with:

$V = S \cup \{x_i, b_i \mid 1 \le i \le n\} \cup \{s^{(i)} \mid s \in S,\ 1 \le i \le 4\} \cup \{b_\star\}$;

$\mu = (Q, E)$ with

$Q = \{0, skin, 1, \ldots, n, state, \star\}$;

$E = \{(0, skin)\} \cup \{(skin, \alpha) \mid \alpha \in Q \setminus \{0, skin\}\}$.

The set R is defined in the following.

For each instruction of the kind $(s, \gamma_i^+, z) \in I$ the set R contains:

$1 : [_{skin}[_i]_i^{b_i}[_{state}]_{state}^s]_{skin}^\phi \to [_{skin}[_i]_i^{b_i s z}]_{skin}$;

$2 : [_{skin}[_i]_i^{b_i s z}]_{skin}^\phi \to [_{skin}[_i]_i^{b_i x_i}[_{state}]_{state}^z]_{skin}^\phi$.

For each instruction of the kind $(s, \gamma_i^-, z, w) \in I$ the set R contains:

$3 : [_{skin}[_{state}]_{state}^s[_\star]_\star^{b_\star}]_{skin}^\phi \to [_{skin}[_{state}]_{state}^{s b_\star}]_{skin}^\phi$;

$4: [_{skin}[_{state}]_{state}^{sb_\star}]_{skin}^{\phi} \rightarrow [_{skin}[_{state}]_{state}^{s^{(1)}s^{(2)}}[_\star]_\star^{b_\star}]_{skin}^{\phi};$

$5: [_{skin}[_{state}]_{state}^{s^{(1)}s^{(2)}}]_{skin}^{\phi} \rightarrow [_{skin}[_{state}]_{state}^{s^{(1)}}[_{state^{(1)}}]_{state^{(1)}}^{s^{(2)}}]_{skin}^{\phi};$

$6: [_{skin}[_i]_i^{x_i}[_{state}]_{state}^{s^{(1)}}]_{skin}^{\phi} \rightarrow [_{skin}[_i]_i^{x_i s^{(1)} s^{(3)}}]_{skin}^{\phi};$

$7: [_{skin}[_i]_i^{x_i s^{(1)} s^{(3)}}]_{skin}^{\phi} \rightarrow [_{skin}[_i]_i^{s^{(1)}}[_{state}]_{state}^{s^{(3)}}]_{skin}^{\phi};$

$8: [_{skin}[_i]_i^{b_i s^{(1)}}[_\star]_\star^{b_\star}]_{skin}^{\phi} \rightarrow [_{skin}[_i]_i^{b_i b_\star}]_{skin}^{\phi};$

$9: [_{skin}[_i]_i^{b_i b_\star}]_{skin}^{\phi} \rightarrow [_{skin}[_i]_i^{b_i}[_\star]_\star^{b_\star}]_{skin}^{\phi};$

$10: [_{skin}[_{state}]_{state}^{s^{(3)}}[_{state^{(2)}}]_{state^{(2)}}^{s^{(4)}}]_{skin}^{\phi} \rightarrow [_{skin}[_{state}]_{state}^{s^{(3)}s^{(4)}z}]_{skin}^{\phi};$

$11: [_{skin}[_{state}]_{state}^{s^{(3)}s^{(4)}z}]_{skin}^{\phi} \rightarrow [_{skin}[_{state}]_{state}^{z}[_{state^{(5)}}]_{state^{(5)}}^{s^{(4)}}]_{skin}^{\phi};$

$12: [_{skin}[_{state^{(5)}}]_{state^{(5)}}^{s^{(4)}}[_\star]_\star^{b_\star}]_{skin}^{\phi} \rightarrow [_{skin}[_\star]_\star^{b_\star}]_{skin}^{\phi};$

$13: [_{skin}[_{state}]_{state}^{s^{(2)}}[_\star]_\star^{b_\star}]_{skin}^{\phi} \rightarrow [_{skin}[_{state}]_{state}^{s^{(2)}b_\star}]_{skin}^{\phi};$

$14: [_{skin}[_{state}]_{state}^{s^{(2)}b_\star}]_{skin}^{\phi} \rightarrow [_{skin}[_{state}]_{state}^{s^{(4)}}[_\star]_\star^{b_\star}]_{skin}^{\phi};$

$15: [_{skin}[_{state}]_{state}^{s^{(1)}}[_{state}]_{state}^{s^{(4)}}]_{skin}^{\phi} \rightarrow [_{skin}[_{state}]_{state}^{s^{(1)}s^{(4)}w}]_{skin}^{\phi};$

$16: [_{skin}[_{state}]_{state}^{s^{(1)}s^{(4)}w}]_{skin}^{\phi} \rightarrow [_{skin}[_{skin}[_{state}]_{state}^{w}[_{state^{(6)}}]_{state^{(6)}}^{s^{(4)}}]_{skin}^{\phi};$

$17: [_{skin}[_{state^{(6)}}]_{state^{(6)}}^{s^{(4)}}[_\star]_\star^{b_\star}]_{skin}^{\phi} \rightarrow [_{skin}[_\star]_\star^{b_\star}]_{skin}^{\phi}.$

Moreover, the set R contains:

$18: [_{skin}[_{state}]_{state}^{s_f}[_\star]_\star^{b_\star}]_{skin}^{\phi} \rightarrow [_{skin}[_{state}]_{state}^{s_f}]_{skin}^{\phi};$

$19: [_{skin}[_{state}]_{state}^{s_f}[_i]_i^{b_i}]_{skin}^{\phi} \rightarrow [_{skin}]_{skin}^{\phi},$ for $1 \le i \le n-1;$

$20: [_{skin}[_n]_n^{b_n}[_{state}]_{state}^{s_f}]_{skin}^{\phi} \rightarrow [_{skin}[_n]_n^{b_n}]_{skin}^{\phi}.$

Each register γ_i, $1 \le i \le n$, has a membrane i associated with it. Each of these membranes has action b_i, moreover, in the initial configurations $val(\gamma_i)$ actions x_i are associated with these membranes. The number of actions x_i associated with membrane i represents the value of register γ_i.

The membrane *state* has the action related to the currently simulated state of the register machine. Initially the action associated with this membrane is s_1. The membrane $\star$, initially present in Π, and $state^{(2)}, state^{(5)}$ and $state^{(6)}$, that can be created during a computation of Π, are ancillary membranes.

The simulation of instructions of the kind $(s, \gamma_i^+, z) \in I$ is performed by rules 1 and 2. Rule 1 allows the i and *state* membrane to mate and to have s and z as actions. Rule 2 allows membrane i with s and z as actions to create a *state* membrane having z as action (simulating the state change), while the i membrane adds x_i to its actions (simulating the addition of 1 to γ_i).

The simulation of instructions of the kind $(s, \gamma_i^-, z, w) \in I$ is performed by the sequence of rules 3, 4, 5, (6, 13), (7, 14), (8, 10), (9, 11), 12 if membrane i contains at least one action x_i or by the sequence of rules 3, 4, 5, 13, 14, 15, 16, 17 is membrane i does not contain any x_i. The rules 6 and 13, 7 and 14, 8 and 10, 9 and 11 are applied in parallel.

In both cases the application of the rules 3, 4 and 5 allows the membrane *state* to change action from s to $s^{(1)}$. Moreover, the membrane $state^{(2)}$ is created with $s^{(2)}$ as action. Also in both cases the next two transitions see the application of rules 13 and 14. The application of these rules allows membrane $state^{(2)}$ to change action from $s^{(2)}$ to $s^{(4)}$.

If membrane i has at least one x_i as action, then the rules 6 and 7 are applied (in parallel with rules 13 and 14, respectively) so that one action x_i associated with membrane i is replaced by $s^{(1)}$ and the membrane state has action $s^{(3)}$ associated with it. In the next two transitions the rules (8, 10) and (9, 11) are applied so that membrane i loses the $s^{(1)}$ action (simulating the subtraction of 1 from register γ_i) and the *state* membrane has action z (simulating the new state of the register machine). The subsequent application of rule 12 allows membrane $state^{(5)}$ to disappear.

If instead membrane i does not have any x_i as action, then the application of rules 15, 16 and 17 allows membrane *state* to have w as action (simulating the new state of the register machine).

When the *state* membrane has s_f as action (associated with the final state of the register machine), then rules 18, 19, and 20 are applied. Rule 18 removes the ancillary membrane $\star$, and rule 19 removes all the i membrane with the exception of the n membrane, that associated with the output register. Rule 20 removes the *state* membrane. It should be clear that the P system has a (final) configuration consisting of the *skin* membrane and one (n) membrane only if the $\star$ membrane and the i, $1 \le i \le n - 1$, membranes have been removed and rule 20 is the last applied rule.

If such a final configuration is not reached, then the P system renders no result. Otherwise, the result is rendered by the cardinality of the actions associated with membrane n. This cardinality is at least 1 because of the presence of b_i. The theorem follows from the fact that at least $n = 2$ registers are needed for a register machine to be computationally complete. $\square$

Using similar techniques it is possible to prove that

Theorem 10.16 $\mathbb{N} OP_5(mate_4, drip_4) = \mathbb{N} RE$.

10.11 Final remarks and research topics

We regard the results in the present chapter as the beginning of a potentially very rich stream of research in Membrane Computing. What is present in the literature of Membrane Computing is far from being a complete description of the computational complexity aspects of membrane systems.

There are very many directions where the research of this stream could go. Here we indicate only a few of them.

In this chapter we saw that some classes of P systems with active membranes are equivalent to **P**, while others can solve **NP-complete** problems.

Suggestion for research 10.1 *Tighten the gap between* **P** *and* **NP** *in terms of P systems with active membranes.*

By this we mean to change the kind of rules used (or the used semantics) by P systems with active membranes in order to minimise the gap between **P** and **NP**.

The rules of type (e) applied according to the semantics (e'') and the rules of type (f) seems very similar: they both divide composite compartments, one using a symbol and the other using the polarities. To the best of our knowledge there is no study on the differences between these two kinds of rules. Can one simulate the other in an efficient way?

Suggestion for research 10.2 *Study the relation between rules of type* (e) *applied according to the semantics* (e'') *and rules of type* (f).

The use of dynamic P/T systems in relation to P systems with active membranes is original to the present monograph. We think that such a link could be of great advantage in the study of both P/T systems and membrane systems. The definition of dynamic P/T systems can be changed so as to include the creation of compartments, that is, the creation of some places and transitions, or even the division of compartments. In particular, concepts such as REACHABILITY and COVERABILITY, well studied in P/T systems, could be linked with the dependency graph used in the proofs presented in Section 10.7. In general we say:

Suggestion for research 10.3 *Investigate further the links between dynamic P/T systems and P systems with active membranes.*

10.12 Bibliographical notes

P systems with active membranes were introduced in [209]. Here they were proved to be computationally complete and a linear time, exponential space

semi-uniform solution to k-SAT was given. The universality result presented
in [209] considered systems with an unbounded number of compartments and
used rules of type *(f)*. This result was improved: in [222] rules of type *(f)* were
no longer considered, in [208] no membrane division rule is used, in [68] no membrane division rule is used and the number of compartments is bounded to be 4,
the number of compartments is decreased to 3 in [165, 253, 255].

Several papers dealt with P systems with active membranes solving **NP**
problems using exponential space and either linear or polynomial time. Here
we list some of these results:

[185, 186] solution to k-SAT using only three polarities and rules of type *(f)*;

[154] polynomial time solution to the VERTEX COVER and HAMILTONIAN
PATH problems using d-division rules for $d \in \mathbb{N}_+$;

[208] polynomial time solution to the HAMILTONIAN PATH problem using 2-
division rules;

[255] linear time solution to the HAMILTONIAN PATH problem using three polarities and rules of the kinds *(a)*, *(b)*, *(c)* and *(e)*;

[187, 188] polynomial time solution to the inverse of one-way functions;

[155] finding keys in the *Data Encryption Standard*;

[97] linear time uniform solution to k-SAT using only rules of the kind *(a)*, *(b)*,
(c) and *(e)* (where Theorem 10.1 was originally proved); in the same paper
a similar result is obtained with systems having a cell-graph as underlying
net;

[12, 229] uniform solutions to **NP** problems using only two polarities and the
proof that k-SAT $\in$ **PMC**$_{\Pi_{(a,b,c,d,e,f)}}$;

[196] P systems with active membranes without polarisations but using bi-stable
catalysts are introduced and proved to solve in a semi-uniform way **NP-
complete** problems;

[234] linear time semi-uniform solution to QSAT using only rules of the kind *(a)*,
(b), *(c)*, *(d)*, *(e)* and *(f)* and polarities;

[13] a uniform solution to QSAT using only rules of the kind *(a)*, *(b)*, *(c)*, *(d)*,
(e) and *(f)* and no polarities;

[57] an efficient uniform solution to SUBSET SUM using only rules of the kind
(a), *(b)*, *(c)*, *(d)* and *(f)* and no polarities.

Several other models of membrane systems having as underlying structure a
dynamic cell-tree were introduced and used to solve **NP** problems. Here we do
not go into the details of these models but simply list the results related to them.

[133, 164] Rules of type *(g)* are introduced. It is proved that such systems are
computationally complete using either multisets [164] or strings [175]. In

[164] it is also proved that these systems can solve in a semi-uniform way the HAMILTONIAN PATH problem.

[43, 156] String replication was introduced and the resulting systems are proved to solve in a semi-uniform way the HAMILTONIAN PATH problem and k-SAT;

[189] P systems using separation, merging and release are introduced and proved to solve **NP-complete** problems;

[190] P systems with separation rules and either polarisations or separations rules are introduced and proved to solve **NP-complete** problems.

The results present in Section 10.5 are based on [194]. This study started in [254]. The definition and use of dynamic P/T systems is original to the present monograph. The result of Theorem 10.5 is slightly more general as it holds also for recogniser basic transition P systems (a model of membrane systems introduced in [207] that we do not treat in this monograph).

Section 10.6 is based on [195], Section 10.7 is based on [102], while Section 10.8 is based on [98].

For more concepts regarding computational complexity, readers can refer to [91, 191].

Brane Calculi was introduced in [40]. Theorem 10.15 is original to this monograph, while in [42] it was proved that $\mathbb{N}\,OP_{11}(\text{mate}_5, \text{drip}_5) = \mathbb{N}\,RE$. Theorem 10.16 was proved in [30]. A similar result but based on cell-graphs is proved in [72]. Other publications bridging Membrane Computing and Brane Calculi are [36, 38].

Another model at the interface of Membrane Computing and Calculi, *mobile ambients* [41], are *mobile membranes*. Mobile membranes [153] describe the movement of membranes using simpler operations than those used in [209]. They were studied in [18, 19, 149–151].

10.12.1 About the notation

In the present chapter we adopted a novel notation to denote classes of P systems. Our notation is of the form $\Pi_{(x_1,\dots,x_n)}$, where $x_i \in \{a, b, c, d, e, f, g\}$, $x_i \neq x_j$ for $i \neq j$, $1 \leq i, j \leq n$, $n \leq 7$.

The notation normally used in Membrane Computing to denote classes of P systems sees an abbreviation, recalling the type of P systems, together with some parameters indicating the kind of rules used. For instance, $\mathcal{AM}_{(+d, +ne)}$ denotes (see, for instance, [97]) the class of recognising P systems with active membranes ($\mathcal{AM}$ stands for active membranes) using division (denoted by $+d$) of composite compartments (denoted by $+ne$), that is, rules of type f.

We consider the notation we adopted to be more immediate as long as the relation between $\{a, b, c, d, e, f, g\}$ and the rules is not changed.

In the bracket representation we used subscripts in each pair of matching square brackets while in the literature of Membrane Computing only the right end bracket, that is the closing one, has a subscript. Also in this case we consider our notation to be clearer.

We used the notation $[_i \ ]_i$ to represent compartments in the bracket representation. In the literature of Membrane Computing the notation $[\ ]_i$ can also be found.

Bibliography

[1] B. Alberts, A. Johnson, J. Lewis, M. Raff, K. Roberts, and P. Walter. *Molecular Biology of the Cell.* Garland Publishing, 4th edition, March 2002.

[2] A. Alhazov and M. Cavaliere. Proton pumping P systems. In Martín-Vide et al. [170], pages 70–88.

[3] A. Alhazov and R. Freund. P systems with one membrane and symport/antiport rules of five symbols are computationally complete. In Gutiérrez-Naranjo et al. [100], pages 19–28. Available from [238].

[4] A. Alhazov, R. Freund, A. Leporati, M. Oswald, and C. Zandron. (Tissue) P systems with unit rules and energy assigned to membranes. *Fundamenta Informaticae*, 74(4):1–18, 2006.

[5] A. Alhazov, R. Freund, and M. Oswald. Symbol/membrane complexity of P systems with symport/antiport. In Freund et al. [67], pages 96–113.

[6] A. Alhazov, R. Freund, and M. Oswald. Cell/symbol complexity of tissue P systems with symport/antiport rules. *International Journal of Foundations of Computer Science*, 17(1):3–25, 2006.

[7] A. Alhazov, R. Freund, M. Oswald, and M. Slavkovik. Extended variants of spiking neural P systems generating strings of vectors of non-negative integers. In Hoogeboom et al. [110], pages 123–134.

[8] A. Alhazov, R. Freund, and Y. Rogozhin. Computational power of symport/antiport: history, advances and open problems. In Freund et al. [67], pages 1–30.

[9] A. Alhazov, R. Freund, and Y. Rogozhin. Some optimal results on communicative P systems with minimal cooperation. In Gutiérrez-Naranjo et al. [99], pages 23–36. Available from [238].

[10] A. Alhazov, M. Margenstern, V. Rogozhin, Y. Rogozhin, and S. Verlan. Communicative P systems with minimal cooperation. In Mauri et al. [177], pages 161–177.

[11] A. Alhazov, C. Martín-Vide, and G. Păun, editors. *Pre-proceedings of the 4th Workshop on Membrane Computing. Tarragona, Spain, July 17–22, 2003*, 2003. Available from [238].

[12] A. Alhazov, L. Pan, and G. Păun. Trading polarizations for labels in P systems with active membranes. *Acta Informatica*, 41(2–3):111–144, 2004.

[13] A. Alhazov and M. J. Pérez-Jiménez. Uniform solution of QSAT using polarizationless active membranes. In J. Durand-Lose and M. Margenstern, editors, *Machines, Computations, and Universality. Proceedings 5th International Conference, MCU 2007, Orléans, France, 2007*, volume 4664 of *Lecture Notes in Computer Science*, pages 122–133. Springer-Verlag, Berlin, Heidelberg, New York, 2007.

[14] A. Alhazov and Y. Rogozhin. Towards a characterisation of P systems with minimal symport/antiport and two membranes. In Hoogeboom et al. [110], pages 135–153.

[15] A. Alhazov and Y. Rogozhin. Skin output in P systems with minimal symport/antiport and two membranes. In Eleftherakis et al. [60], pages 97–112.

[16] A. Alhazov, Y. Rogozhin, and S. Verlan. Symport/antiport tissue P systems with minimal cooperation. In Gutiérrez-Naranjo et al. [99], pages 37–52. Available from [238].

[17] A. Alhazov, Y. Rogozhin, and S. Verlan. Minimal cooperation in symport/antiport tissue P systems. *International Journal of Foundations of Computer Science*, 18(1):163–180, 2007.

[18] B. Aman and G. Ciobanu. Decidability results for mobile membranes derived from mobile ambients. *Pre-proceedings of the Fourth Conference on Computability in Europe, CiE 2008, Athens (Greece)*.

[19] B. Aman and G. Ciobanu. On the reachability problem in P systems with mobile membranes. In Freund et al. [67], pages 113–123.

[20] M. Amos. *Theoretical and Experimental DNA Computation*. Natural computing series. Springer-Verlag, Berlin, Heidelberg, New York, 2005.

[21] M. Amos. *Genesis Machines: The New Science of Biocomputation*. Atlantic Books, 2006.

[22] I. Antoniou, C. S. Calude, and M. J. Dinneen, editors. *Unconventional Models of Computation, UMC'2K, Proceedings on the Second International Conference on Unconventional Models of Computation*. Springer-Verlag, Berlin, Heidelberg, New York, 2001.

[23] B. S. Baker and R. V. Book. Reversal-bounded multipushdown machines. *Journal of Computer and System Science*, 8:315–332, 1974.

[24] H. G. Baker. *Rabin's Proof of the Undecidability of the Reachability Set Inclusion Problem of Vector Addition Systems*. Cambridge, Mass.: MIT, Project MAC, Computation Structures Group Memo 79, 1973.

[25] Ja. M. Barzin'. On a certain class of Turing machines (Minsky machines). *Algebra i Logika*, 1(6):42–51, 1962/1963. (in Russian) MR 27 #2415.

[26] F. Bernardini and R. Freund. Tissue P systems with communication modes. In Hoogeboom et al. [110], pages 170–182.

[27] F. Bernardini and M. Gheorghe. On the power of minimal symport/antiport. In Martín-Vide et al. [170], pages 72–83.

[28] F. Bernardini and M. Gheorghe. Cell communication in tissue P systems and cell division in population P systems. *Soft Computing*, 9(9):640–649, 2005.

[29] F. Bernardini and A. Păun. Universality of minimal symport/antiport: five membranes suffice. In Martín-Vide et al. [170], pages 43–45.

[30] D. Besozzi, N. Busi, G. Franco, R. Freund, and G. Păun. Two universality results for (mem)brane systems. In M. A. Gutiérrez-Naranjo, G. Păun, A. Riscos-Núñez, and F. R. Romero-Campero, editors, *Proceedings of the Fourth Brainstormin Week on Membrane Computing, Seville, Spain, January 30–February 3, 2006*, pages 49–62. Fénix Editoria, Sevilla, 2006.

[31] A. Binder, R. Freund, M. Oswald, and L. Vock. Extended spiking neural P systems with excitatory and inhibitory astrocytes. In Gutírrez-Naranjo et al. [103], pages 63–72. Available from [238].

[32] P. Bonizzoni, C. Ferretti, G. Mauri, and R. Zizza. Separating some splicing models. *Information Processing Letters*, 79(6):255–259, 2001.

[33] R. Borrego-Ropero, D. Diaz-Pernil, and M. J. Pérez-Jiménez. Tissue simulator: A graphical tool for tissue P systems. In *Pre-proceedings of the International Workshop on Automata for Cellular and Molecular Computing*, pages 23–34, 2007. Budapest, Hungary, August 31, 2007.

[34] P. Brodal. *The central nervous system. Structure and function.* Oxford University Press, 3rd edition, 2004.

[35] H.-D. Burkhard. Ordered firing in Petri nets. *Journal of Information Processing and Cybernetics*, 17(2–3):71–86, 1981.

[36] N. Busi. On the computational power of the mate/bud/drip brane calculus: interleaving vs. maximal parallelism. In Freund et al. [67], pages 144–158.

[37] N. Busi. Towards a causal semantics for brane calculi. In Gutírrez-Naranjo et al. [103], pages 97–112. Available from [238].

[38] N. Busi. Using well-structured transition systems to decide divergence for catalytic P systems. *Theoretical Computer Science*, 372(2–3):125–135, 2007.

[39] C. S. Calude, G. Păun, G. Rozenberg, and A. Salomaa, editors. *Multiset Processing: Mathematical, Computer Science, and Molecular Computing Points of View [Workshop on Multiset Processing, WMP 2000, Curtea de Arges, Romania, August 21–25, 2000]*, volume 2235 of *Lecture Notes in Computer Science*. Springer-Verlag, Berlin, Heidelberg, New York, 2001.

[40] L. Cardelli. Brane calculi. In V. Danos and V. Schächter, editors, *Computational Methods in Systems Biology, International Conference CMSB 2004,*

Paris, France, May 26–28, 2004, Revised Selected Papers, volume 3082 of *Lecture Notes in Computer Science*, pages 257–278. Springer-Verlag, Berlin, Heidelberg, New York, 2004.

[41] L. Cardelli and A. D. Gordon. Mobile ambients. In M. Nivat, editor, *Foundations of Software Science and Computation Structure, First International Conference, FoSSaCS'98, Held as Part of the European Joint Conferences on the Theory and Practice of Software, ETAPS'98, Lisbon, Portugal, March 28–April 4, 1998, Proceedings*, volume 1378 of *Lecture Notes in Computer Science*, pages 140–155. Springer-Verlag, Berlin, Heidelberg, New York, 1998.

[42] L. Cardelli and G. Păun. An universality result for a (mem)brane calculus based on mate/drip operations. *International Journal of Foundations of Computer Science*, 2005.

[43] J. Castellanos, G. Păun, and A. Rodriguez-Paton. P systems with worm-objects. In *IEEE 7th. International Conference on String Processing and Information Retrieval, SPIRE 2000*, pages 64–74, 2000. La Coruna, Spain.

[44] M. Cavaliere, O. Egecioglu, O. H. Ibarra, S. Woodworth, M. Ionescu, and G. Păun. Asynchronous spiking neural P systems. Technical Report 9-2007, Microsoft Research—University of Trento, 2007.

[45] M. Cavaliere and D. Genova. P systems with symport/antiport of rules. *The Journal of Universal Computer Science*, 10(5):540–558, 2004.

[46] M. Cavaliere, C. Martín-Vide, and G. Păun. Brainstorming week on membrane computing. Technical Report 26/03, Universitat Rovira i Virgili, Tarragona, Spain, 2003.

[47] H. Chen, R. Freund, M. Ionescu, G. Păun, and M. J. Pérez-Jiménez. On string languages generated by spiking neural P systems. *Fundamenta Informaticae*, 75(1–4):141–162, 2007.

[48] H. Chen, M. Ionescu, T.-O. Ishdorj, A. Păun, G. Păun, and M. J. Pérez-Jiménez. Spiking neural P systems with extended rules: Universality and languages. *International Journal on Natural Computing*, 7(2):147–166, 2008.

[49] H. Chen, M. Ionescu, A. Păun, G. Păun, and B. Popa. On trace languages generated by spiking neural P systems. In *Eighth International Workshop on Descriptional Complexity of Formal Systems, (DCFS 2006)*, pages 94–105, 2006. June 21–23, 2006, Las Cruces, New Mexico, USA.

[50] H. Chen, T.-O. Ishdorj, G. Păun, and M. J. Pérez-Jiménez. Handling languages with spiking neural p systems with extended rules. *Romanian Journal of Information Science and Technology*, 9(3):151–162, 2006.

[51] G. M. Cooper and R. E. Hausman. *The Cell. A Molecular Approach*. ASM press, 4th edition, 2007.

[52] D. W. Corne and P. Frisco. Dynamics of HIV infection studied with cellular automata and conformon-P systems. *BioSystems*, 91(3):531–544, 2008. Special issue: P-systems applications to systems biology.

[53] D. W. Corne, P. Frisco, G. Păun, G. Rozenberg, and A. Salomaa. Pre-proceedings Membrane Computing. 9th International Workshop, WMC 2008, Edinburgh, UK, July 2008, 2007. Available at [238].

[54] E. Csuhaj-Varjú, L. Kari, and G. Păun. Test tube distributed systems based on splicing. *Computers and AI*, 15(2–3):211–232, 1996.

[55] E. Csuhaj-Varjú, G. Păun, and G. Vaszil. Tissue-like P systems communicating by request. In K. Kamala and R. Rama, editors, *Formal Language Aspects of Natural Computing*, volume 3, pages 143–153. Ramanujan Mathematical Society, 2006.

[56] E. Csuhaj-Varjú and G. Vaszil. P automata or purely communicating accepting P systems. In Păun et al. [221], pages 219–233.

[57] D. Díaz-Perin, M. A. Gutiérrez-Narajo, M. J. Pérez-Jiménez, and A. Riscos-Núñez. A logaritmic bound for solving subset sum with P systems. In Eleftherakis et al. [60], pages 301–315.

[58] S. Donatelli and J. Kleijn, editors. *Application and Theory of Petri Nets*, volume 1639 of *Lecture Notes in Computer Science*. Springer-Verlag, Berlin, Heidelberg, New York, 1999.

[59] A. Ehrenfeucht, T. Harju, I. Petre, D. M. Prescott, and G. Rozenberg. *Computation in Living Cells. Gene Assembly in Ciliates*. Springer-Verlag, Berlin, Heidelberg, New York, 2003.

[60] G. Eleftherakis, P. Kefalas, G. Păun, G. Rozenberg, and A. Salomaa, editors. *Membrane Computing. 8th International Workshop, WMC 2007, Thessaloniki, Greece, June 2007, Revised Selected and Invited Papers*, volume 4860 of *Lecture Notes in Computer Science*. Springer-Verlag, Berlin, Heidelberg, New York, 2007.

[61] J. Esparza. Petri nets, commutative context-free grammars, and basic parallel processes. *Fundamenta Informaticae*, 31(1):13–25, 1997.

[62] F. Freund, R. Freund, M. Margenstern, M. Oswald, Y. Rogozhin, and S. Verlan. P systems with cutting/recombination rules or splicing rules assigned to membranes. In Martín-Vide et al. [170], pages 191–202.

[63] F. Freund, R. Freund, and M. Oswald. Splicing test tube systems and their relation to splicing membrane systems. In Jonoska et al. [139], pages 139–151.

[64] R. Freund. Special variants of P systems inducing an infinite hierarchy with respect to the number of membranes. *Bulletin of EATCS*, 75:209–219, 2001.

[65] R. Freund, L. Kari, M. Oswald, and P. Sosík. Computationally universal P systems without priorities: two catalysts are sufficient. *Theoretical Computer Science*, 330(2):251–266, 2005.

[66] R. Freund, L. Kari, and G. Păun. DNA computation based on splicing: The existence of universal computers. *Theory of Computing Systems*, 32:69–112, 1999.

[67] R. Freund, G. Lojka, M. Oswald, and G. Păun, editors. *Membrane Computing. 6th International Workshop, WMC 2005, Vienna, Austria, July 18–21, 2005, Revised Selected and Invited Papers*, volume 3850 of *Lecture Notes in Computer Science*. Springer-Verlag, Berlin, Heidelberg, New York, 2006.

[68] R. Freund, C. Martin-Vide, and G. Paun. From regulated rewriting to computing with membranes: collapsing hierarchies. *Theoretical Computer Science*, 312:143–188, 2004.

[69] R. Freund and M. Oswald. P systems with activated/prohibited membrane channels. In Păun et al. [221], pages 261–269.

[70] R. Freund and M. Oswald. A short note on analysing P systems with antiport rules. *Bulletin of the European Association for Theoretical Computer Science*, 78:231–236, October 2002.

[71] R. Freund and M. Oswald. Tissue P systems with symport/antiport rules of one symbols are computationally complete, 2005. Available from [238].

[72] R. Freund and M. Oswald. Tissue P systems and (mem)brane systems with mate and drip operations working on strings. *Electronic notes in Theoretical Computer Science*, 171:105–115, 2007.

[73] R. Freund, M. Oswald, and P. Sosík. Reducing the number of catalysts needed in computationally universal P systems without priorities. In E. Csuhaj-Varjú, C. Kintala, D. Wotschke, and G. Vaszil, editors, *5th International Workshop of Formal Systems*, pages 102–113. MTA SZTAKI, Budhapest, 2003. Budapest, Hungary.

[74] R. Freund and G. Păun. On deterministic P systems, 2003. Available from [238].

[75] R. Freund, G. Păun, and M. J. Pérez-Jiménez. Tissue P systems with channels states. *Theoretical Computer Science*, 330:101–116, 2005.

[76] R. Freund and S. Verlan. A formal framework for static (tissue) P systems. In Eleftherakis et al. [60], pages 271–284.

[77] P. Frisco. On two variants of splicing super-cell systems. *Romanian Journal of Information Science and Technology*, 4(1–2):89–100, 2001.

[78] P. Frisco. The conformon-P system: A molecular and cell biology-inspired computability model. *Theoretical Computer Science*, 312(2–3):295–319, 2004.

[79] P. Frisco. About P systems with symport/antiport. *Soft Computing*, 9(9): 664–672, 2005.

[80] P. Frisco. Infinite hierarchies of conformon-P systems. In Hoogeboom et al. [110], pages 395–408.

[81] P. Frisco. P systems, Petri nets, and Program machines. In Freund et al. [67], pages 209–223.

[82] P. Frisco. Conformon-P systems with negative values. In Eleftherakis et al. [60], pages 331–344.

[83] P. Frisco. An hierarchy of recognising computational processes. Technical report, Heriot-Watt University, 2007. HW-MACS-TR-0047.

[84] P. Frisco. On s-sum vectors. Technical report, Heriot-Watt University, 2008. HW-MACS-TR-0060.

[85] P. Frisco and D. W. Corne. Modeling the dynamics of HIV infection with conformon-P systems and cellular automata. In Eleftherakis et al. [60], pages 21–32.

[86] P. Frisco and R. T. Gibson. A simulator and an evolution program for conformon-P systems. In *SYNASC 2005, 7th International Symposium on Simbolic and Numeric Algorithms for Scientific Computing*, pages 427–430. IEEE Computer Society, 2005. Workshop on Theory and Applications of P Systems, TAPS, Timisoara, Romania, September 26–27, 2005.

[87] P. Frisco and H. J. Hoogeboom. Simulating counter automata by P systems with symport/antiport. In Păun et al. [221], pages 288–301.

[88] P. Frisco and H. J. Hoogeboom. P systems with symport/antiport simulating counter automata. *Acta Informatica*, 41(2–3):145–170, 2004.

[89] P. Frisco, H. J. Hoogeboom, and P. Sant. A direct construction of a universal P system. *Fundamenta Informaticae*, 49(1–3):103–122, 2002.

[90] P. Frisco and G. Păun. No cycles in compartments. Starting from conformon-P systems. In Gutiérrez-Naranjo and Păun [101], pages 157–170. Available from [238].

[91] M. R. Garey and D. S. Johnson. *Computers and intractability: A Guide to the Theory of NP-Completeness*. W. H. Freeman and Co., San Francisco, 1979.

[92] M. H. Garzon and H. Yan, editors. *DNA Computing*, volume 4848 of *Lecture Notes in Computer Science*, 2008. 13th International Meeting on DNA Computing, DNA13, Memphis, TN, USA, Revised Selected Papers.

[93] R. Gershoni, E. Keinan, G. Păun, R. Piran, T. Ratner, and S. Shoshani. Research topics arising from the (planned) P systems implementation experiment in Technion. In Gutiérrez-Naranjo and Păun [101], pages 183–192. Available from [238].

[94] D. E. Green and S. Ji. *Molecular Basis of Electron Transport*, chapter The electromechanical model of mitochondrial structure and function, pages 1–44. Academic Press, New York, 1972. J. Schultz, B. F. Cameron (eds).

[95] S. A. Greibach. Remarks on blind and partially blind one-way multicounter machines. *Theoretical Computer Science*, 7:311–324, 1978.

[96] M. A. Gutiérrez-Naranjo, M. J. Pérez-Jiménez, and D. Ramírez-Martínez. A software tool for verification of spiking neural P systems. *International Journal on Natural Computing*, 7(4):485–497, 2008.

[97] M. A. Gutiérrez-Naranjo, M. J. Pérez-Jiménez, A. Riscos-Nùñez, and F. J. Romero-Campero. On the efficiency of cell-like and tissue-like recognizer membrane systems. Presented at the Workshop on Nature Inspired Cooperative Strategies for Optimization—NICSO 2006, Granada, June 29–30 2006.

[98] M. A. Gutiérrez-Naranjo, M. J. Pérez-Jiménez, and F. J. Romero-Campero. A linear solution for QSAT with membrane creation. In Freund et al. [67], pages 241–252.

[99] M. A. Gutiérrez-Naranjo, G. Păun, and M. J. Pérez-Jiménez, editors. *Cellular Computing. Complexity Aspects. Proceedings of the ESF Exploratory Workshop on Cellular Computing (Complexity Aspects)*. Fénix Editoria, Sevilla, 2005. Available from [238].

[100] M. A. Gutiérrez-Naranjo, A. Riscos-Núñez, F. R. Romero-Campero, and D. Sburlan, editors. *Proceedings of the Third Brainstorming week on Membrane Computing. Sevilla, Spain, January 31–February 4, 2005*. Fénix Editoria, Sevilla, 2005. Available from [238].

[101] M. A. Gutiérrez-Naranjo and G. Păun, editors. *Proceedings of the Sixth Brainstorming Week on Membrane Computing. Sevilla, February 4–8, 2008*. Fenix Editora, Sevilla, 2008. Available from [238].

[102] M. A. Gutírrez-Naranjo, M. J. Pérez-Jiménez, A. Riscos-Núñez, and F. J. Romero-Campero. On the power of dissolution in P systems with active membranes. In Freund et al. [67], pages 226–242.

[103] M. A. Gutírrez-Naranjo, G. Păun, A. Romero-Jiménez, and A. Riscos-Núñez, editors. *Proceedigns of the Fifth Brainstorming Week on Membrane Computing. Sevilla, Spain, January 29–February 2, 2007*. Fenix Editora, Sevilla, 2007. Available from [238].

[104] M. Hack. The quality problem for vector addition systems is undecidable. *Theoretical Computer Science*, 2(1):77–95, 1976.

[105] M. Hagiya and A. Ohuchi, editors. *DNA Computing: 8th International Workshop on DNA-Based Computers, DNA8, Sapporo, Japan, June 10–13, 2002. Revised Papers*, volume 2568 of *Lecture Notes in Computer Science*. Hokkaido University, Springer Verlag, Berlin, Heidelberg, New York, 2002.

[106] T. Head. Formal language theory and DNA: an analysis of the generative capacity of specific recombinant behaviors. *Bulletin of Mathematical Biology*, 49(6):737–759, 1987.

[107] T. Head. Splicing schemes and DNA. In G. Rozenberg and A. Salomaa, editors, *Lindenmayer Systems: Impact on Theoretical Computer Science*

and Developmental Biology, pages 371–383. Springer-Verlag, Berlin, Heidelberg, New York, 1992.

[108] T. Head, G. Păun, and D. Pixton. *Handbook of formal languages, vol. 2: linear modeling: background and application,* chapter Language theory and molecular genetics: generative mechanisms suggested by DNA recombination, pages 295–360. Springer-Verlag, Berlin, Heidelberg, New York, 1997.

[109] H. J. Hoogeboom. Carriers and counters. P-systems with carriers vs. (blind) counter automata. In M. Ito and M. Toyama, editors, *Developments in Language Theory. 6th International Conference, DLT 2002, Kyoto, Japan, September 18-21, 2002, Revised Papers*, volume 2450 of *Lecture Notes in Computer Science*, pages 140–151. Springer-Verlag, Berlin, Heidelberg, New York, 2003.

[110] H. J. Hoogeboom, G. Păun, G. Rozenberg, and A. Salomaa, editors. *Membrane Computing. 7th International Workshop, WMC 2006, Leiden, Netherlands, July 17-21, 2006, Revised, Selected, and Invited Papers*, volume 4361 of *Lecture Notes in Computer Science*. Springer-Verlag, Berlin, Heidelberg, New York, 2006.

[111] P. W. Hoogersa, H. C. M. Kleijn, and P. S. Thiagarajan. A trace semantics for Petri nets. *Information and Computation*, 117:98–114, 1995.

[112] J. E. Hopcroft and J.-J. Pansiot. On the reachability problem for 5-dimensional vector addition systems. *Theoretical Computer Science*, 8:135–159, 1979.

[113] J. E. Hopcroft and D. Ullman. *Introduction to Automata Theory, Languages, and Computation*. Addison-Wesley, 1979.

[114] D. T. Huynh. The complexity of equivalence problems for commutative grammars. *Information and Control*, 66(1–2):103–121, 1985.

[115] O. H. Ibarra. Reversal-bounded multicounter machines and their decision problems. *Journal of the ACM*, 25:116–133, 1978.

[116] O. H. Ibarra. On determinism versus nondeterminism in P systems. *Theoretical Computer Science*, 344:120–133, 2005.

[117] O. H. Ibarra. On membrane hierarchy in P systems. *Theoretical Computer Science*, 334:115–129, 2005.

[118] O. H. Ibarra. Some recent results concerning deterministic P systems. In Gutiérrez-Naranjo et al. [99], pages 49–54. Available from [238].

[119] O. H. Ibarra, Z. Dang, and O. Egecioglu. Catalytic P systems, semilinear sets, and vector addition systems. *Theoretical Computer Science*, 312:379–399, 2004.

[120] O. H. Ibarra and G. Păun. Characterizations of context-sensitive languages and other languages classes in terms of symport/antiport P systems. *Theoretical Computer Science*, 358(1):88–103, 2006.

[121] O. H. Ibarra and S. Woodworth. Characterizations of some classes of spiking neural P systems. *International Journal on Natural Computing*, 7(4):499–517, 2008.

[122] O. H. Ibarra and S. Woodworth. On bounded symport/antiport P systems. In A. Carbone and N. Pierce, editors, *DNA computing. 11th International Workshop on DNA Computing, DNA11, London, ON, Canada, June 6–9, 2005. Revised Selected Papers*, volume 3892 of *Lecture Notes in Computer Science*, pages 129–143. Springer-Verlag, Berlin, Heidelberg, New York, 2005.

[123] O. H. Ibarra and S. Woodworth. Characterizations of some restricted spiking neural P systems. In Hoogeboom et al. [110], pages 424–442.

[124] O. H. Ibarra and S. Woodworth. On symport/antiport P systems with small number of objects. *International Journal of Computer Mathematics*, 83(7):613–629, 2006.

[125] O. H. Ibarra, S. Woodworth, H.-C. Yen, and Z. Dang. On symport/antiport P systems and semilimear sets. In Freund et al. [67], pages 253–271.

[126] O. H. Ibarra and H.-C. Yen. Deterministic catalytic systems are not universal. *Theoretical Computer Science*, 363(2):149–161, 2006.

[127] K. Culik II and T. Harju. Dominoes and the regularity of DNA splicing languages. In G. Ausiello, M. Dezani-Ciancaglini, and S. Ronchi Della Rocca, editors, *Automata, Languages and Programming*, volume 372 of *Lecture Notes in Computer Science*, pages 222–233. Springer-Verlag, Berlin, Heidelberg, New York, 1989. 6th International Colloquium, ICALP89, Stresa, Italy, July 11–15, 1989, Proceedings.

[128] K. Culik II and T. Harju. Splicing semigroups of dominoes and DNA. *Discrete Applied Mathematics*, 31:261–277, 1991.

[129] Essential Science Indicators: http://esi-topics.com.

[130] M. Ionescu, C. Martín-Vide, A. Păun, and G. Păun. Unexpected universality results for three classes of P systems with symport/antiport. In Hagiya and Ohuchi [105], pages 281–290.

[131] M. Ionescu, A. Păun, G. Păun, and M. J. Pérez-Jiménez. Computing with spiking neural P systems: Traces and small universal systems. In C. Mao and T. Yokomori, editors, *DNA Computing. 12th International Meeting on DNA Computing, (DNA12), Revised Selected Papers*, volume 4287 of *Lecture Notes in Computer Science*, pages 1–16. Springer-Verlag, Berlin, Heidelberg, New York, 2007.

[132] M. Ionescu, G. Păun, and T. Yokomori. Spiking neural P systems. *Fundamenta Informaticae*, 71(2–3):279–308, 2006.

[133] M. Ito, C. Martin-Vide, and G. Păun. A characterization of Parikh sets of ET0L languages in terms of P systems. In M. Ito, G. Păun, and S. Yu, editors, *Words, Semigroups, and Transducers*, pages 239–254. World Scientific Publishing Co., 2001.

[134] M. Jantzen. Language theory of Petri nets. In *Advances in Petri nets 1986, part I on Petri nets: central models and their properties*, pages 397–412. Springer-Verlag, Berlin, Heidelberg, New York, 1987.

[135] S. Ji. The Bhopalator: a molecular model of the living cell based on the concepts of conformons and dissipative structures. *Journal of Theoretical Biology*, 116:395–426, 1985.

[136] S. Ji. Free energy and information contents of conformons in proteins and DNA. *BioSystems*, 54:107–214, 2000.

[137] S. Ji. The Bhopalator: an information/energy dual model of the living cell (II). *Fundamenta Informaticae*, 49(1–3):147–165, 2002.

[138] J. P. Jones and Y. V. Matijasevič. Register machine proof of the theorem on exponential Diophantine representation of enumerable sets. *The Journal of Symbolic Logic*, 49(3):818–829, 1984.

[139] N. Jonoska, G. Păun, and G. Rozenberg, editors. *Aspects of Molecular Computing. Essay dedicated to Tom Head on the occasion of his 70th birthday*, volume 2950 of *Lecture Notes in Computer Science*. Springer-Verlag, Berlin, Heidelberg, New York, 2004.

[140] E. R. Kandel, J. H. Schwartz, and T. M. Jessell. *Principles of Neural Science*. McGraw-Hill Medical, 4th edition, 2000.

[141] L. Kari, C. Martín-Vide, and A. Păun. On the universality of P systems with minimal symport/antiport rules. In Jonoska et al. [139], pages 254–265.

[142] P. Kefalas, G. Eleftherakis, M. Holcombe, and M. Gheorghe. Simulation and verification of P systems through communicating X-machines. *BioSystems*, 70(2):135–148, 2003.

[143] G. Kemeny and I. M. Goklany. Polarons and conformons. *Journal of Theoretical Biology*, 40:107–123, 1973.

[144] G. Kemeny and I. M. Goklany. Quantum mechanical model for conformons. *Journal of Theoretical Biology*, 48:23–38, 1974.

[145] J. Kleijn and M. Koutny. Processes of membrane systems with promoters and inhibitors. Technical Report CS-TR: 986, School of Computing Science, University of Newcastle (UK), 2006.

[146] J. Kleijn, M. Koutny, and G. Rozenberg. Towards a Petri net semantics for membrane systems. In Freund et al. [67], pages 292–309.

[147] J. Kleijn, M. Koutny, and G. Rozenberg. Process semantics for membrane systems. *Journal of Automata, Languages and Combinatorics*, 11:321–340, 2008.

[148] S. Krishna. *Languages of P systems: computability and complexity*. PhD thesis, Indian institute of technology Madras, 2001. India.

[149] S. N. Krishna. The power of mobility: four membranes suffice. In S. B. Cooper and L. Torenvliet B. Löwe and, editors, *New Computational Paradigms. First Conference on Computability in Europe, CiE 2005,*

Amsterdam, The Netherlands, June 8–12, Proceedings, volume 3526 of Lecture Notes in Computer Science, pages 242–251. Springer-Verlag, Berlin, Heidelberg, New York, 2005.

[150] S. N. Krishna. Upper and lower bounds for the computational power of P systems with mobile membranes. In A. Beckmann, U. Berger, B. Löwe, and J. V. Tucker, editors, Logical Approaches to Computational Barriers. Second Conference on Computability in Europe, CiE 2006, Swansea, UK, June 30–July 5, 2006, Proceedings, volume 3988 of Lecture Notes in Computer Science, pages 526–535. Springer-Verlag, Berlin, Heidelberg, New York, 2006.

[151] S. N. Krishna and G. Ciobanu. On the computational power of enhanced mobile membranes. Pre-proceedings of the Fourth Conference on Computability in Europe, CiE 2008, Athens (Greece).

[152] S. N. Krishna, K. Lakshmanan, and R. Rama. Tissue P systems with contextual and rewriting rules. In Păun et al. [221], pages 339–351.

[153] S. N. Krishna and G. Păun. P systems with mobile membranes. International Journal on Natural Computing, 4(3):255–274, 2005.

[154] S. N. Krishna and R. Rama. A variant of P systems with active membranes: solving NP-complete problems. Romanian Journal of Information Science and Technology, 2(4):357–367, 1999.

[155] S. N. Krishna and R. Rama. Breaking DES using P systems. Theoretical Computer Science, 299(1–3):495–508, 2003.

[156] S. N. Krishna and R. Rama. P systems with replicated rewriting. Journal of Automata, Languages and Combinatorics, 6(3):345–350, 2001.

[157] K. Lakshmanan and R. Rama. On the power of tissue P systems with insertion and deletion rules. In Alhazov et al. [11], pages 304–318. Available from [238].

[158] J. Van Leeuwen, editor. Handbook of Theoretical Computer Science. Elsevier and MIT Press, 1990.

[159] A. Leporati, G. Mauri, C. Zandron, G. Păun, and M. J. Pérez-Jiménez. Uniform solutions to SAT and Subset Sum by spiking neural P systems, 2008. submitted.

[160] A. Leporati, C. Zandron, C. Ferretti, and G. Mauri. On the computational power of spiking neural P systems, 2007. in press.

[161] A. Leporati, C. Zandron, C. Ferretti, and G. Mauri. Solving numerical NP-complete problems with spiking neural P systems. In Eleftherakis et al. [60], pages 336–352.

[162] C. Li, Z. Dang, O. H. Ibarra, and Hsu-Chun Yen. Signaling P systems and verification problems. In L. Caires, G. F. Italiano, L. Monteiro, C. Palamidessi, and M. Yung, editors, ICALP, volume 3580 of Lecture Notes in Computer Science, pages 1462–1473. Springer-Verlag, Berlin, Heidelberg, New York, 2005.

[163] H. Lodish, A. Berk, C. A. Kaiser, M. Krieger, M. P. Scott, A. Bretscher, H. Ploegh, and P. T. Matsudaira. *Molecular Cell Biology*. W. H. Freeman & Co Ltd, 6th edition, 2007.

[164] M. Madhu and K. Krithivasan. P systems with membrane creation: universality and efficiency. In M. Margenstern and Y. Rogozhin, editors, *Machines, Computations, and Universality, Third International Conference, MCU 2001, Chisinau, Moldavia, May 23–27, 2001, Proceedings*, volume 2055 of *Lecture Notes in Computer Science*, pages 276–287. Springer-Verlag, Berlin, Heidelberg, New York, 2001.

[165] M. Madhu and K. Krithivasan. Improved results about universality of P systems. *Bulletin of the European Association for Theoretical Computer Science*, 76:162–168, September 2002.

[166] V. Manca. A proof of regularity for finite splicing. In Jonoska et al. [139], pages 309–317.

[167] M. Margenstern, V. Rogozhin, J. Rogozhin, and S. Verlan. About P systems with minimal symport/antiport rules and four membranes. In Mauri et al. [176], pages 283–294. Available from [238].

[168] M. Margenstern and Y. Rogozhin. About time-varying distributed H systems. In A. Condon and G. Rozenberg, editors, *DNA Computing*, volume 2054 of *Lecture Notes in Computer Science*, pages 53–62. Springer Verlag, Berlin, Heidelberg, New York, 2000. 6th International Workshop on DNA-Based Computers, DNA 2000, Leiden, The Netherlands, June 13–17, 2000, Revised Papers.

[169] M. Margenstern, Y. Rogozin, and S. Verlan. Time-varying distributed H systems of degree 2 can carry out parallel computations. Poster paper at 8th International Workshop on DNA-Based Computers, DNA 2002, Sapporo, Japan, 10–13 June 2002.

[170] C. Martín-Vide, G. Mauri, G. Păun, G. Rozenberg, and A. Salomaa, editors. *Membrane Computing. International Workshop, WMC 2003, Tarragona, Spain, July 17–22, 2003, Revised Papers*, volume 2933 of *Lecture Notes in Computer Science*. Springer-Verlag, Berlin, Heidelberg, New York, 2003.

[171] C. Martín-Vide, A. Păun, and G. Păun. On the power of P systems with symport rules. *The Journal of Universal Computer Science*, 8:317–331, 2002.

[172] C. Martín-Vide and G. Păun, editors. *Fundamenta Informaticae*, volume 49(1–3). IOS Press, 2002. Special Issue on Membrane Computing.

[173] C. Martín-Vide, G. Păun, and A. Rodríguez Patón. P systems with immediate communication. *Romanian Journal of Information Science and Technology*, 4(1–2):171–182, 2001.

[174] C. Martin-Vide, G. Păun, J. Pazos, and A. Rodríguez-Patón. Tissue P systems. *Theoretical Computer Science*, 296:295–326, 2003.

[175] C. Martin-Vide, G. Păun, and A. Rodríguez-Patón. On P systems with membrane creation. *Computer Science Journal of Moldavia*, 9(2):134–145, 2001.

[176] G. Mauri, G. Păun, M. J. Pérez-Jímenez, G. Rozenberg, and A. Salomaa, editors. *Pre-proceedings of the Fifth Workshop on Membrane Computing WMC5, held in Milan, Italy, June 14–16, 2004*, 2004. Available from [238].

[177] G. Mauri, G. Păun, M. J. Pérez-Jímenez, G. Rozenberg, and A. Salomaa, editors. *Membrane Computing. 5th International Workshop, WMC 2004, Milan, Italy, June 14–16, 2004, Revised Selected and Invited Papers*, volume 3365 of *Lecture Notes in Computer Science*. Springer-Verlag, Berlin, Heidelberg, New York, 2005.

[178] E. W. Mayr. Persistence of vector replacement systems is decidable. *Acta Informatica*, 15:309–318, 1981.

[179] R. Milner. *Communication and Concurrency*. Prentice-Hall, Englewood, N.J., 1989.

[180] R. Milner. *Communicating and Mobile Systems: the π-calculus*. Cambridge University press, 1999.

[181] M. L. Minsky. Recursive unsolvability of Post's problem of "tag" and other topics in theory of Turing machines. *Annals of Mathematics*, 74(3):437–455, 1961.

[182] M. L. Minsky. *Computation: Finite and Infinite Machines*. Automatic computation. Prentice-Hall, 1967.

[183] M. Mutyam, V. J. Prakash, and K. Krithivasan. Rewriting tissue P systems. *The Journal of Universal Computer Science*, 10(9):1250–1271, 2004.

[184] H. Nagda, A. Păun, and A. Rodríguez-Patón. P systems with symport/antiport and time. In Hoogeboom et al. [110], pages 463–476.

[185] A. Obtulowicz. Note on some recursive family of P systems with active membranes. Available from [238].

[186] A. Obtulowicz. Deterministic P systems for solving SAT problem. *Romanian Journal of Information Science and Technology*, 4(1–2):195–202, 2001.

[187] A. Obtulowicz. Membrane computing and one-way functions. *International Journal of Foundations of Computer Science*, 12(4):551–558, 2001.

[188] A. Obtulowicz. On P systems with active membranes solving integer factorizing problem in polynomial time. In Calude et al. [39], pages 267–286.

[189] L. Pan, A. Alhazov, and T.-O. Ishdorj. Further remarks on P systems with active membranes, separation, merging, and release rules. *Soft Computing*, 9(9):686–690, 2005.

[190] L. Pan and T.-O. Ishdorj. P systems with active membranes and separation rules. *The Journal of Universal Computer Science*, 10(5):630–649, 2004.

[191] C. H. Papadimitriou. *Computational Complexity*. Addison-Wesley Pub. Co., 1994.

[192] A. Pérez-Jímenez, M. J. Pérez-Jímenez, and F. Sancho-Caparrini. Computing a partial mapping by P systems: Design and verification. In *BWC2003* [46], pages 247–260.

[193] A. Pérez-Jímenez, M. J. Pérez-Jímenez, and Fernando Sancho-Caparrini. Formal verification of a transition P system generating the set 2#2. In *BWC2003* [46], pages 261–269.

[194] M. J. Pérez-Jiménez, A. Romero-Jiménez, and F. Sancho-Caparrini. The P versus NP problem through cellular computing with membranes. In Jonoska et al. [139], pages 338–352.

[195] M. J. Pérez-Jiménez and A. Riscos-Núñez. A linear-time solution to the Knapsack problem using P systems with active membranes. In Martín-Vide et al. [170], pages 250–268.

[196] M. J. Pérez-Jiménez and F. J. Romero-Campero. Trading polarizations for bi-stable catalysts in P systems with active membranes. In Mauri et al. [177], pages 373–388.

[197] C. A. Petri. *Kommunikation mit Automaten*. PhD thesis, Rheinisch-Westfälisches Institut für Instrumentelle Mathematik an der Universität Bonn, 1962. Germany, Schrift Nr. 2 (in German).

[198] D. Pixton. Regularity of splicing languages. *Discrete Applied Mathematics*, 69(1–2):101–124, August 1996.

[199] D. Pixton. Splicing in abstract families of languages. *Theoretical Computer Science*, 234:135–166, 2000.

[200] V. J. Prakash. On the power of tissue P systems working in the maximal-one mode. In Alhazov et al. [11], pages 356–364. Available from [238].

[201] V. J. Prakash and K. Krithivasan. Simulating boolean circuits with tissue P systems. In Mauri et al. [176], pages 343–359. Available from [238].

[202] A. Păun. On P systems with global rules. In N. Jonoska and N. C. Seeman, editors, *Proceedings of DNA Computing: 7th International Workshop on DNA-Based Computers, DNA7, Tampa, FL, USA, June 10–13, 2001. Revised Papers*, volume 2340, pages 329–339. Springer Verlag, Berlin, Heidelberg, New York, 2001.

[203] A. Păun and G. Păun. The power of communication: P systems with symport/antiport. *New Generation Computing*, 20(3):295–306, 2002.

[204] A. Păun and M. Păun. On the membrane computing based on splicing. In C. Martíin-Vide and V. Mitrana, editors, *Where Do Mathematics, Computer Science and Biology Meet*, pages 409–422. Kluwer Academic, Dortrecht, 2001.

[205] G. Păun. Regular extended H systems are computationally universal. *Journal of Automata, Languages, Combinatorics*, 1(1):27–36, 1996.

[206] G. Păun. *Structures in Logic and Computer Science. A Selection of Essays in Honor of A. Ehrenfeucht*, volume 1261 of *Lecture Notes in Computer Science*, chapter DNA computing: distributed splicing systems, pages 351–370. Springer Verlag, Berlin, Heidelberg, New York, 1997. J. Mycielsky, G. Rozenberg and A. Salomaa (eds).

[207] G. Păun. Computing with membranes. *Journal of Computer and System Science*, 1(61):108–143, 2000.

[208] G. Păun. On P systems with membrane division. In Antoniou et al. [22], pages 187–201.

[209] G. Păun. P systems with active membranes: attacking NP-complete problems. *Journal of Automata, Languages and Combinatorics*, 6(1):75–90, 2001.

[210] G. Păun. *Membrane Computing. An Introduction*. Springer-Verlag, Berlin, Heidelberg, New York, 2002.

[211] G. Păun. Further twentysix open problems in membrane computing. In Gutiérrez-Naranjo et al. [100], pages 249–262. Available from [238].

[212] G. Păun. Spiking neural P systems with astrocyte-like control. *The Journal of Universal Computer Science*, 13(11):1707–1721, 2007.

[213] G. Păun, J. Pazos, M. J. Pérez-Jiménez, and A. Rodriguez-Paton. Symport/antiport P systems with three objects are universal. *Fundamenta Informaticae*, 64:1–4, 2005.

[214] G. Păun and M. J. Pérez-Jiménez. Spiking neural P systems. an overview. In A. B. Porto, A. Pazos, and W. Buno, editors, *Advancing Artificial Intelligence through Biological Process Applications*. Idea Group Publishing, London, 2008. in press.

[215] G. Păun, M. J. Pérez-Jímenez, and G. Rozenberg. Infinite spike trains in spiking neural P systems. Submitted, 2005.

[216] G. Păun, M. J. Pérez-Jímenez, and G. Rozenberg. Spike trains in spiking neural P systems. *International Journal of Foundations of Computer Science*, 17(4):975–1002, 2006.

[217] G. Păun, M. J. Pérez-Jiménez, and G. Rozenberg. Computing morphisms by spiking neural P systems. *International Journal of Foundations of Computer Science*, 18(6):1371–1382, 2007.

[218] G. Păun, G. Rozenberg, and A. Salomaa. Computing by splicing. *Theoretical Computer Science*, 168(2):321–336, 1996.

[219] G. Păun, G. Rozenberg, and A. Salomaa. Restricted use of the splicing operation. *International Journal of Computer Mathematics*, 60:17–32, 1996.

[220] G. Păun, G. Rozenberg, and A. Salomaa. *DNA Computing: New Computing Paradigms*. Springer Verlag, Berlin, Heidelberg, New York, September 1998.

[221] G. Păun, G. Rozenberg, A. Salomaa, and C. Zandron, editors. *Membrane Computing. International Workshop, WMC-CdeA2002, Curtea de Arges, Romania, August 19–23, 2002, Revised Papers*, volume 2597 of *Lecture Notes in Computer Science*. Springer-Verlag, Berlin, Heidelberg, New York, 2002.

[222] G. Păun, Y. Suzuki, H. Tanaka, and T. Yokomori. On the power of membrane division in P systems. *Theoretical Computer Science*, 324(1):61–85, 2004.

[223] G. Păun and T. Yokomori. Membrane computing based on splicing. In E. Winfree and D. K. Gifford, editors, *DNA Based Computers V*, volume 54 of *DIMACS Series in Discrete Mathematics and Theoretical Computer Science*, pages 217–232. American Mathematical Society, 1999.

[224] G. Păun and T. Yokomori. Simulating H systems by P systems. *Journal of Universal Computer Science*, 6(2):178–193, 2000.

[225] Gheorghe Păun. Computing with membranes. Technical Report 208, Turku Center for Computer Science-TUCS, 1998.

[226] D. Ramírez-Martínez and M. A. Gutiérrez-Naranjo. A software tool for dealing with spiking neural P systems. In Gutírrez-Naranjo et al. [103], pages 299–314. Available from [238].

[227] W. Reisig. *Petri Nets: An Introduction*, volume 4 of *Monographs in Theoretical Computer Science. An EATCS Series*. Springer-Verlag, Berlin, Heidelberg, New York, 1985.

[228] W. Reisig and G. Rozenberg, editors. *Lectures on Petri Nets I: Basic Models*, volume 1491 of *Lecture Notes in Computer Science*. Springer-Verlag, Berlin, Heidelberg, New York, 1998.

[229] A. Riscos-Núñez. *Cellular programming: efficient resolution of NP-complete numerical problems*. PhD thesis, University of Seville, 2004. Spain.

[230] Y. Rogozhin and S. Verlan. On the rule complexity of universal tissue P systems. In Freund et al. [67], pages 356–362.

[231] G. Rozenberg and J. Engelfriet. *Elementary net systems*, pages 12–121. Volume 1491 of Reisig and Rozenberg [228], 1998.

[232] G. Rozenberg and A. Salomaa. *Handbook of Formal Languages*. Springer-Verlag, Berlin, Heidelberg, New York, 1996.

[233] P. Sosík. P systems versus register machines: two universality proofs. In *Pre-Proceedings of the Workshop on Membrane Computing (WMC-CdeA2002), Curtea de Arges, Romania*, pages 371–382. MolCoNet (Molecular Computing Network) Publication, 2002.

[234] P. Sosík. The computational power of cell division in P systems: Beating down parallel computers? *International Journal on Natural Computing*, 2(3):287–298, 2003.

[235] P. Sosík. The power of catalysts and priorities in membrane systems. *Grammars*, 6(1):13–24, 2003.

[236] P. Sosík and J. Matysek. Membrane computing: when communication is enough. In C. Calude, M. J. Dinneen, and F. Peper, editors, *Unconventional Models of Computation, Third International Conference, UMC 2002, Kobe, Japan, October 15–19, 2002, Proceedings*, volume 2509 of *Lecture Notes in Computer Science*, pages 264–275. Springer-Verlag, Berlin, Heidelberg, New York, 2002.

[237] About splicing P systems with immediate communication and non-extended splicing P systems. S. verlan. In Martín-Vide et al. [170], pages 369–382.

[238] The P Systems Webpage. http://ppage.psystems.eu/.

[239] Manca V. Splicing normalization and regularity. In C. Calude and G. Păun, editors, *Finite Versus Infinite. Contribution to an Eternal Dilemma*, Discrete Mathematics and Theoretical Computer Science, pages 199–215. Springer Verlag, Berlin, Heidelberg, New York, 2000.

[240] J. van Leeuwen. A partial solution to the reachability-problem for vector-addition systems. In *STOC '74: Proceedings of the sixth annual ACM symposium on Theory of computing*, pages 303–309, New York, NY, USA, 1974. ACM.

[241] G. Vaszil. On the size of P systems with minimal symport/antiport. In Mauri et al. [177], pages 404–413.

[242] S. Verlan. *Head systems and application to bio-informatics*. PhD thesis, LITA, Universite' de Metz, 2004. France.

[243] S. Verlan. Tissue P systems with minimal symport/antiport. In C. Calude, E. Calude, and M. J. Dinneen, editors, *Developments in Language Theory, 8th International Conference, DLT 2004, Auckland, New Zealand, December 13–17, 2004, Proceedings*, volume 3340 of *Lecture Notes in Computer Science*, pages 418–429. Springer-Verlag, Berlin, Heidelberg, New York, 2004.

[244] S. Verlan. A boundary result on enhanced time-varying distributed H systems with parallel computations. *Theoretical Computer Science*, 344(2–3):226–242, 2005.

[245] S. Verlan, F. Bernardini, M. Gheorghe, and M. Margenstern. Computational completeness of tissue P systems with conditional uniport. In Hoogeboom et al. [110], pages 521–535.

[246] S. Verlan and M. Margenstern. About splicing P systems with one membrane. *Fundamenta Informaticae*, 65(3):279–290, 2005.

[247] S. Verlan and R. Zizza. 1-splicing vs. 2-splicing: separating results. In *WORDS 2003, Turku, Finland, 1–14 September 2003*, 2003. Pre-proceedings TUCS Report, University of Turku.

[248] C. Versari. Encoding catalytic P systems in π. *Electron. Notes Theor. Comput. Sci.*, 171(2):171–186, 2007.

[249] D. Voet and J. G. Voet. *Biochemistry*. John Wiley and Sons, 3th edition, December 2003.

[250] M. V. Volkenstein. The conformon. *Journal of Theoretical Biology*, 34:193–195, 1972.

[251] Wikipedia. `http://www.wikipedia.org`, accessed on November 2007.

[252] X. Xu. Tissue P systems with parallel rules on channels. In *Pre-proceedings of the 1st International Conference on Bio-Inspired Computing—Theory and Applications*, pages 168–177, 2006. Wuhan, China.

[253] C. Zandron. *A Model for Molecular Computing: Membrane Systems*. PhD thesis, Università degli Studi di Milano, Italy, 2002.

[254] C. Zandron, C. Ferretti, and G. Mauri. Solving NP-complete problems using P systems with active membranes. In Antoniou et al. [22], pages 289–301.

[255] C. Zandron, G. Mauri, and C. Ferretti. Universality and normal forms on membrane systems. In R. Freund and A. Kelemenova, editors, *Proceedings of the International Workshop Grammar Systems 2000. Bad Ischl, Austria, July 2000*, pages 61–74. Silesian University, Opava, Czech Republic, 2000.

Index